SOLUTIONS AND PROBLEM-SOLVING MANUAL FOR

GENETICS: A CONCEPTUAL APPROACH

THIRD EDITION

JUNG H. CHOI
MARK E. MCCALLUM

W. H. FREEMAN AND COMPANY
NEW YORK

ISBN-13: 978-1-4292-0353-1
ISBN-10: 1-4292-0353-6

Printed in the United States of America

Third printing

W. H. Freeman and Company
41 Madison Avenue
New York, NY 10010
Houndmills, Basingstoke RG21 6XS, England
www.whfreeman.com

Contents

Chapter One: Introduction to Genetics

COMPREHENSION QUESTIONS

Section 1.1

*1. How does the Hopi culture contribute to the high incidence of albinism among members of the Hopi tribe?
In the Hopi culture, albino individuals were considered special and awarded special status in the village. Hopi male albinos were not required to work the fields, thus avoiding extensive exposure to sunlight that could prove damaging or deadly. Albinism was a positive trait and not a negative physical condition, and that allowed albinos to have more children, increasing the frequency of the albino allele. Finally, the small population size of the Hopi tribe may have helped increase the allele frequency of the albino gene due to chance.

2. Outline some of the ways in which genetics is important to each of us.
Genetics directly influences our lives and is fundamental to what and who we are. For example, genes affect our appearance (e.g., eye color, height, weight, skin pigmentation, and hair color). Our susceptibility to diseases and disorders is affected by our genetic makeup. Genes also influence our intelligence and personality.

*3. Give at least three examples of the role of genetics in society today.
Genetics plays important roles in the diagnosis and treatment of hereditary diseases: in breeding plants and animals for improved production and disease resistance; and in producing pharmaceuticals and novel crops through genetic engineering.

4. Briefly explain why genetics is crucial to modern biology.
Genetics is crucial to modern biology in that it provides unifying principles: All organisms use nucleic acid as their genetic material, and all organisms encode genetic information in the same manner. The study of many other biological disciplines, such as developmental biology, ecology, and evolutionary biology, is supported by genetics.

*5. List the three traditional subdivisions of genetics and summarize what each covers.
Transmission genetics: inheritance of genes from one generation to the next, gene-mapping.
Molecular genetics: structure, organization, and function of genes at a molecular level.
Population genetics: genes and changes in genes in populations.

6. What are some characteristics of model genetic organisms that make them useful for genetic studies?
Model genetic organisms have relatively short generation times, produce numerous progeny, are amenable to laboratory manipulations, and can be maintained and propagated inexpensively.

Section 1.2

7. When and where did agriculture first arise? What role did genetics play in the development of the first domesticated plants and animals?
Agriculture first arose 10,000 to 12,000 years ago in the area now referred to as the Middle East (i.e., Turkey, Iran, Iraq, Syria, Jordan, and Israel). Early farmers selectively bred individual wild plants or animals that had useful characteristics with others that had similar useful traits. The farmers then selected for offspring that contained those useful features. Early farmers did not completely understand genetics, but they clearly understood that breeding individual plants or animals with desirable traits would lead to offspring that contained these same traits. This selective breeding led to the development of domesticated plants and animals.

*8. Outline the notion of pangenesis and explain how it differs from the germ-plasm theory.
Pangenesis theorizes that information for creating each part of the offspring's body originates from each part of the parent's body and is passed through the reproductive organs to the embryo at conception. Pangenesis suggests that changes in parts of the parent's body may be passed to the offspring's body. The germ-plasm theory, in contrast, states that the reproductive cells possess all of the information required to make the complete body; the rest of the body contributes no information to the next generation.

9. What does the concept of the inheritance of acquired characteristics propose and how is it related to the theory of pangenesis?
The theory of inheritance of acquired characteristics postulates that traits acquired during one's lifetime can be transmitted to offspring. It developed from pangenesis, which postulates that information from all parts of one's body is transmitted to the next generation. Thus, for example, learning acquired in the brain or larger arm muscles developed through exercise could be transmitted to offspring.

*10. What is preformationism? What does it have to say about how traits are inherited?
Preformationism is the theory that the offspring results from a miniature adult form that is already preformed in the sperm or the egg. All traits would thus be inherited from only one parent, either the father or the mother, depending on whether the homunculus (the preformed miniature adult) resided in the sperm or the egg.

11. Define blending inheritance and contrast it with preformationism.
The theory of blending inheritance proposes that the egg and sperm from two parents contains material that blends upon conception, influencing the development of the offspring. This theory indicates that the offspring is an equal blend of the two parents. In preformationism, the offspring inherits all of its traits from one parent.

12. How did developments in botany during the seventeenth and eighteenth centuries contribute to the rise of modern genetics?

Botanists of the seventeenth and eighteenth centuries developed new techniques for crossing plants and creating plant hybrids. These early experiments provided essential background work for Mendel's plant crosses. Mendel's work laid the foundation for the study of modern genetics.

*13. Who first discovered the basic principles that laid the foundation for our modern understanding of heredity?
Gregor Mendel

14. List some of advances in genetics that have occurred in the twentieth century.

1902 *Proposal that genes are located on chromosomes by Walter Sutton*
1910 *Discovery of the first genetic mutation in a fruit fly by Thomas Hunt Morgan*
1930 *The foundation of population genetics by Ronald A. Fisher, John B. S. Haldane, and Sewall Wright*
1940s *The use of viral and bacterial genetic systems*
1953 *Three-dimensional structure of DNA described by Watson and Crick*
1966 *Deciphering of the genetic code*
1973 *Recombinant DNA experiments*
1977 *Chemical and enzymatic methods for DNA sequencing developed by Walter Gilbert and Frederick Sanger*
1986 *PCR developed by Kary Mullis*
1990 *Gene therapy*

Section 1.3

15. What are the two basic cell types (from a structural perspective) and how do they differ?
The two basic cell types are prokaryotic and eukaryotic. Prokaryotic cells do not have a nucleus, and their chromosomes are found within the cytoplasm. They do not possess membrane-bound cell organelles. Eukaryotic cells possess a nucleus and membrane-bound cell organelles.

*16. Outline the relations between genes, DNA, and chromosomes.
Genes are composed of DNA nucleotide sequences that are located at specific positions in chromosomes.

APPLICATION QUESTIONS AND PROBLEMS

Section 1.1

17. What is the relation between genetics and evolution?
In essence, evolution is change in the genetic composition of a population over generations. Mutation generates new genetic variants, recombination generates new combinations of genetic variants, and natural selection or other evolutionary processes cause a change in the proportions of specific genetic variants in the population.

*18. For each of the following genetic topics, indicate whether it focuses on transmission genetics, molecular genetics, or population genetics.

a. Analysis of pedigrees to determine the probability of someone inheriting a trait
Transmission genetics

b. Study of the genetic history of people on a small island to determine why a genetic form of asthma is so prevalent on the island
Population genetics

c. The influence of nonrandom mating on the distribution of genotypes among a group of animals.
Population genetics

d. Examination of the nucleotide sequences found at the ends of chromosomes
Molecular genetics

e. Mechanisms that ensure a high degree of accuracy during DNA replication
Molecular genetics

f. Study of how the inheritance of traits encoded by genes on sex chromosomes (sex-linked traits) differs from the inheritance of traits encoded by genes on nonsex chromosomes (autosomal traits)
Transmission genetics

Section 1.2

*19. Genetics is said to be both a very old science and a very young science. Explain what is meant by this statement.
Genetics is old in the sense that humans have been aware of hereditary principles for thousands of years and have applied them since the beginning of agriculture and the domestication of plants and animals. It is very young in the sense that the fundamental principles were not uncovered until Mendel's time, and the discovery of the structure of DNA and the principles of recombinant DNA have occurred within the last 60 years.

20. Match the description (a–d) with the correct theory or concept listed below.

a. Each reproductive cell contains a complete set of genetic information.
Germ-plasm theory

b. All traits are inherited from one parent.
Preformationism

c. Genetic information may be altered by use of a feature.
Inheritance of acquired characteristics

d. Different genetic information occurs in cells of different tissues.
Pangenesis

*21. Compare and contrast the following ideas about inheritance.

a. Pangenesis and germ-plasm theory
Pangenesis theorizes that units of genetic information (pangenes) from all parts of the body are carried through the reproductive organs to the embryo where each unit directs the formation of its own specific part of the body. According to the germ-plasm theory, the germ-line tissue or gamete producing cells found within the reproductive organs contain the complete set of genetic information that is passed to the gametes. Both theories are similar in that each predicts the passage of genetic information occurs through the reproductive organs.

b. Preformationism and blending inheritance
Preformationism predicts that the sperm or egg contains a miniature preformed adult called the homunculus. During development, the homunculus would grow to produce the offspring. Only one parent would contribute genetic traits to the offspring. Blending inheritance requires contributions of genetic material from both parents. According to the theory of blending inheritance, genetic contributions from the parents blend to produce the genetic material of the offspring. Once blended, the genetic material could not be separated for future generations.

c. The inheritance of acquired characteristics and our modern theory of heredity
The theory of inheritance of acquired characteristics postulates that traits acquired during one's lifetime alter the genetic material and can be transmitted to offspring. Our modern theory of heredity indicates that offspring inherit genes located on chromosomes from their parents. These chromosomes segregate during meiosis in the germ cells and are passed into the gametes.

Section 1.3

*22. Compare and contrast the following terms:

a. Eukaryotic and prokaryotic cells
Both cell types have lipid bilayer membranes, DNA genomes, and machinery for DNA replication, transcription, translation, energy metabolism, response to stimuli, growth, and reproduction. Eukaryotic cells have a nucleus containing chromosomal DNA and possess internal membrane-bound organelles.

b. Gene and allele
A gene is a basic unit of hereditary information, usually encoding a functional RNA or polypeptide. An allele is a variant form of a gene, arising through mutation.

c. Genotype and phenotype
The genotype is the set of genes or alleles an organism has inherited from its parent(s). The expression of the genes of a particular genotype, through interaction with environmental factors, produces the phenotype, the observable trait.

d. DNA and RNA
Both are nucleic acid polymers. RNA contains ribose, whereas DNA contains deoxyribose. RNA also contains uracil as one of the four bases, whereas DNA contains thymine. The other three bases are common to both DNA and RNA. Finally, DNA is usually double-stranded, consisting of two complementary strands

and very little secondary structure, whereas RNA is single-stranded with regions of internal base-pairing to form complex secondary structures.

e. DNA and chromosome
Chromosomes are structures formed of DNA and associated proteins. The DNA contains the genetic information.

CHALLENGE QUESTIONS

Section 1.1

23. We now know as much or more about the genetics of humans as any other organism, and humans are the foscus of many genetic studies. Do you think humans should be considered a model genetic organism? Why or why not?
Arguments against considering humans as a model genetic organism:
Although human genetics has been intensively studied, humans should not be considered a model genetic organism because they lack the characteristics of model genetic organisms. Humans have a long time between generations, usually bear only one offspring per mating, and are expensive to maintain! Most importantly, humans are not amenable to laboratory manipulation; the ethical barriers to controlling human matings and subjecting humans to experiments are insurmountable. Ultimately, the purpose of model organisms is to enable experimental investigations that are not possible with humans.

Arguments in favor of humans as a model genetic organism:
Human genetics has been extensively studied, with vast accumulated literature and knowledge. Extensive records of marriages, births, deaths, medical histories, and migrations provide a wealth of data. The human genome has been sequenced. Some unique features of humans, such as long life-span, diverse habitats and behaviors, varied diets, and relatively large size make animal models inadequate for study of many aspects of human biology and genetics. Ultimately, we are our own model organism.

24. Describe some of the ways in which your own genetic makeup affects you as a person. Be as specific as you can.
Answers will vary but should include observations similar to those in the following example: Genes affect my physical appearance; for example, they probably have largely determined the fact that I have brown hair and brown eyes. Undoubtedly, genes have affected my height of five feet, seven inches, which is quite close to the height of my father and mother, and my slim build. My dark complexion mirrors the skin color of my mother. I have inherited susceptibilities to certain diseases and disorders that tend to run in my family; these include asthma, a slight tremor of the hand, and vertigo.

25. Describe at least one trait that appears to run in your family (appears in multiple members of the family). Do you think this trait runs in your family because it is an inherited trait or because it is caused by environmental factors that are common to family members? How might you distinguish between these possibilities?

My two brothers and I share two traits: we are all three taciturn (we don't speak much) and smart (just don't ask my teenage daughter). Although the literature provides evidence for a genetic component for intelligence, I'm not aware of any studies on the heritability of being taciturn. If I were to investigate to what extent these traits are determined by the environment or by heredity, I would look at studies of twins who had been separated at birth and lived in different environments to adulthood. Such studies would separate environmental factors from genetic factors, whereas studies of family members reared in the same household are confounded by the fact that the family members experienced similar environments. If the trait had a strong genetic component, we would expect identical twins reared apart to be similarly taciturn or similarly intelligent. One would have to devise some objective measure of these traits–degrees of being taciturn or smart.

Section 1.3

*26. Suppose that life exists elsewhere in the universe. All life must contain some type of genetic information, but alien genomes might not consist of nucleic acids and have the same features as those found in the genomes of life on Earth. What do you think might be the common features of all genomes, no matter where they exist?
All genomes must have the ability to store complex information and must have the capacity to vary. The blueprint for the entire organism must be contained within the genome of each reproductive cell. The information has to be in the form of a code that can be used as a set of instructions for assembling the components of the cells. The genetic material of any organism must be stable, be replicated precisely, and be transmitted faithfully to the progeny, but must be capable of mutating.

27. Pick one of the following ethical or social issues and give your opinion on this issue. For background information, you might read one of the articles on ethics marked with an asterisk and listed in the Suggested Readings at the end of the chapter.

a. Should a person's genetic makeup be used in determining his or her eligibility for life insurance?
Arguments pro: *Genetic susceptibility to certain types of diseases or conditions is relevant information regarding consequences of exposure to certain occupational hazards. Genes that will result in neurodegenerative diseases, such as Huntington disease, Alzheimer's, or breast cancer, could logically be considered preexisting conditions. Insurance companies have a right, and arguably a duty to their customers, to exclude people with genetic preconditions so that insurance rates can be lowered for the general population.*
Arguments con: *The whole idea of insurance is to spread the risk and pool assets. Excluding people based on their genetic makeup would deny insurance to people who need it most. Indeed, as information about various genetic risks accumulates, more people would become excluded until only a small fraction of the population is insurable.*

b. Should biotechnology companies be able to patent newly sequenced genes?
Pro: *Patenting genes provides companies with protection for their investment in research and development of new drugs and therapies. Without such patent protection, companies would have less incentive to expend large amounts of money in genetic research and thus would slow the pace of advancement of medical research. Such a result would be detrimental to everyone.*
Con: *Patents on human genes would be like allowing companies to patent a human arm. Genes are integral parts of our selves, so how can a company patent something that every human has?*

c. Should gene therapy be used on people?
Pro: *Gene therapy can be used to cure previously incurable or intractable genetic disorders and to relieve the suffering of millions of people.*
Con: Gene therapy may lead to genetic engineering of people for unsavory ends. Who determines what is a genetic defect? Is short stature a genetic defect?

d. Should genetic testing be made available for inherited conditions for which there is no treatment or cure?
Pro: *Information will provide relief from unnecessary anxiety (if the test is negative). Even if the test result is positive for a genetic disorder, it provides the individual, the family, and friends with information and time to prepare. Information about one's own genetic makeup is a right; every person should be able to make his or her own choice as to whether he or she wants this information.*
Con: *If there is no treatment or cure, a positive test result can have no good consequences. It's like receiving a death sentence or sentence of extended punishment. It will only engender feelings of hopelessness and depression and may cause some people to terminate their own lives prematurely. Applied to the unborn as a prenatal test, it may lead to abortion.*

e. Should governments outlaw the cloning of people?
Pro (for outlawing human cloning): *There is no medical necessity for human cloning. There is no good research objective that can be attained solely via human cloning. Human cloning creates human beings for a purpose and may be a form of slavery. As such, it violates our sense that all humans should have free will. The risk of birth defects and complicated pregnancies is too great to justify these experiments with humans.*
Con (against outlawing human cloning): *Cloning people is just another method of assisted reproduction for infertile couples. People should be free to do whatever they want with their own bodies. Research on human cloning may lead to stem cell lines and methods that can be used medically to grow organ replacements.*

*28. Suppose that you could undergo genetic testing at age 18 for susceptibility to a genetic disease that would not appear until middle age and has no available treatment.

a. What would be some of the possible reasons for having such a genetic test and some of the possible reasons for not having the test?
Having the genetic test removes doubt about the potential for the disorder—either you are susceptible or not. By knowing about the potential of a genetic disorder lifestyle changes could possibly be made to lessen the impact of the disease or lessen the risk.

The types and nature of future medical tests could be positively impacted by the genetic testing, thus allowing for early warning and screening for the disease. The knowledge could also have impact on future family plans and allow for informed decisions regarding future offspring and the potential of passing the trait to your offspring. Also by knowing the future, one could plan one's life accordingly.

Reasons for not having the test typically revolve around the potential for testing positive for the susceptibility to the genetic disease. If the susceptibility was detected, the potential for discrimination could exist. For example, your employer (or possibly future employer) might see you as a long-term liability, thus affecting employment options. Insurance companies may not want to insure you for that condition or symptoms of the disorder, and potentially social stigmatism associated with the disease could be a factor. Knowledge of the potential future condition could lead to psychological difficulties in coping with the anxiety of waiting for the disease to manifest.

b. Would you personally want to be tested? Explain your reasoning.
There is no "correct" answer, but for me, yes, I would personally want to be tested. The test would remove doubt about the susceptibility particularly if the genetic disease had been demonstrated to occur in my family. Either a positive or negative result would allow for informed planning of lifestyle, medical testing, and family choices in the future.

Chapter Two: Chromosomes and Cellular Reproduction

COMPREHENSION QUESTIONS

Section 2.1

*1. Give some genetic differences between prokaryotic and eukaryotic cells.

Prokaryotic cell	*Eukaryotic cell*
No nucleus *No paired chromosomes (haploid)* *Typically single circular chromosome containing a single origin of replication* *Single chromosome is replicated with each copy moving to opposite sides of the cell* *No histone proteins bound to DNA*	*Nucleus present* *Paired chromosomes common (diploid)* *Typically multiple linear chromosomes containing centromeres, telomeres, and multiple origins of replication* *Chromosomes are replicated and segregate during mitosis or meiosis to the proper location* *Histone proteins are bound to DNA*

2. Why are the viruses that infect mammalian cells useful for studying the genetics of mammals?
It is thought that viruses must have evolved after their host cells because a host is required for viral reproduction. Viral genomes are closely related to their host genomes. The close relationship between a mammalian virus and its mammalian cell host, along with the simpler structure of the viral particle, makes it useful in studying the genetics of mammals. The viral genome will have a similar structure to the mammalian cell host, but because it has fewer genes, it will be easier to decipher the interactions and regulation of the viral genes.

Section 2.2

*3. List three fundamental events that must take place in cell reproduction.
(1) A cell's genetic information must be copied.
(2) The copies of the genetic information must be separated from one another.
(3) The cell must divide.

4. Outline the process by which prokaryotic cells reproduce.
(1) Replication of the circular chromosome takes place.
(2) The two replicated chromosomal copies attach to the plasma membrane.
(3) The plasma membrane grows, which results in the separation of the two chromosomes.
(4) A new cell wall is formed between the two chromosomes, producing two cells, each with its own chromosome.

5. Name three essential structural elements of a functional eukaryotic chromosome and describe their functions.
   (1) Centromere: serves as the point of attachment for the spindle fibers (microtubules).
   (2) Telomeres, or the natural ends of the linear eukaryotic chromosome: serve to stabilize the ends of the chromosome; may have a role in limiting cell division.
   (3) Origins of replication: serve as the starting place for DNA synthesis.

*6. Sketch and label four different types of chromosomes based on the position of the centromere.

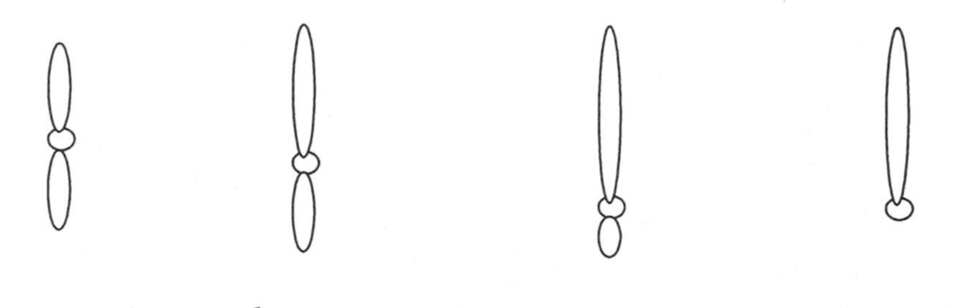

7. List the stages of interphase and the major events that take place in each stage.
   Three predominant stages are found in interphase of cells active in the cell cycle.
   (1) G_1 (Gap 1). In this phase, the cell grows and synthesizes proteins necessary for cell division. During G_1, the G_1/S checkpoint takes place. Once the cell has passed this checkpoint, it is committed to divide.
   (2) S phase. During S phase, DNA replication takes place.
   (3) G_2 (Gap 2). In G_2, additional biochemical reactions take place that prepare the cell for mitosis. A major checkpoint in G_2 is the G_2/M checkpoint. Once the cell has passed this checkpoint, it enters into mitosis.
   A fourth stage is frequently found in cells prior to the G_1/S checkpoint. Cells may exit the active cell cycle and enter into a nondividing stage called G_0.

*8. List the stages of mitosis and the major events that take place in each stage.
   (1) Prophase: The chromosomes condense and become visible, the centrosomes move apart, and microtubule fibers form from the centrosomes.
   (2) Prometaphase: The nucleoli disappear and the nuclear envelope begins to disintegrate, allowing for the cytoplasm and nucleoplasm to join. The sister chromatids of each chromosome are attached to microtubles from the opposite centrosomes.
   (3) Metaphase: The spindle microtubules are clearly visible and the chromosomes arrange themselves on the equatorial plane of the cell.
   (4) Anaphase: The sister chromatids separate at the centromeres after the breakdown of cohesin protein, and the newly formed daughter chromosomes move to the opposite poles of the cell.
   (5) Telophase: The nuclear envelope reforms around each set of daughter chromosomes. Nucleoli reappear. Spindle microtubules disintegrate.

9. Briefly describe how the chromosomes move toward the spindle poles during anaphase.
Due to the actions of the microtubule subunits attached to the kinetochores of the chromosome and motor proteins such as kinesin, the chromosomes are pulled toward the spindle poles during anaphase. The spindle fibers are composed of tubulin protein subunits. As the tubulin subunits are removed from the "–" end of the microtubule, the chromosome is pulled (or "reeled in") toward the spindle pole as the microtubule is shortened. While at the "+" end, the kinetochore is removing tubulin subunits of the microtubule attached to the kinetochore with the net effect being the movement of the chromosome closer to the spindle pole. Molecular motor proteins, such as kinesin, are responsible for removing the subunits at the "+" and "–" ends of the microtubules and thus generate the force needed to move the chromosomes.

*10. What are the genetically important results of the cell cycle?
In the mitotic cell cycle, the genetic material is precisely copied so that the two resulting cells contain the same genetic information. In other words, the cells have genomes identical to each other and to the mother cell.

11. Why are the two cells produced by the cell cycle genetically identical?
The two cells are genetically identical because during S phase an exact copy of each DNA molecule was created. These exact copies give rise to the two identical sister chromatids. Mitosis ensures that each new cell receives one copy of the two identical sister chromatids. Thus, the newly formed cells will contain identical daughter chromosomes.

12. What are checkpoints? List some of the important checkpoints in the cell cycle. What two general classes of compounds regulate progression through the cell cycle?
Checkpoints function to ensure that all the cellular components, such as important proteins and chromosomes, are present and functioning before the cell moves to the next stage of the cell cycle. If components are missing or not functioning, the checkpoint will prevent the cell from moving to the next stage. The checkpoints prevent defective cells from replicating and malfunctioning.

These checkpoints occur throughout the various stages of the cell cycle. Important checkpoints include the G_1/S checkpoint, which occurs during G_1 prior to the S phase; the G_2/M checkpoint, which occurs in G_2 prior to mitosis; and the spindle-assembly checkpoint, which occurs during mitosis.
Two types of proteins are responsible for movement through the cell cycle: cyclin proteins and cyclin-dependent kinases.

Section 2.3

13. What are the stages of meiosis and what major events take place in each stage?
Meiosis I: Separation of homologous chromosomes

Prophase I: The chromosomes condense and homologous pairs of chromosomes undergo synapsis. While the chromosomes are synapsed, crossing over occurs. The nuclear membrane disintegrates and the meiotic spindle begins to form.

Metaphase I: The homologous pairs of chromosomes line up on the equatorial plane of the metaphase plate.

Anaphase I: Homologous chromosomes separate and move to opposite poles of the cell. Each chromosome possesses two sister chromatids.

Telophase I: The separated homologous chromosomes reach the spindle poles and are at opposite ends of the cell.

Meiosis I is followed by cytokinesis, resulting in the division of the cytoplasm and the production of two haploid cells. These cells may skip directly into meiosis II or enter interkinesis, where the nuclear envelope reforms and the spindle fibers break down.

Meiosis II: Separation of sister chromatids

Prophase II: Chromosomes condense, the nuclear envelope breaks down, and the spindle fibers form.

Metaphase II: Chromosomes line up at the equatorial plane of the metaphase plate.

Anaphase II: The centromeres split, which results in the separation of sister chromatids.

Telophase II: The daughter chromosomes arrive at the poles of the spindle. The nuclear envelope reforms, and the spindle fibers break down. Following meiosis II, cytokinesis takes place.

*14. What are the major results of meiosis?

Meiosis involves two cell divisions, thus producing four new cells (in many species). The chromosome number of a haploid cell produced by meiosis I (haploid) is half the chromosome number of the original diploid cell. Finally, the cells produced by meiosis are genetically different from the original cell and genetically different from each other.

15. What two processes unique to meiosis are responsible for genetic variation? At what point in meiosis do these processes take place?

(1) Crossing over, which begins during the zygotene stage of prophase I and is completed near the end of prophase I

(2) The random distribution of chromosomes to the daughter cells, which takes place in anaphase I of meiosis.

*16. List similarities and differences between mitosis and meiosis. Which differences do you think are most important and why?

Mitosis	*Meiosis*
A single cell division produces two genetically identical progeny cells.	*Two cell divisions usually result in four progeny cells that are not genetically identical.*
Chromosome number of progeny cells and the original cell remain the same.	*Daughter cells are haploid and have half the chromosomal complement of the original diploid cell as a result of the separation of homologous pairs during anaphase I.*
Daughter cells and the original cell are genetically identical. No separation of homologous chromosomes or crossing over takes place.	*Crossing over in prophase I and separation of homologous pairs during anaphase I produce daughter cells that are genetically different from each other and from the original cell.*
Homologous chromosomes do not synapse.	*Synapsis of homologous chromosomes takes place during prophase I.*
In metaphase, individual chromosomes line up on the metaphase plate.	*In metaphase I, homologous pairs of chromosomes line up on the metaphase plate. Individual chromosomes line up in metaphase II.*
In anaphase, sister chromatids separate.	*In anaphase I, homologous chromosomes separate. Separation of sister chromatids takes place in anaphase II.*

The most important difference is that mitosis produces cells genetically identical to each other and to the original cell, resulting in the orderly passage of information from one cell to its progeny. In contrast, by producing progeny that do not contain pairs of homologous chromosomes, meiosis results in the reduction of chromosome number from the original cell. Meiosis also allows for genetic variation through crossing over and the random assortment of homologous chromosomes.

17. Briefly explain why sister chromatids remain together in anaphase I but separate in anaphase of mitosis and anaphase II of meiosis.
During the S phase, the cohesin protein complex forms and holds together the sister chromatids throughout the early stages of mitosis or meiosis I. In mitosis at the end of metaphase, the cohesin molecules that connect the sister chromatids at the centromeres are cleaved by the protein separase, which allows the sister chromatids to separate. Prior to this point, separase is kept inactive by a protein called securin. Securin is broken down at the end of metaphase, allowing separase to become active.

In meiosis, a similar process occurs. The cohesin complexes form at the centromeres of the sister chromatids during the S phase. At the beginning of meiosis, cohesin molecules are also found along the entire length of the chromosome arms assisting in the formation of the synaptonemal complex and holding together the two homologs. During anaphase I of meiosis, the cohesin molecules along the arms are cleaved by activated separase allowing the

homologs to separate. However, the cohesin complexes at the centromeres of the sister chromatids are protected from the action of separase and are unaffected. The result is that sister chromatids remained attached during anaphase I. At the end of metaphase II, the protection of the cohesin molecules at the centromeres is lost, and the separase proteins can now cleave the cohesin complex, which allows the sister chromatids to separate.
Also during metaphase I, proteins called monopolins function to allow only the two kinetochores of sister chromatids to orient toward the same spindle pole and attach to microtubules of the same pole. The end result is that the attached sister chromatids move toward the same pole during meiosis.

18. Outline the process of spermatogenesis in animals. Outline the process of oogenesis in animals.
In animals, spermatogenesis occurs in the testes. Primordial diploid germ cells divide mitotically to produce diploid spermatogonia that can either divide repeatedly by mitosis or enter meiosis. A spermatogonium that has entered prophase I of meiosis is called a primary spermatocyte and is diploid. Upon completion of meiosis I, two haploid cells, called secondary spermatocytes, are produced. Upon completing meiosis II, the secondary spermatocytes produce a total of four haploid spermatids.

Female animals produce eggs through the process of oogenesis. Similar to what takes place in spermatogenesis, primordial diploid cells divide mitotically to produce diploid oogonia that can divide repeatedly by mitosis, or enter meiosis. An oogonium that has entered prophase I is called a primary oocyte and is diploid. Upon completion of meiosis I, the cell divides, but unequally. One of the newly produced haploid cells receives most of the cytoplasm and is called the secondary oocyte. The other haploid cell receives only a small portion of the cytoplasm and is called the first polar body. Ultimately, the secondary oocyte will complete meiosis II and produce two haploid cells. One cell, the ovum, will receive most of the cytoplasm from the secondary oocyte. The smaller haploid cell is called the second polar body. Typically, the polar bodies disintegrate, and only the ovum is capable of being fertilized.

19. Outline the process by which male gametes are produced in plants. Outline the process of female gamete formation in plants.
Plants alternate between a multicellular haploid stage called the gametophyte and a multicellular diploid stage called the sporophyte. Meiosis in the diploid sporophyte stage of plants produces haploid spores that develop into the gametophyte. The gametophyte produces gametes by mitosis.

In flowering plants, the microsporocytes found in the stamen of the flower undergo meiosis to produce four haploid microspores. Each microspore divides by mitosis to produce the pollen grain, or the microgametophyte. Within the pollen grain are two haploid nuclei. One of the haploid nuclei divides by mitosis to produce two sperm cells. The other haploid nucleus directs the formation of the pollen tube.

Female gamete production in flowering plants takes place within the megagametophyte. Megasporocytes found within the ovary of a flower divide by meiosis to produce four megaspores. Three of the megaspores disintegrate, while the remaining megaspore divides mitotically to produce eight nuclei that form the embryo sac (or female gametophyte). Of the eight nuclei, one will become the egg.

APPLICATION QUESTIONS AND PROBLEMS

Section 2.2

20. A certain species has three pairs of chromosomes: an acrocentric pair, a metacentric pair, and a submetacentric pair. Draw a cell of this species as it would appear in metaphase of mitosis.

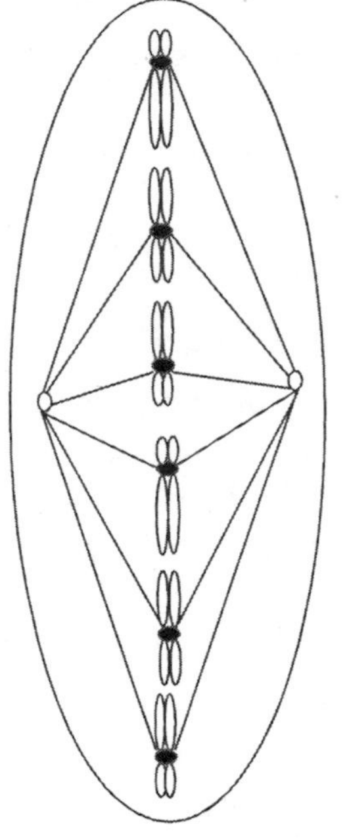

21. A biologist examines a series of cells and counts 160 cells in interphase, 20 cells in prophase, 6 cells in prometaphase, 2 cells in metaphase, 7 cells in anaphase, and 5 cells in telophase. If the complete cell cycle requires 24 hours, what is the average duration of M phase in these cells? Of metaphase?
To determine the average duration of M phase in these cells, the proportion of cells in interphase, or in each stage of M phase, should be calculated by dividing the number of cells in each stage by the total number of cells counted. To calculate the time required for a given phase, multiply 24 hours by the proportion of cells at that stage. This will give the average duration of each stage in hours.

Stage	*Number of cells counted*	*Proportion of cells at each stage*	*Average duration (hours)*
Interphase	*160*	*0.80*	*19.2*
Prophase	*20*	*0.10*	*2.4*
Prometaphase	*6*	*0.03*	*0.72*
Metaphase	*2*	*0.01*	*0.24*
Anaphase	*7*	*0.035*	*0.84*
Telophase	*5*	*0.025*	*0.6*
Totals	*200*	*1.0*	*24*

The average duration of M phase can be determined by adding up the hours spent in each stage of mitosis. In these cells, M phase lasts 4.8 hours. The table shows that metaphase requires 0.24 hours, or 14.4 minutes.

Section 2.3

*22. A cell in G_1 of interphase has 12 chromosomes. How many chromosomes and DNA molecules will be found per cell when this original cell progresses to the following stages?
The number of chromosomes and DNA molecules depends on the stage of the cell cycle. Each chromosome contains only one centromere, but after the completion of S phase, and prior to anaphase of mitosis or anaphase II of meiosis, each chromosome will consist of two DNA molecules.

a. *G_2 of interphase*
G_2 of interphase occurs after S phase, when the DNA molecules are replicated. Each chromosome now consists of two DNA molecules. So a cell in G_2 will contain 12 chromosomes and 24 DNA molecules.

b. *Metaphase I of meiosis*
Neither homologous chromosomes nor sister chromatids have separated by metaphase I of meiosis. Therefore, the chromosome number is 12, and the number of DNA molecules is 24.

c. *Prophase of mitosis*
This cell will contain 12 chromosomes and 24 DNA molecules.

d. *Anaphase I of meiosis*
During anaphase I of meiosis, homologous chromosomes separate and begin moving to opposite ends of the cell. However, sister chromatids will not separate until anaphase II of meiosis. The number of chromosomes is still 12, and the number of DNA molecules is 24.

e. *Anaphase II of meiosis*
Homologous chromosomes were separated and migrated to different daughter cells at the completion of meiosis I. However, in anaphase II of meiosis, sister chromatids separate, resulting in a temporary doubling of the chromosome number in the now haploid daughter cell. The number of chromosomes and the number of DNA molecules present will both be 12.

f. *Prophase II of meiosis*
The daughter cells in prophase II of meiosis are haploid. The haploid cells will contain six chromosomes and 12 DNA molecules.

g. *After cytokinesis following mitosis*
After cytokinesis following mitosis the daughter cells will enter G_1. Each cell will contain 12 chromosomes and 12 DNA molecules.

h. *After cytokinesis following meiosis II*
After cytokinesis following meiosis II, the haploid daughter cells will contain six chromosomes and six DNA molecules.

23. How are the events that take place in spermatogenesis and oogenesis similar? How are they different?
Both spermatogenesis and oogenesis begin similarly in that the diploid primordial cells (spermatogonia and oogonia) can undergo multiple rounds of mitosis to produce more primordial cells, or both types of cells can enter into meiotic division. In spermatogenesis, cytokinesis is equal, resulting in haploid cells of similar sizes. Upon completion of meiosis II, four haploid spermatids have been produced for each spermatogonium that began meiosis. In oogenesis, cytokinesis is unequal. At the completion of meiosis I in oogenesis, a secondary oocyte is produced, which is much larger and contains more cytoplasm than the other haploid cell produced, called the first polar body. At the completion of meiosis II, the secondary oocyte divides, producing the ovum and the second polar body. Again, the division of the cytoplasm in cytokinesis is unequal, with the ovum receiving most of the cytoplasmic material. Usually, the polar bodies disintegrate, leaving the ovum as the only product of meiosis.

*24. All of the following cells, shown in various stages of mitosis and meiosis, come from the same rare species of plant. What is the diploid number of chromosomes in this plant? Give the names of each stage of mitosis or meiosis shown.

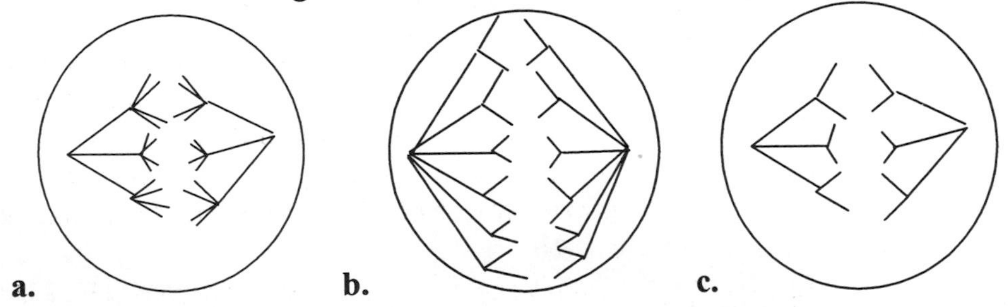

To determine the diploid chromosome number in this plant, the number of centromeres present within a cell that contains homologous pairs of chromosomes must be determined. Remember, each chromosome possesses a single centromere. The location and presence of a centromere are determined by the attachment of the spindle fibers to the chromosome, which occurs at the centromere in the above diagram. Only the cell in stage (a) clearly has homologous pairs of chromosomes. So the diploid chromosome number for cells of this species of plant is six.

a. *This cell is undergoing anaphase of meiosis I, as indicated by the separation of the homologous pairs of chromosomes.*

b. *In the diagram, the cell contains six chromosomes, the diploid chromosome number for this species. Also in this cell, sister chromatids have separated, resulting in a doubling of the chromosome number within the cell from six to 12. Based on the number of chromosomes, the separation of sister chromatids in this cell must be occurring during anaphase of mitosis.*

c. *Again, sister chromatids are being separated, but the number of chromosomes present in the cell is only six. This indicates that no homologs are present within the cell, so in this cell the separation of sister chromatids is occurring in anaphase II of meiosis.*

25. The amount of DNA per cell of a particular species is measured in cells found at various stages of meiosis, and the following amounts are obtained. Match the amounts of DNA on the left with the corresponding stages of the cell cycle on the right. You may use more than one stage for each amount of DNA.

Stage of mitosis

a. G_1
b. Prophase I
c. G_2
d. Following telophase II and cytokinesis
e. Anaphase I
f. Metaphase II

Amount of DNA per cell	
d	3.7 pg
a, f	7.3 pg
b, c, e	14.6 pg

The amount of DNA in the cell will be doubled after the completion of S phase in the cell cycle and prior to cytokinesis in either mitosis or meiosis I. At the completion of cytokinesis following meiosis II, the amount of DNA will be halved.

a. *G_1 occurs prior to S phase and the doubling of the amount of DNA and prior to the completion of the meiosis II and cytokinesis, which will result in a haploid cell containing one-half the amount of DNA that was contained in the cell in G_1.*

b. *During prophase I of meiosis, the amount of DNA in the cell is two times the amount in G_1. The homologus chromosomes are still located within a single cell, and there are two sister chromatids per chromosome.*

c. *G_2 takes place directly after the completion of S phase, so the amount of DNA is two times the amount prior to the S phase.*

d. *Following cytokinesis associated with meiosis II, each daughter cell will contain only one-half the amount of DNA of a mother cell found in G_1 of interphase. By the completion of cytokinesis associated with meiosis II, both homologous pairs of chromosomes and sister chromatids have been separated into different daughter cells. Therefore, each daughter cell will contain only one-half the amount of DNA of the original cell in G_1.*

e. *During anaphase I of meiosis, the amount of DNA in the cell is two times the amount in G_1. The homologus chromosomes are still located within a single cell, and there are two sister chromatids per chromosome.*

f. *Metaphase II takes place after the cytokinesis associated with meiosis I and results in the daughter cells receiving only one-half the DNA found in their mother cell. In metaphase II of meiosis, the amount of DNA in each cell is the same as G_1 because each chromosome still consists of two DNA molecules (two sister chromatids per chromosome).*

*26. Fill in the following table.

Event	Mitosis	Meiosis I	Meiosis II
Does crossing over occur?	No	Yes	No
What separates in anaphase?	Sister chromatids	Homologous pairs of chromosomes	Sister chromatids
What lines up on the metaphase plate?	Individual chromosomes	Homologous pairs of chromosomes	Individual chromosomes
Does cell division usually take place?	Yes	Yes	Yes
Do homologs pair?	No	Yes	No
Is genetic variation produced?	No	Yes	No

*27. What would be the consequences of each of the following events to the outcome of mitosis or meiosis?

a. Mitotic cohesin fails to form early in mitosis.
Cohesin is necessary to hold the sister chromatids together until anaphase of mitosis. If cohesin fails to form early in mitosis, the sister chromatids could separate prior to anaphase. The result would be improper segregation of chromosomes to daughter cells.

b. Shugoshin is absent during meiosis.
Shugoshin protects cohesin proteins from degradation at the centromere during meiosis I. Cohesin at the arms of the homologous chromosomes is not protected by shugoshin and is broken in anaphase I, allowing for the two homologs to separate. If shugosin is absent during meiosis, then the cohesin at the centromere may be broken, allowing for the separation of sister chromatids along with the homologs during anaphase I, leading to improper segregation of chromosomes to daughter cells.

c. Shugoshin does not break down after anaphase I of meiosis.
If shugoshin is not broken down, then the cohesins at the centromere will remain protected from degradation. The intact cohesins will prevent the sister chromatids from separating during anaphase II of meiosis, resulting in an improper separation of sister chromatids and daughter cells with too many or too few chromosomes.

28. A cell in prophase II of meiosis contains 12 chromosomes. How many chromosomes would be present in a cell from the same organism if it were in prophase of mitosis? Prophase I of meiosis?

A cell in prophase II of meiosis will contain the haploid number of chromosomes. For this organism, 12 chromosomes represent the haploid chromosome number of a cell, or one complete set of chromosomes.

A cell from the same organism that is undergoing prophase of mitosis would contain a diploid number of chromosomes, or two complete sets of chromosomes, which means that homologous pairs of chromosomes are present. So a cell in this stage should contain 24 chromosomes.

Homologous pairs of chromosomes have not been separated by prophase I of meiosis. During this stage, a cell of this organism will contain 24 chromosomes.

29. A cell has eight chromosomes in G_1 of interphase. Draw a picture of this cell with its chromosomes at the following stages. Indicate how many DNA molecules are present at each stage.

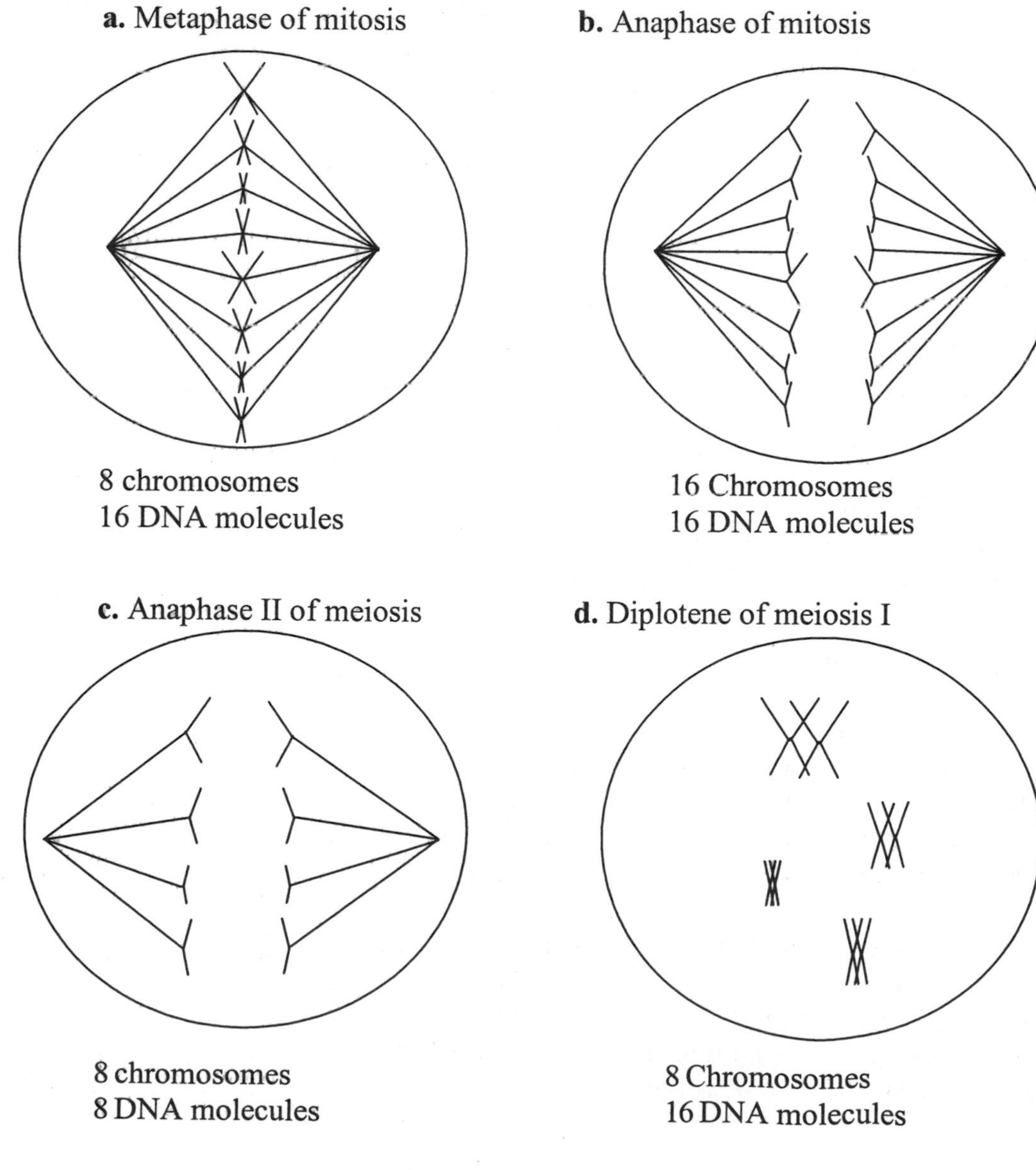

*30. The fruit fly *Drosophila melanogaster* has four pairs of chromosomes, whereas the house fly *Musca domestica* has six pairs of chromosomes. All things being equal, in which species would you expect to see more genetic variation among the progeny of a cross? Explain your answer.

The progeny of an organism whose cells contain more homologous pairs of chromosomes should be expected to exhibit more variation. The number of different combinations of chromosomes that are possible in the gametes is 2^n, *where* n *is equal to the number of homologous pairs of chromosomes. For the fruit fly with four pairs of chromosomes, the number of possible combinations is* $2^4 = 16$. *For* Musca domestica *with six pairs of chromosomes, the number of possible combinations is* $2^6 = 64$.

*31. A cell has two pairs of submetacentric chromosomes, which we will call chromosomes I_a, I_b, II_a, and II_b (chromosomes I_a and I_b are homologs, and chromosomes II_a and II_b are homologs). Allele *M* is located on the long arm of chromosome I_a, and allele *m* is located at the same position on chromosome I_b. Allele *P* is located on the short arm of chromosome I_a, and allele *p* is located at the same position on chromosome I_b. Allele *R* is located on chromosome II_a, and allele *r* is located at the same position on chromosome II_b.

a. Draw these chromosomes, labeling genes *M, m, P, p, R,* and *r,* as they might appear in metaphase I of meiosis. Assume that there is no crossing over.

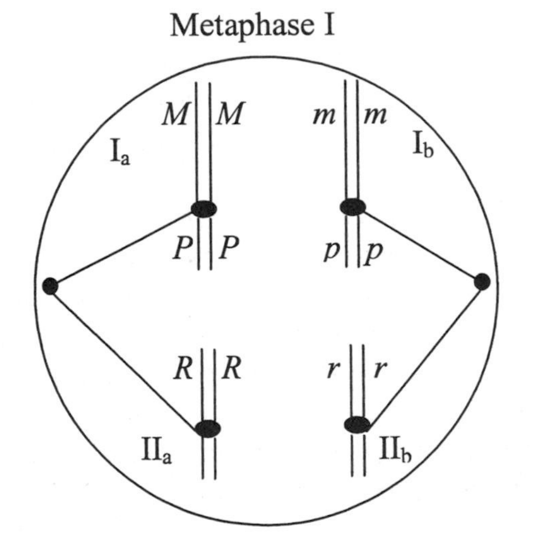

b. Considering the random separation of chromosomes in anaphase I, draw the chromosomes (with labeled genes) present in all possible types of gametes that might result from this cell going through meiosis. Assume that there is no crossing over.

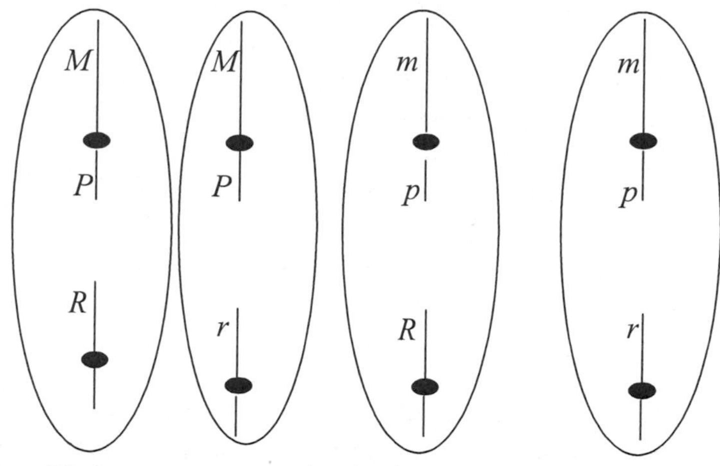

32. A horse has 64 chromosomes and a donkey has 62 chromosomes. A cross between a female horse and a male donkey produces a mule, which is usually sterile. How many chromosomes does a mule have? Can you think of any reasons for the fact that most mules are sterile?
The haploid egg produced by the female horse contains 32 chromosomes. The haploid sperm produced by the male donkey contains 31 chromosomes. The union of the horse and donkey gametes will produce a zygote containing 63 chromosomes. From the zygote, the adult mule will develop and will contain cells with a chromosome number of 63. Because an odd number of chromosomes in the mule's cells are present, at least one chromosome will not have a homolog. During the production of gametes by meiosis when pairing and separation of homologous chromosomes occurs, the odd chromosome will be unable to pair up. Furthermore, the mule's chromosomes, which are contributed by the horse and donkey, are from two different species. Not all of the mule's chromosomes may be able to find a suitable homolog during meiosis I and thus may not synapse properly during prophase I of meiosis. If improper synapsis or no synapsis occurs during prophase I, this will result in faulty segregation of chromosomes to the daughter cells produced at the conclusion of meiosis I. This leads to gametes that have abnormal numbers of chromosomes. When these abnormal gametes unite, the resulting zygote has an abnormal number of chromosomes and will be nonviable.

33. Normal somatic cells of horses have 64 chromosomes ($2n = 64$). How many chromosomes and DNA molecules will be present in the following types of horse cells?

Cell type	Number of chromosomes	Number of DNA molecules
a. Spermatogonium	*64*	*64*

Assuming the spermatogonium is in G_1 prior to the production of sister chromatids in S phase, the chromsome number will be the diploid number of chromosomes.

Cell type	Number of chromosomes	Number of DNA molecules
b. First polar body	*32*	*64*

The first polar body is the product of meiosis I, so it will be haploid;, but the sister chromatids have not separated, so each chromosome will consist of two sister chromatids.

Cell type	Number of chromosomes	Number of DNA molecules
c. Primary oocyte	*64*	*128*

The primary oocyte has stopped in prophase I of meiosis. So the homologs have not yet separated, and each chromosome consists of two sister chromatids.

Cell type	Number of chromosomes	Number of DNA molecules
d. Secondary spermatocyte	*32*	*64*

The secondary spermatocyte is a product of meiosis I and has yet to enter meiosis II. So the secondary spermatocyte will be haploid because the homologous pairs were separated in meiosis I; but each chromosome is still composed of two sister chromatids.

*34. A primary oocyte divides to give rise to a secondary oocyte and a first polar body. The secondary oocyte then divides to give rise to an ovum and a second polar body.

a. Is the genetic information found in the first polar body identical with that found in the secondary oocyte? Explain your answer.
No, the information is not identical with that found in the secondary oocyte. The first polar body and the secondary oocyte are the result of meiosis I. In meiosis I, homologous chromosomes segregate and thus both the first polar body and secondary oocyte will contain only one member of each original chromosome pair, and these will have different alleles of some of the genes. Also, the recombination that took place in prophase I will have generated new and different arrangements of genetic material for each member of the pair.

b. Is the genetic information found in the second polar body identical with that in the ovum? Explain your reasoning.
No, the information is not identical. The second polar body and the ovum will contain the same members of the homologous pairs of chromosomes that were separated during meiosis I and produced by the separation of sister chromatids during anaphase II. However, the sister chromatids are no longer identical. The sister chromatids have undergone recombination during prophase I and thus contain genetic information that is not identical to the other sister chromatids.

CHALLENGE QUESTIONS

Section 2.3

35. Eighty to ninety percent of the most common chromosome abnormalities in humans arise because the chromosomes fail to divide properly in *female* oogenesis. Can you think of a reason why failure of chromosome division might be more common in female gametogenesis than male gametogenesis?
Male gametogenesis, or spermatogenesis in human males, occurs regularly. Once the spermatogonium begins meiosis, the process quickly goes to completion, resulting in the formation of four spermatids that can mature into sperm cells. Female gametogenesis, or oogenesis in human females, is more complicated. Each oogonium enters meiosis I but stops at prophase I, generating a primary oocyte. This primary oocyte remains frozen in prophase I until ovulation begins and continues through meiosis I. Only if the egg is fertilized will meiosis II be completed. Because the primary oocyte is present at birth, the completion of meiosis I by a primary oocyte may not occur for many years (35 to 40 years or more). The length of time could lead to degradation or damaging of the meiotic machinery (such as the meiotic spindle fibers or cohesin complex). The damaged meiotic machinery could result in an improper separation of homologous pairs or of sister chromatids during the meiotic process. The spermatogenesis process does not have this time delay, which may protect the process from age-induced damage to the meiotic machinery.

36. On average, what proportion of the genome in the following pairs of humans would be exactly the same if no crossing over occurred? (For the purposes of this question only, we will ignore the special case of the X and Y sex chromosomes and assume that all genes are located on nonsex chromosomes.)

 a. Father and child
 The father will donate one-half of his chromosomes to his child. Therefore, the father and child will have one-half of their genomes that are similar.

 b. Mother and child
 The mother will donate one-half of her chromosomes to her child. Therefore, the mother and child will have one-half of their genomes that are similar.

 c. Two full siblings (offspring who have the same two biological parents)
 The parents can contribute only one-half of their genome to each offspring. So it is likely that the siblings share one-fourth of their genes from one parent. Because each sibling would share one-fourth of their genes from each parent, their total relatedness is one-half (or ¼ + ¼).

 d. Half siblings (offspring that have only one biological parent in common)
 Half siblings share only one-fourth of their genomes with each other because they have only one parent in common.

 e. Uncle and niece
 An uncle would share one-half of his genomes with his sibling, who would share one-half of his or her genome with his or her child. So, an uncle and niece would share one-fourth of their genomes (½ × ½).

f. Grandparent and grandchild
The grandparent and grandchild would share one-fourth of their genomes because the grandchild would share one-half of her genome with her parent and the parent would share one-half of her genome with the child's grandparent.

*37. Female bees are diploid and male bees are haploid. The haploid males produce sperm and can successfully mate with diploid females. Fertilized eggs develop into females and unfertilized eggs develop into males. How do you think the process of sperm production in male bees differs from sperm production in other animals?
Most male animals produce sperm by meiosis. In haploid male bees, meiosis will not occur, since meiosis can only occur in diploid cells. Male bees can still produce sperm, but only through mitosis. Haploid cells that divide mitotically produce more haploid cells.

Chapter Three: Basic Principles of Heredity

COMPREHENSION QUESTIONS

Section 3.1

*1. Why was Mendel's approach to the study of heredity so successful?
Mendel was successful for several reasons. He chose to work with a plant, Pisum sativum, *that was easy to cultivate, grew relatively rapidly, and produced many offspring whose phenotype was easy to determine, which allowed Mendel to detect mathematical ratios of progeny phenotypes. The seven characteristics he chose to study were also important because they exhibited only a few distinct phenotypes and did not show a range of variation. Finally, by looking at each trait separately and counting the numbers of the different phenotypes, Mendel adopted a reductionist experimental approach and applied the scientific method. From his observations, he proposed hypotheses that he was then able to test empirically.*

2. What is the difference between genotype and phenotype?
Genotype refers to the genes or the set of alleles found within an individual. Phenotype refers to the manifestation of a particular character or trait.

Section 3.2

*3. What is the principle of segregation? Why is it important?
The principle of segregation, or Mendel's first law, states that an organism possesses two alleles for any one particular trait and that these alleles separate during the formation of gametes. In other words, one allele goes into each gamete. The principle of segregation is important because it explains how the genotypic ratios in the haploid gametes are produced.

4. How are Mendel's principles different from the concept of blending inheritance discussed in Chapter 1?
Mendel's principles assert that the genetic factors or alleles are discrete units that remain separate in an individual organism with a trait encoded by the dominant allele being the only one observed if two different alleles are present. According to Mendel's principles, if an individual contains two different alleles, then the individual's gametes could contain either of these two alleles (but not both). Blending inheritance proposes that offspring are the result of blended genetic material from the parent and the genetic factors are not discrete units. Once blended, the combined genetic material could not be separated from each other in future generations.

5. What is the concept of dominance? How does dominance differ from incomplete dominance?
The concept of dominance states that when two different alleles are present in a genotype, only the dominant allele is expressed in the phenotype. Incomplete dominance occurs when different alleles are expressed in a heterozygous individual, and the resulting phenotype is intermediate to the phenotypes of the two homozygotes.

6. What are the addition and multiplication rules of probability and when should they be used?
The addition and multiplication rules are two rules of probability used by geneticists to predict the ratios of offspring in genetic crosses. The multiplication rule allows for predicting the probability of two or more independent events occurring together. According to the multiplication rule, the probability of two independent events occurring together is the product of their probabilities of occurring independently. The addition rule allows for predicting the likelihood of a single event that can happen in two or more ways. It states that the probability of a single mutually exclusive event can be determined by adding the probabilities of the two or more different ways in which this single event could take place. The multiplication rule allows us to predict how alleles from each parent can combine to produce offspring, while the addition rule is useful in predicting phenotypic ratios once the probability of each type of progeny can be determined.

7. Give the genotypic ratios that may appear among the progeny of simple crosses and the genotypes of the parents that may give rise to each ratio.

Genotypic ratio	*Parental genotype*
1:2:1	Aa × Aa
1:1	Aa × aa
Uniform progeny	AA × AA aa × aa AA × aa

*8. What is the chromosome theory of inheritance? Why was it important?
Walter Sutton developed the chromosome theory of inheritance. The theory states that genes are located on the chromosomes. The independent segregation of pairs of homologous chromosomes in meiosis provides the biological basis for Mendel's two rules of heredity.

Section 3.3

*9. What is the principle of independent assortment? How is it related to the principle of segregation?
According to the principle of independent assortment, genes for different characteristics that are at different loci segregate independently of one another. The principle of segregation indicates that the two alleles at a locus separate; the principle of independent assortment indicates that the separation of alleles at one locus is independent of the separation of alleles at other loci.

10. In which phases of mitosis and meiosis are the principles of segregation and independent assortment at work?
In anaphase I of meiosis, each pair of homologous chromosomes segregate independently of all other pairs of homologous chromosomes. The assortment is dependent on how the homlogs line up during metaphase I. This assortment of

homologs explains how genes located on different pairs of chromosomes will separate independently of one another. Anaphase II results in the separation of sister chromatids and subsequent production of gametes carrying single alleles for each gene locus as predicted by Mendel's principle of segregation.

Section 3.4

11. How is the goodness-of-fit chi-square test used to analyze genetic crosses? What does the probability associated with a chi-square value indicate about the results of a cross?
The goodness-of-fit chi-square test is a statistical method used to evaluate the role of chance in causing deviations between the observed and the expected numbers of offspring produced in a genetic cross. The probability value obtained from the chi-square table refers to the probability that random chance produced the deviations of the observed numbers from the expected numbers.

APPLICATION QUESTIONS AND PROBLEMS

Section 3.1

12. What characteristics of an organism would make it well suited for studies of the principles of inheritance? Can you name several organisms that have these characteristics?

Useful characteristics

- *Are easy to grow and maintain*
- *Grow rapidly, producing many generations in a short period*
- *Produce large numbers of offspring*
- *Have distinctive phenotypes that are easy to recognize*

Examples of organisms that meet these criteria

- *Neurospora, a fungus*
- *Saccharomyces cerevisiae, a yeast*
- *Arabidopsis, a plant*
- *Caenorhabditis elegans, a nematode*
- Drosophilia melanogaster, *a fruit fly*

Section 3.2

*13. In cucumbers, orange fruit color (*R*) is dominant over cream fruit color (*r*). A cucumber plant homozygous for orange fruits is crossed with a plant homozygous for cream fruits. The F_1 are intercrossed to produce the F_2.

a. Give the genotypes and phenotypes of the parents, the F_1, and the F_2.
The cross of a homozygous cucumber plant that produces orange fruit (RR) *with a homozygous cucumber plant that produces cream fruit* (rr) *will result in an F_1 generation heterozygous for the orange fruit phenotype.*

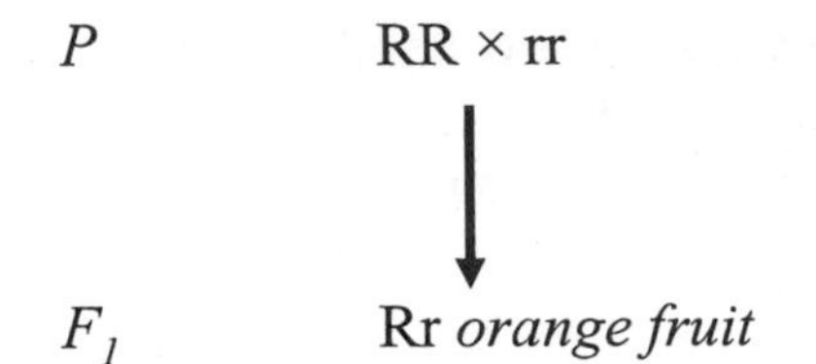

Intercrossing the F_1 will produce F_2 that are expected to show a 3:1 orange- to-cream-fruit phenotypic ratio.

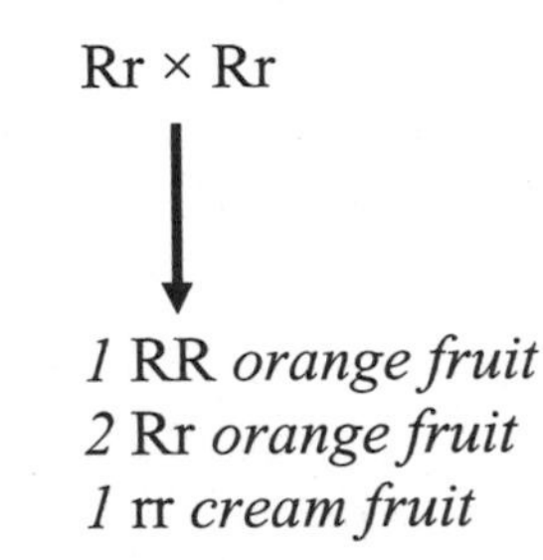

b. Give the genotypes and phenotypes of the offspring of a backcross between the F_1 and the orange parent.
The backcross of the F_1 orange offspring (Rr) *with homozygous orange parent* (RR) *will produce progeny that all have the orange fruit phenotype. However, one-half of the progeny will be expected to be homozygous for orange fruit and one-half of the progeny will be expected to be heterozygous for orange fruit.*

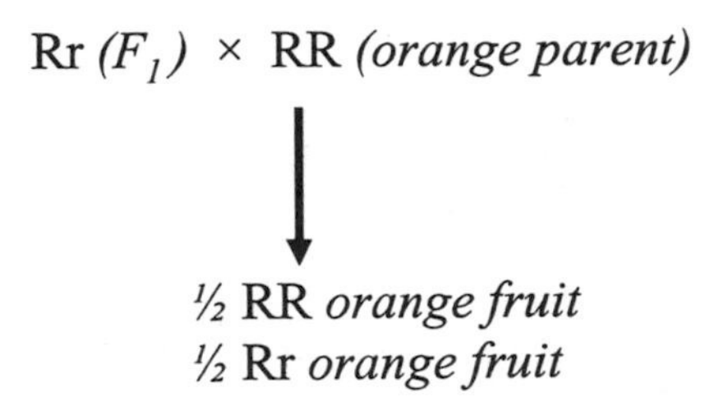

c. Give the genotypes and phenotypes of a backcross between the F_1 and the cream parent.
The backcross of the F_1 offspring (Rr) *with the cream parent* (rr) *is also a testcross. The product of this testcross should produce progeny, one-half of which are heterozygous for orange fruit and one-half of which are homozygous for cream fruit.*

Rr *(F_1)* × rr *(cream parent)*

↓

½ Rr *orange fruit*
½ rr *cream fruit*

14. J. W. McKay crossed a stock melon plant that produced tan seeds with a plant that produced red seeds and obtained the following results (J. W. McKay. 1936. *Journal of Heredity* 27:110–112).

Cross	F_1	F_2
tan ♀ × red ♂	13 tan seeds	93 tan, 24 red seeds

a. Explain the inheritance of tan and red seeds in this plant.
The F_1 generation contains all tan seed producing progeny and is the result of crossing a tan-seed-producing plant with a red-seed-producing plant. The F_1 result suggests that the tan phenotype is dominant to red. In the F_2 generation, the ratio of tan- to red-seed-producing plants is about 3.9 to 1, which is similar but not identical to a 3 to 1 ratio expected for monohybrid cross involving dominant and recessive alleles. The F_2 ratio suggests that the F_1 parents are heterozygous dominant for tan color.

b. Assign symbols for the alleles in this cross and give genotypes for all the individual plants.
We will define the tan allele as "R" and the recessive red allele as "r."
Tan-seed-producing ♀ parent: RR
Red-seed-producing ♂ parent: rr
F_1 tan-seed-producing offspring: Rr
F_2 tan-seed-producing offspring: RR *or* Rr
F_2 red-seed-producing offspring: rr

*15. White (*w*) coat color in guinea pigs is recessive to black (*W*). In 1909, W. E. Castle and J. C. Phillips transplanted an ovary from a black guinea pig into a white female whose ovaries had been removed. They then mated this white female with a white male. All the offspring from the mating were black in color (W. E. Castle and J. C. Phillips. 1909. *Science* 30:312–313).

a. Explain the results of this cross.
Although the white female gave birth to the offspring, her eggs were produced by the ovary from the black female guinea pig. The transplanted ovary produced only eggs containing the allele for black coat color. Like most mammals, guinea pig females produce primary oocytes early in development, and thus the transplanted ovary already contained primary oocytes produced by the black female guinea pig.

b. Give the genotype of the offspring of this cross.
The white male guinea pig contributed a "w" allele, while the white female guinea pig contributed the "W" allele from the transplanted ovary. The offspring are thus Ww.

c. What, if anything, does this experiment indicate about the validity of the pangenesis and the germ-plasm theories discussed in Chapter 1?
The transplant experiment supports the germ-plasm theory. According to the germ-plasm theory, only the genetic information in the germ-line tissue in the reproductive organs is passed to the offspring. The production of black guinea

pig offspring suggests that the allele for black coat color was passed to the offspring from the transplanted ovary in agreement with the germ-plasm theory. According to the pangenesis theory, the genetic information passed to the offspring originates at various parts of the body and travels to the reproductive organs for transfer to the gametes. If pangenesis were correct, then the guinea pig offspring should have been white. The white coat alleles would have traveled to the transplanted ovary and then to into the white female's gametes. The absence of any white offspring indicates that pangenesis did not occur.

*16. In cats, blood type A results from an allele (I^A) that is dominant over an allele (i^B) that produces blood type B. There is no O blood type. The blood types of male and female cats that were mated and the blood types of their kittens follow. Give the most likely genotypes for the parents of each litter.

	Male parent	Female parent	Kittens
a.	blood type A	blood type B	4 kittens with blood type A, 3 with blood type B
b.	blood type B	blood type B	6 kittens with blood type B
c.	blood type B	blood type A	8 kittens with blood type A
d.	blood type A	blood type A	7 kittens with blood type A, 2 with blood type B
e.	blood type A	blood type A	10 kittens with blood type A
f.	blood type A	blood type B	4 kittens with blood type A, 1 with blood type B

a. Male with blood type A × Female with blood type B
Because the female parent has blood type B, she must have the genotype i^Bi^B The male parent could be either I^AI^A or I^Ai^B. However, as some of the offspring are kittens with blood type B, the male parent must have contributed an i^B allele to these kittens. Therefore, the male must have the genotype of I^Ai^B.

b. Male with blood type B × Female with blood type B
Because blood type B is caused by the recessive allele i^B, both parents must be homozygous for the recessive allele or i^Bi^B. Each contributes only the i^B allele to the offspring.

c. Male with blood type B × Female with blood type A
Again, the male with type B blood must be i^Bi^B. A female with type A blood could have either the I^AI^A or I^Ai^B genotypes. Because all of her kittens have type A blood, this suggests that she is homozygous for the for I^A allele (I^AI^A) and contributes only the I^A allele to her offspring. It is possible that she is heterozygous for type A blood, but if so it is unlikely that chance alone would have produced eight kittens with blood type A.

d. Male with blood type A × Female with blood type A
Because kittens with blood type A and blood type B are found in the offspring, both parents must be heterozygous for blood type A, or I^Ai^B. With both parents being heterozygous, the offspring would be expected to occur in a 3:1 ratio of blood type A to blood type B, which is close to the observed ratio.

e. Male with blood type A × Female with blood type A
Only kittens with blood type A are produced, which suggests that each parent is homozygous for blood type A (I^AI^A), or that one parent is homozygous for blood type A (I^AI^A), and the other parent is heterozygous for blood type A (I^Ai^B). The

data from the offspring will not allow us to determine the precise genotype of either parent.

f. Male with blood type A × Female with blood type B
On the basis of her phenotype, the female will be $i^B i^B$*. In the offspring, one kitten with blood type B is produced. This kitten would require that both parents contribute an* i^B *to produce its genotype. Therefore, the male parent's genotype is* $I^A i^B$*. From this cross, the number of kittens with blood type B would be expected to be similar to the number of kittens with blood type A. However, due to the small number of offspring produced, random chance could have resulted in more kittens with blood type A than kittens with blood type B.*

17. Joe has a white cat named Sam. When Joe crosses Sam with a black cat, he obtains one-half white kittens and one-half black kittens. When the black kittens are interbred, they produce all black kittens. On the basis of these results, would you conclude that white or black coat color in cats is a recessive trait? Explain your reasoning.
The black coat color is likely recessive. When Sam was crossed with a black cat, one-half the offspring were white and one-half were black. This ratio potentially indicates that one of the parental cats is heterozygous dominant while the other parental cat is homozygous recessive—a testcross. The interbreeding of the black kittens produced only black kittens, indicating that the black kittens are likely to be homozygous, and thus the black coat color is the recessive trait.

If the black allele was dominant, we would have expected the black kittens to be heterozygous, containing a black coat color allele and a white coat color allele. Under this condition, we would expect one-fourth of the progeny from the interbred black kittens to have white coats. Because this did not happen, we can conclude that the black coat color is recessive.

18. In sheep, lustrous fleece (*L*) results from an allele that is dominant over an allele for normal fleece (*l*). A ewe (adult female) with lustrous fleece is mated with a ram (adult male) with normal fleece. The ewe then gives birth to a single lamb with normal fleece. From this single offspring, is it possible to determine the genotypes of the two parents? If so, what are their genotypes? If not, why not?
Yes, it is possible to determine the genotype of each parent, assuming that the dominant lustrous allele (L) *exhibits complete penetrance. The ram and the single lamb must be homozygous for the normal allele* (l) *because both have the normal fleece phenotype. Because the lamb receives only a single allele* (l) *from the ram, the ewe must have contributed the other recessive* l *allele. Therefore, the ewe must be heterozygous for lustrous fleece.*
In summary:

Lustrous fleece ewe × *Normal fleece ram*
(Ll) (ll)
↓
normal fleece lamb
(ll)

*19. In humans, alkaptonuria is a metabolic disorder in which affected persons produce black urine (see the introduction to this chapter). Alkaptonuria results from an allele (*a*) that is recessive to the allele for normal metabolism (*A*). Sally has normal metabolism, but her brother has alkaptonuria. Sally's father has alkaptonuria, and her mother has normal metabolism.

a. Give the genotypes of Sally, her mother, her father, and her brother.

Sally's father, who has alkaptonuria, must be aa. *Her brother, who also has alkaptonuria, must be* aa *as well. Because both parents must have contributed one* a *allele to her brother, Sally's mother, who is phenotypically normal, must be heterozygous* (Aa). *Sally, who is normal, received the* A *allele from her mother but must have received an* a *allele from her father.*

The genotypes of the individuals are: Sally (Aa), *Sally's mother* (Aa), *Sally's father* (aa), *and Sally's brother* (aa).

b. If Sally's parents have another child, what is the probability that this child will have alkaptonuria?

Sally's father (aa) × *Sally's mother* (Aa)

Sally's mother has a ½ chance of contributing the a *allele to her offspring. Sally's father can contribute only the* a *allele. The probability of an offspring with genotype* aa *and alkaptonuria is therefore ½ × 1 = ½.*

c. If Sally marries a man with alkaptonuria, what is the probability that their first child will have alkaptonuria?

Since Sally is heterozygous (Aa), *she has a ½ chance of contributing the* a *allele. Her husband with alkaptonuria* (aa) *can only contribute the* a *allele. The probability of their first child having alkaptonuria* (aa) *is ½ × 1 = ½.*

20. Suppose that you are raising Mongolian gerbils. You notice that some of your gerbils have white spots, whereas others have solid coats. What type of crosses could you carry out to determine whether white spots are due to a recessive or a dominant allele?

If white spots are recessive, then any gerbil with white spots must be homozygous for white spots (ww), *and a cross between two white-spotted gerbils* (ww × ww) *should produce offspring with only white spots. If white spots are dominant to solid, then a cross between a gerbil with white spots and a gerbil with a solid coat should produce either progeny all having solid coats* (WW × ww → Ww) *or progeny where one-half have solid coats and the other half have white spots* (Ww × ww → ½ Ww ½ ww).

*21. Hairlessness in American rat terriers is recessive to the presence of hair. Suppose that you have a rat terrier with hair. How can you determine whether this dog is homozygous or heterozygous for the hairy trait?

We will use h *for the hairless allele and* H *for the dominant. Because* H *is dominant to* h, *a rat terrier with hair could be either homozygous* (HH) *or heterozygous* (Hh). *To determine which genotype is present in the rat terrier with hair, cross this dog with a hairless rat terrier* (hh). *If the terrier with hair is homozygous* (HH), *then no hairless offspring will be produced. However, if the terrier is heterozygous* (Hh) *then we would expect one-half of the offspring to be hairless.*

22. In snapdragons, red flower color (*R*) is incompletely dominant over white flower color (*r*); the heterozygotes produce pink flowers. A red snapdragon is crossed with a white snapdragon, and the F_1 are intercrossed to produce the F_2.

 a. Give the genotypes and phenotypes of the F_1 and F_2, along with their expected proportions.

 Because the red flower color (R) *is not completely dominant to the white flower color* (r), *the* F_1 *heterozygotes* (Rr) *will all be pink.*

 Crossing the F_1*:* Rr × Rr

 ↓

 F_2*:* ¼ RR *red*
 ½ Rr *pink*
 ¼ rr *white*

 b. If the F_1 are backcrossed to the white parent, what are the genotypes and phenotypes of the offspring?

 Crossing the F_1 *to the white parent:* Rr × rr
 Offspring: ½ Rr *pink and* ½ rr *white*

 c. If the F_1 are backcrossed to the red parent, what are the genotypes and phenotypes of the offspring?

 Crossing the F_1 *to the red parent:* Rr × RR
 Offspring: ½ Rr *pink and* ½ RR *red*

23. What is the probability of rolling one six-sided die and obtaining the following numbers?

 a. 2

 Because 2 is only found on one side of a six-sided die, then there is a 1/6 chance of rolling a two.

 b. 1 or 2

 The probability of rolling a 1 on a six-sided die is 1/6. Similarly, the probability of rolling a 2 on a six-sided die is 1/6. Because the question asks what is the probability of rolling a 1 or *a 2, and these are mutually exclusive events, we should use the additive rule of probability to determine the probability of rolling a 1 or a 2:*

 (p *of rolling a 1)* + (p *of rolling a 2)* = p *of rolling either a 1 or a 2*
 1/6 + 1/6 = 2/6 = 1/3 probability of rolling either a 1 or a 2

 c. An even number

 The probability of rolling an even number depends on the number of even numbers found on the die. A single die contains three even numbers (2, 4, 6). The probability of rolling any one of these three numbers on a six-sided die is 1/6. To determine the probability of rolling either a 2, a 4, or a 6, we apply the additive rule: 1/6 + 1/6 + 1/6 = 3/6 = ½.

d. Any number but a 6
The number 6 is found only on one side of a six-sided die. The probability of rolling a 6 is therefore 1/6. The probability of rolling any number but 6 is (1 – 1/6) = 5/6.

*24. What is the probability of rolling two six-sided dice and obtaining the following numbers?

a. 2 and 3
To calculate the probability of rolling two six-sided dice and obtaining a 2 and a 3, we will need to use the product and additive rules. There are two possible ways in which to obtain the 2 and the 3 on the dice.
There is a 1/6 chance of rolling a 2 on the first die and a 1/6 chance of rolling a 3 on the second die. The probability of this taking place is therefore 1/6 × 1/6 = 1/36.
There is also a 1/6 chance of rolling a 3 on the first die and a 1/6 chance of rolling a 2 of the second die. Again, the probability of this taking place is 1/6 × 1/6 = 1/36. So the probability of rolling a 2 and a 3 would be 1/36 + 1/36 = 2/36 or 1/18.

b. 6 and 6
There is only one way to roll two 6's on a pair of dice: the first die must be a 6 and the second die must be a 6. The probability is 1/6 × 1/6 = 1/36.

c. At least one 6
To determine the probability of rolling at least one 6, we need to know the probability of rolling a 6 on a given die and the probability of not rolling a 6 on that same die. The probability of rolling a 6 is 1/6, while the probability of not rolling a 6 is 1 – 1/6 = 5/6.
(probability of a 6 on first die) + [(probability of no 6 on first die) × (probability of 6 on second die)] = the probability of at least one 6.
(1/6) + (5/6)(1/6) = 6/36 + 5/36 = 11/36 chance of rolling at least one 6

d. Two of the same number (two 1's, or two 2's, or two 3's, etc.)
There are several ways to roll two of the same number. You could roll two 1's, two 2's, two 3's, two 4's, two 5's, or two 6's. Using the multiplication rule, the probability of rolling two 1s is 1/6 × 1/6 = 1/36. The same is true of two 2's, two 3's, two 4's, two 5's, and two 6's. Using the addition rule, the probability of rolling either two 1's, two 2's, two 3's, two 4's, two 5's, and two 6's is 1/36 + 1/36 +1/36 +1/36 +1/36 + 1/36 = 6/36 = 1/6.

e. An even number on both dice
Three out of the six sides of a die are even numbers, so there is a 3/6 probability of rolling an even number on each of the dice. The chance of having an even number on both dice is (3/6)(3/6) = 9/36, or ¼.

f. An even number on at least one die
Three out of the six sides of a die are even numbers, so the probability of rolling an even number on the one die is 3/6. The probability of not rolling an even number is 3/6. An even number on at least one die could be obtained by rolling (a) an even on the first but not on the second die (3/6 × 3/6 = 9/36), (b) an even on the second die but not on the first (3/6 × 3/6 = 9/36), or (c) an even on both

dice ($3/6 \times 3/6 = 9/36$). Using the addition rule to obtain the probability of either a or b or c, we obtain $9/36 + 9/36 + 9/36 = 27/36 = 3/4$.

*25. In a family of seven children, what is the probability of obtaining the following numbers of boys and girls?

a. All boys

$(1/2)^7 = 1/128$

b. All children of the same sex

The children could be all boys or all girls:

$(1/2)^7$ chance of being all boys and $(1/2)^7$ chance of being all girls

$1/128 + 1/128 = 2/128$ or $1/64$ chance of being either all boys or all girls

Parts c–e require the use of the binomial expansion. Let a equal the probability of being a girl and b equal the probability of being a boy. The probabilities of a and b are ½.

$$(a + b)^7 = a^7 + 7a^6b + 21a^5b^2 + 35a^4b^3 + 35a^3b^4 + 21a^2b^5 + 7ab^6 + b^7$$

c. Six girls and one boy

The probability for part (c) is provided for by the term $7a^6b$. Because the probabilities of a and b are ½, then the overall probability is $7(1/2)^6(1/2) = 7/128$.

d. Four boys and three girls

This probability is provided for by the term $35a^3b^4$. The overall probability is $35(1/2)^3(1/2)^4 = 35/128$.

e. Four girls and three boys

Using the term $35a^4b^3$, we see that the overall probability is $35(1/2)^4(1/2)^3 = 35/128$.

26. Phenylketonuria (PKU) is a disease that results from a recessive gene. Two normal parents produce a child with PKU.

Because the two normal parents have a child with PKU, each parent must be heterozygous. We will define the recessive PKU allele as p and the dominant normal allele as P. Therefore, both parents have the genotype Pp.

a. What is the probability that a sperm from the father will contain the PKU allele?

The father has a ½ chance of donating a sperm with the PKU allele.

b. What is the probability that an egg from the mother will contain the PKU allele?

The mother's egg has a ½ chance of containing the PKU allele.

c. What is the probability that their next child will have PKU?

Each parent has a ½ chance of donating the p allele to the child. So, the child has a $1/2 \times 1/2 = 1/4$ chance of having PKU.

d. What is the probability that their next child will be heterozygous for the PKU gene?

Each parent has a ½ chance of donating the P allele or a ½ chance of donating the p allele to the child. Therefore, the child has a $(1/2 \times 1/2) + (1/2 \times 1/2) = 1/2$ chance of being heterozygous.

*27. In German cockroaches, curved wing (cv) is recessive to normal wing ($cv+$). A homozygous cockroach having normal wings is crossed with a homozygous cockroach having curved wings. The F_1 are intercrossed to produce the F_2. Assume that the pair of chromosomes containing the locus for wing shape is metacentric.

Draw this pair of chromosomes as it would appear in the parents, the F_1, and each class of F_2 progeny at metaphase I of meiosis. Assume that no crossing over takes place. At each stage, label a location for the alleles for wing shape (*cv* and *cv*+) on the chromosomes.

Parents:

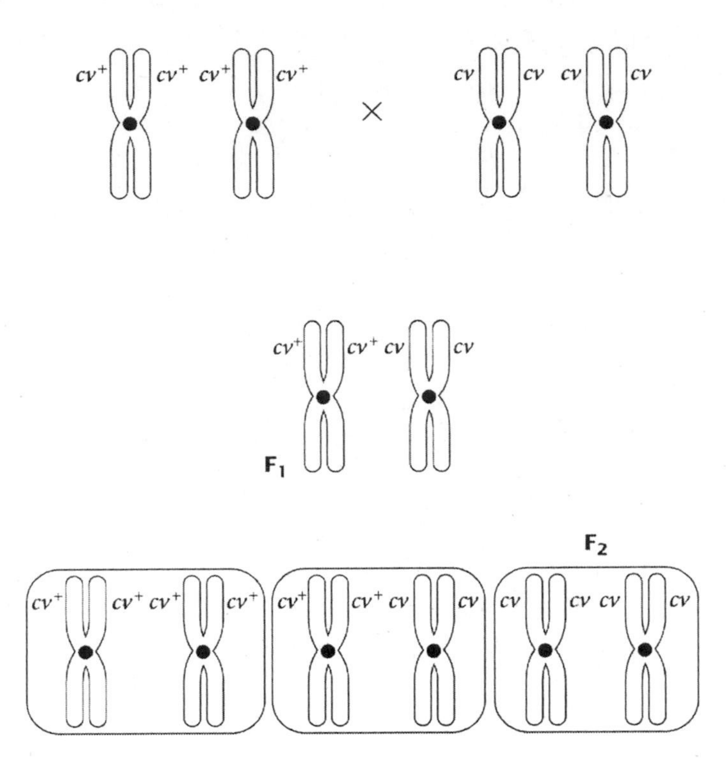

*28. In guinea pigs, the allele for black fur (*B*) is dominant over the allele for brown (*b*) fur. A black guinea pig is crossed with a brown guinea pig, producing five F_1 black guinea pigs and six F_1 brown guinea pigs.

a. How many copies of the black allele (*B*) will be present in *each* cell from an F_1 black guinea pig at the following stages: G_1, G_2, metaphase of mitosis, metaphase I of meiosis, metaphase II of meiosis, and after the second cytokinesis following meiosis? Assume that no crossing over takes place.

The cross of a black guinea pig with a brown guinea pig produced black and brown guinea pigs in the offspring. Because the brown guinea pig is homozygous (bb), *the black guinea pig must be heterozygous* (Bb).

Black guinea pigs (Bb) × *Brown guinea pigs* (bb) → *F_1 five black guinea pigs* (Bb) *and six brown guinea pigs* (bb).

To determine the number of copies of the B *allele or the* b *allele at the different stages of the cell cycle, we need to remember that following the completion of S phase and prior to anaphase of mitosis and anaphase II of meiosis, each chromosome will consist of two sister chromatids.*

In the F_1 black guinea pigs (Bb), *only one chromosome possesses the black allele, so we would expect in G_1 one black allele; G_2, two black alleles; metaphase of mitosis, two black alleles; metaphase I of meiosis, two black alleles; after* *cytokinesis of meiosis, one black allele but only in one-half of the cells produced by meiosis. (The remaining one-half will not contain the black allele.)*

b. How many copies of the brown allele (*b*) will be present in each cell from an F_1 brown guinea pig at the same stages? Assume that no crossing over takes place.
In the F_1 brown guinea pigs (bb), *both homologs possess the brown allele, so we would expect in G_1 two brown alleles; G_2, four brown alleles; metaphase of mitosis, four brown alleles; metaphase I of meiosis, four brown alleles; metaphase II, two brown alleles; and after cytokinesis of meiosis, one brown allele.*

Section 3.3

29. In watermelons, bitter fruit (*B*) is dominant over sweet fruit (*b*), and yellow spots (*S*) are dominant over no spots (*s*). The genes for these two characteristics assort independently. A homozygous plant that has bitter fruit and yellow spots is crossed with a homozygous plant that has sweet fruit and no spots. The F_1 are intercrossed to produce the F_2.

a. What will be the phenotypic ratios in the F_2?
P: homozygous bitter fruit, yellow spots (BB SS) × *homozygous sweet fruit and no spots* (bbss)

F_1: All progeny have bitter fruit and yellow spots (Bb Ss).
The F_1 are intercrossed to produce the F_2: Bb Ss × Bb Ss.
The F_2 phenotypic ratios are as follows:

9/16 bitter fruit and yellow spots
3/16 bitter fruit and no spots
3/16 sweet fruit and yellow spots
1/16 sweet fruit and no spots

b. If an F_1 plant is backcrossed with the bitter, yellow-spotted parent, what phenotypes and proportions are expected in the offspring?
The backcross of an F_1 plant (Bb Ss) *with the bitter, yellow-spotted parent* (BB SS) *will produce all bitter, yellow-spotted offspring.*

c. If an F_1 plant is backcrossed with the sweet, nonspotted parent, what phenotypes and proportions are expected in the offspring?
The backcross of a F_1 plant (Bb Ss) *with the sweet, nonspotted parent* (bb ss) *will produce the following phenotypic proportions in the offspring:*

¼ bitter fruit and yellow spots
¼ bitter fruit and no spots
¼ sweet fruit and yellow spots
¼ sweet fruit and no spots

30. In cats, curled ears (*Cu*) result from an allele that is dominant over an allele for normal ears (*cu*). Black color results from an independently assorting allele (*G*) that is dominant over an allele for gray (*g*). A gray cat homozygous for curled ears is mated with a homozygous black cat with normal ears. All the F_1 cats are black and have curled ears.

   a. If two of the F_1 cats mate, what phenotypes and proportions are expected in the F_2?

   If F_1 cats mated, GgCucu × GgCucu, *then the following proportions and phenotypes are expected in the F_2:*

   9/16 black with curly ears
   3/16 black with normal ears
   3/16 gray with curly ears
   1/16 gray with normal ears

   b. An F_1 cat mates with a stray cat that is gray and possesses normal ears. What phenotypes and proportions of progeny are expected from this cross?

   The mating of an F_1 cat (GgCucu) *with a gray cat with normal ears* (ggcucu) *is a testcross in which we would expect to produce equal numbers of all the different progeny classes:*

   ¼ black with curly ears
   ¼ black with normal ears
   ¼ gray with curly ears
   ¼ gray with normal ears

*31. The following two genotypes are crossed: *AaBbCcddEe* × *AabbCcDdEe*. What will the proportion of the following genotypes be among the progeny of this cross?

The simplest procedure for determining the proportion of a particular genotype in the offspring is to break the cross down into simple crosses and consider the proportion of the offspring for each cross.

AaBbCcddEe × AabbCcDdEe
Locus 1: Aa × Aa = ¼ AA, ½ Aa, ¼ aa
Locus 2: Bb × bb = ½ Bb, ½ bb
Locus 3: Cc × Cc = ¼ CC, ½ Cc, ¼ cc
Locus 4: dd × Dd = ½ Dd, ½ dd
Locus 5: Ee × Ee = ¼ EE, ½ Ee, ¼ ee

a. AaBbCcDdEe: ½ (Aa) × ½ (Bb) × ½ (Cc) × ½ (Dd) × ½ (Ee) = *1/32*

b. AabbCcddee: ½ *(Aa)* × ½ *(bb)* × ½ *(Cc)* × ½ *(dd)* × ¼ *(ee)* = *1/64*

c. aabbccddee: ¼ *(aa)* × ½ *(bb)* × ¼ *(cc)* × ½ *(dd)* × ¼ *(ee)* = *1/256*

d. AABBCCDDEE: *Will not occur. The* AaBbCcddEe *parent cannot contribute a* D *allele, and the* AabbCcDdEe *parent cannot contribute a* B *allele. Therefore, their offspring cannot be homozygous for the* BB *and* DD *gene loci.*

32. In mice, an allele for apricot eyes (a) is recessive to an allele for brown eyes (a^+). At an independently assorting locus, an allele for tan (t) coat color is recessive to an allele for black (t^+) coat color. A mouse that is homozygous for brown eyes and black coat color is crossed with a mouse having apricot eyes and a tan coat. The resulting F_1 are intercrossed to produce the F_2. In a litter of eight F_2 mice, what is the probability that two will have apricot eyes and tan coats?
The F_1 will have brown eyes and tan coats, and the genotype $a^+a\ t^+t$.
The F_2 will be produced by intercrossing the F_1: $a^+a\ t^+t \times a^+a\ t^+t$. *By considering each locus individually with a simple cross, we can easily calculate the proportion of any offspring class in the F_2.*

Locus 1: $a^+a \times a^+a = \frac{1}{4}\ a^+a^+,\ \frac{1}{2}\ a^+a,\ \frac{1}{4}\ aa$

Producing the phenotypic ratio:	*¾ brown eyes* (a^+a^+ *or* a^+a)
	¼ apricot eyes (aa)

Locus 2: $t^+t \times t^+t = \frac{1}{4}\ t^+t^+,\ \frac{1}{2}\ t^+t,\ \frac{1}{4}\ tt$

Producing the phenotypic ratio:	*¾ black coat* (t^+t^+, t^+t)
	¼ tan coat (tt)

To determine the probability that, out of a litter of eight mice, two will have apricot eyes and a tan coat, we first need to determine the likelihood of an apricot mouse with a tan coat being produced from the mating of the F_1:
aa tt: ¼ (aa) × ¼ (tt) = *1/16.*

The probability of two mice with this phenotype can then be determined using the binomial expansion defining "a" as the probability that 1/16 of the mice will have apricot eyes and tan coats, while defining "b" as the probability that 15/16 will have another phenotype.

$$(a + b)^8 = a^8 + 8a^7b + 28a^6b^2 + 56a^5b^3 + 70a^4b^4 + 56a^3b^5 + 28a^2b^6 + 8ab^7 + b^8$$

The probability of having two apricot mice with tan coats is provided by the term $28a^2b^6$ *in the binomial. The probability of "a" is 1/16, while the probability of "b" is 15/16. So the overall probability is* $28(1/16)^2(15/16)^6 = 0.074$.

33. In cucumbers, dull fruit (D) is dominant over glossy fruit (d), orange fruit (R) is dominant over cream fruit (r), and bitter cotyledons (B) are dominant over nonbitter cotyledons (b). The three characters are encoded by genes located on different pairs of chromosomes. A plant homozygous for dull, orange fruit and bitter cotyledons is crossed with a plant that has glossy, cream fruit and nonbitter cotyledons. The F_1 are intercrossed to produce the F_2.
All of the F_1 plants have dull, orange fruit and bitter cotyledons (DdRrBb). *By intercrossing the F_1, the F_2 are produced. The expected phenotypic ratios in the F_2 can be calculated more easily by examining the phenotypic ratios produced by the individual crosses of each gene locus.*

F_1 are intercrossed: DdRrBb × DdRrBb

Locus 1: Dd × Dd = *¾ dull (*DD *and* Dd*); ¼ glossy (*dd*)*
Locus 2: Rr × Rr = *¾ orange (*RR *and* Rr*); ¼ cream (*rr*)*
Locus 3: Bb × Bb = *¾ bitter (*BB *and* Bb*); ¼ nonbitter (*bb*)*

a. Give the phenotypes and their expected proportions in the F_2.
dull, orange, bitter: *¾ dull × ¾ orange × ¾ bitter = 27/64*
dull, orange, nonbitter: *¾ dull × ¾ orange × ¼ nonbitter = 9/64*
dull, cream, bitter: *¾ dull × ¼ cream × ¾ bitter = 9/64*
dull, cream, nonbitter: *¾ dull × ¼ cream × ¼ nonbitter = 3/64*
glossy, orange, bitter: *¼ glossy × ¾ orange × ¾ bitter = 9/64*
glossy, orange, nonbitter: *¼ glossy × ¾ orange × ¼ nonbitter = 3/64*
glossy, cream, bitter: *¼ glossy × ¼ cream × ¾ bitter = 3/64*
glossy, cream, nonbitter: *¼ glossy × ¼ cream × ¼ nonbitter = 1/64*

b. An F_1 plant is crossed with a plant that has glossy, cream fruit and nonbitter cotyledons. Give the phenotypes and expected proportions among the progeny of this cross.
Intercrossing the F_1 with a plant that has glossy, cream fruit and nonbitter cotyledons is an example of a testcross. All progeny classes will be expected in equal proportions because the phenotype of the offspring will be determined by the alleles contributed by the F_1 parent.
DdRrCc *(F_1)* × ddrrcc *(tester)*
*F_1 Locus 1 (*Dd*): ½* D *and ½* d
*F_1 Locus 2 (*Rr*): ½* R *and ½* r
*F_1 Locus 3 (*Cc*): ½* C *and ½* c
dull, orange, bitter: ½ dull × ½ orange × ½ bitter = 1/8
dull, orange, nonbitter: ½ dull × ½ orange × ½ nonbitter = 1/8
dull, cream, bitter: ½ dull × ½ cream × ½ bitter = 1/8
dull, cream, nonbitter: ½ dull × ½ cream × ½ nonbitter = 1/8
glossy, orange, bitter: ½ glossy × ½ orange × ½ bitter = 1/8
glossy, orange, nonbitter: ½ glossy × ½ orange × ½ nonbitter = 1/8
glossy, cream, bitter: ½ glossy × ½ cream × ½ bitter = 1/8
glossy, cream, nonbitter: ½ glossy × ½ cream × ½ nonbitter = 1/8

*34. *A* and *a* are alleles located on a pair of metacentric chromosomes. *B* and *b* are alleles located on a pair of acrocentric chromosomes. A cross is made between individuals having the following genotypes: *Aa Bb* × *aa bb*.

a. Draw the chromosomes as they would appear in each type of gamete produced by the individuals of this cross.

Gametes from *Aa Bb* individual:

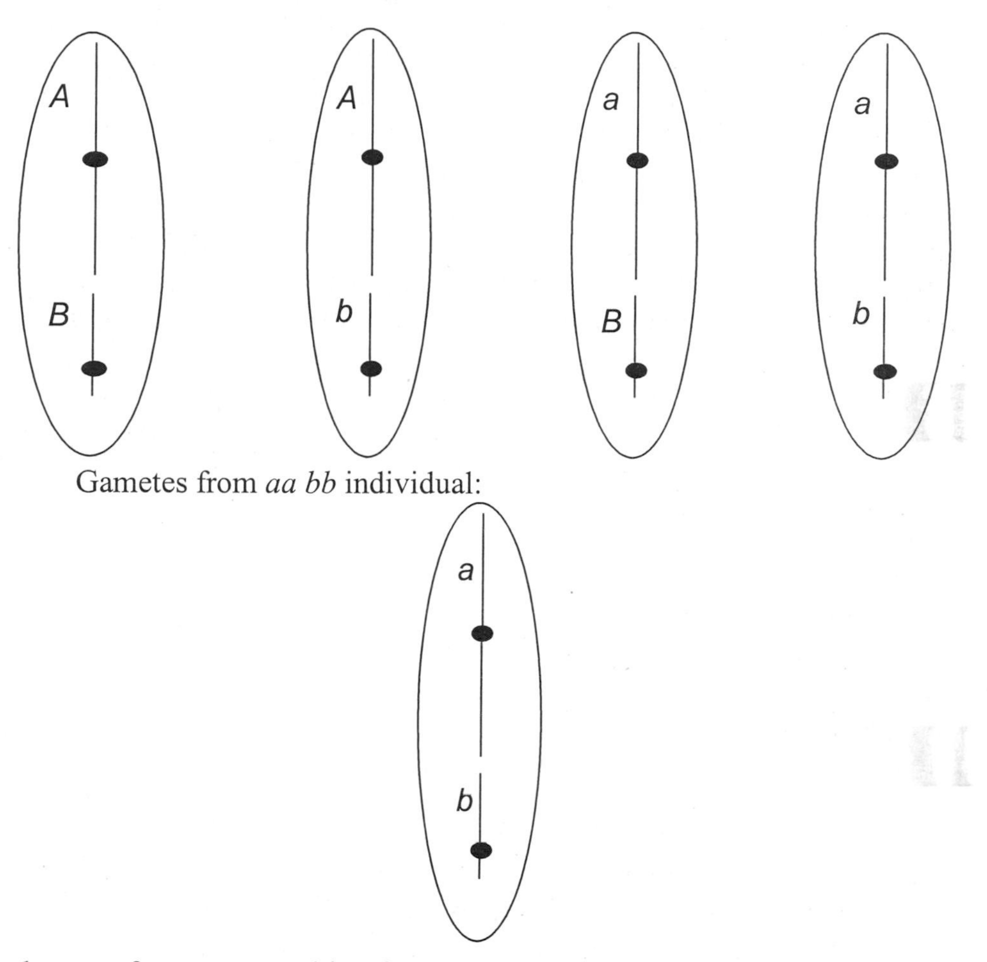

Gametes from *aa bb* individual:

b. For each type of progeny resulting from this cross, draw the chromosomes as they would appear in a cell at G_1, G_2, and metaphase of mitosis.

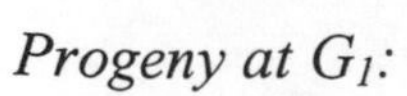
Progeny at G_1:

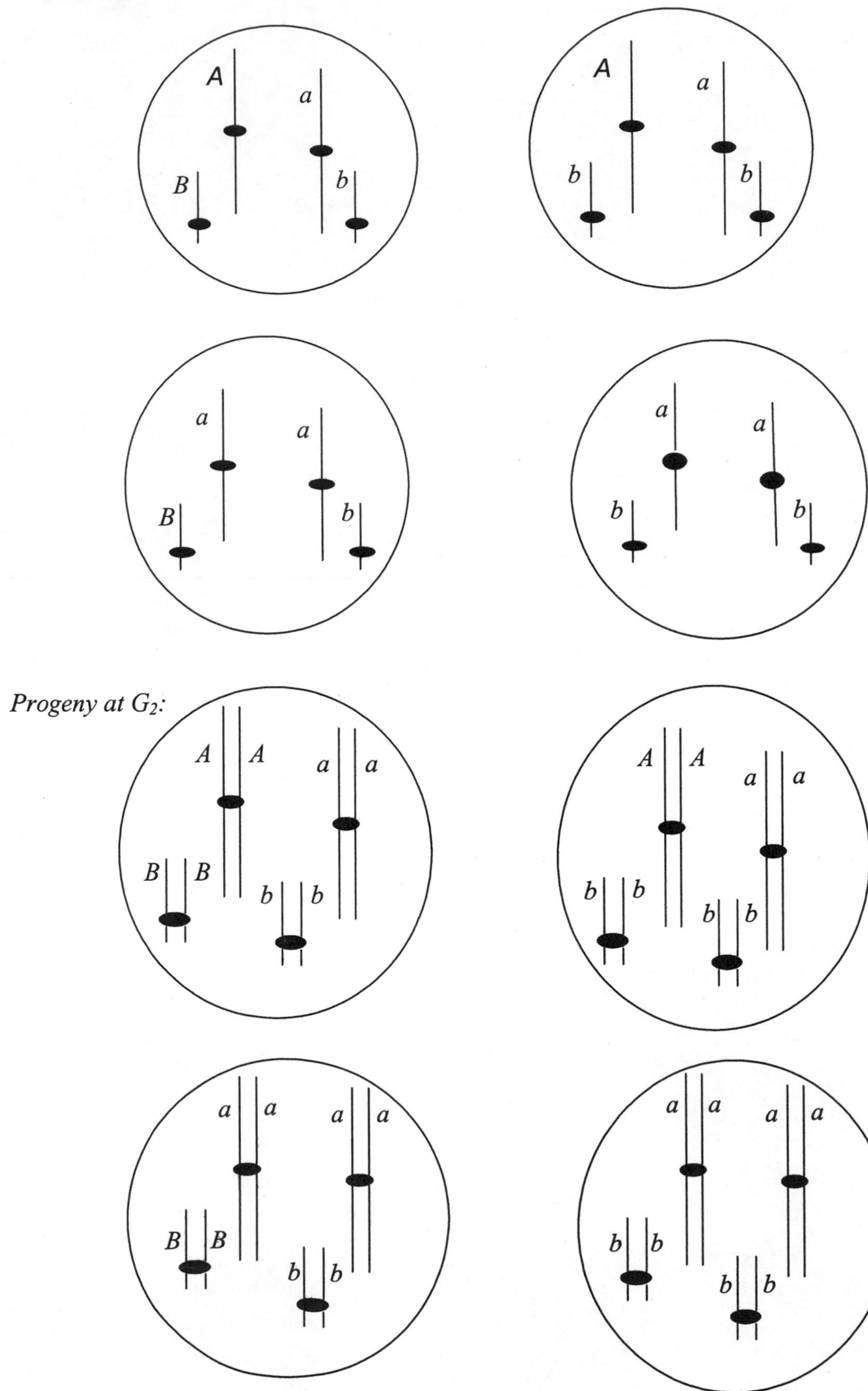
A
a
B
b
A
a
b
b
a
a
B
b
a
a
b
b
Progeny at G_2:
A A
a a
B B
b b
A A
a a
b b
b b
a a
a a
B B
b b
a a
a a
b b
b b

Progeny at metaphase of mitosis:

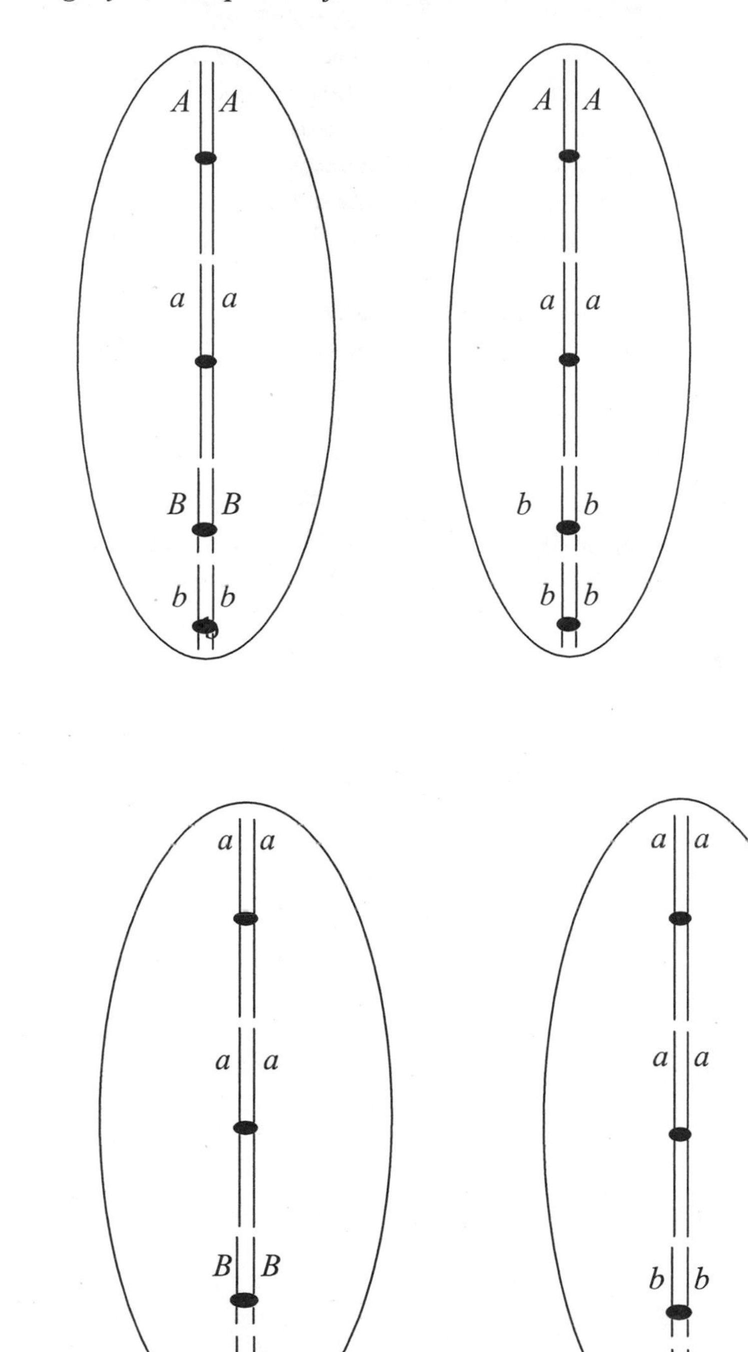

The order of chromosomes on metaphase plate can vary.

Section 3.4

*35. J. A. Moore investigated the inheritance of spotting patterns in leopard frogs (J. A. Moore. 1943. *Journal of Heredity* 34:3–7). The pipiens phenotype had the normal spots that give leopard frogs their name. In contrast, the burnsi phenotype lacked spots on its back. Moore carried out the following crosses, producing the progeny indicated.

Parent phenotypes	**Progeny phenotypes**
burnsi × burnsi	39 burnsi, 6 pipiens
burnsi × pipiens	23 burnsi, 33 pipiens
burnsi × pipiens	196 burnsi, 210 pipiens

a. On the basis of these results, what is the most likely mode of inheritance of the burnsi phenotype?
The cross of two burnsi individuals produced both burnsi and pipiens offspring. The result suggests that these individuals were heterozygous with each possessing a burnsi allele and a pipiens allele. The cross also suggests that the burnsi allele is dominant to the pipiens allele. The two crosses of burnsi and pipiens individuals suggest the crosses were between homozygous recessive individuals (pipiens) and heterozygous dominant individuals (burnsi).

b. Give the most likely genotypes of the parent in each cross.
We will define the burnsi allele as "P" *and the pipiens allele as* "p."
burnsi (Pp) × *burnsi* (Pp)
burnsi (Pp) × *pipiens* (pp)
burnsi (Pp) × *pipiens* (pp)

c. Use a chi-square test to evaluate the fit of the observed numbers of progeny to the number expected on the basis of your proposed genotypes.
In each of the crosses, we expect that the burnsi allele is dominant to the pipiens allele. By making that assumption, we can make predictions regarding the phenotypic ratios of the offspring and the genotypes of the parents.
For the cross of burnsi × burnsi (Pp × Pp), *we would expect a phenotypic ratio of 3:1 in the offspring.*

Phenotype	*Observed (O)*	*Expected (E)*	*$(O–E)^2/E$ or (X^2)*
burnsi	*39*	*33.75*	*.81*
pipiens	*6*	*11.25*	*2.45*
Total	*45*	*45*	*3.26*

Degrees of freedom (df) = (number of phenotypic classes) – 1.
*Because there are two phenotypic classes, the degrees of freedom (df) are 1. From the chi-square table, we can see that the calculated chi-square value falls between 2.706 (*P *of .1) and 3.841 (*P *of .05). The probability is sufficient that differences between what we expected and what we observed could have been generated by chance and that our parents are as predicted,* (Pp × Pp).

For the cross of burnsi × pipiens (Pp × pp), *we would expect a phenotypic ratio of 1:1 in the offspring.*

Phenotype	*Observed (O)*	*Expected (E)*	*$(O–E)^2/E$ or (X^2)*
burnsi	*23*	*28*	*.89*
pipiens	*33*	*28*	*.89*
Total	*56*	*56*	*1.78*

Because there are two phenotypic classes, the degrees of freedom (df) are 1. From the chi-square table, we can see that the calculated chi-square value falls between 0.455 (P *of .5) and 2.706 (* P *of .1). The probability is sufficient that differences between what we expected and what we observed could have been generated by chance and that our parents are as predicted,* (Pp × pp).
For the second cross of burnsi × pipiens (Pp × pp), *we would expect a phenotypic ratio of 1:1 in the offspring.*

Phenotype	*Observed (O)*	*Expected (E)*	*$(O–E)^2/E$ or (X^2)*
burnsi	*196*	*203*	*.24*
pipiens	*210*	*203*	*.24*
Total	*406*	*406*	*.48*

Again because there are two phenotypic classes, the degrees of freedom (df) are 1. The calculated chi-square value falls between 0.455 (P *of .5) and 2.706 (* P *of .1). The probability is sufficient that differences between what we expected and what we observed could have been generated by chance and that our parents are as predicted,* (Pp × pp).

36. In the 1800s, a man with dwarfism who lived in Utah produced a large number of descendants: 22 children, 49 grandchildren, and 250 great-grandchildren (see the illustration of a family pedigree), many of whom were also dwarfs (F. F. Stephens. 1943. *Journal of Heredity* 34:229–235). The type of dwarfism found in this family is called Schmid-type metaphyseal chondrodysplasia, although it was originally thought to be achondroplastic dwarfism. Among the families of this kindred, dwarfism appeared only in members who had one parent with dwarfism. When one parent was a dwarf, the following numbers of children were produced.

Family in which one parent had dwarfism	**Children with normal stature**	**Children with dwarfism**
A	15	7
B	4	6
C	1	6
D	6	2
E	2	2
F	8	4
G	4	4
H	2	1
I	0	1
J	3	1
K	2	3
L	2	1
M	2	0
N	1	0
O	0	2
Total	52	40

a. With the assumption that Schmid-type metaphyseal chondrodysplasia is rare, is this type of dwarfism inherited as a dominant or recessive trait? Explain your reasoning?
The cumulative data from the above crosses suggest that the crosses are possibly the result of a homozygous individual mating with a heterozygous individual (potentially 1:1 ratio among the offspring). Assuming that the allele for Schmid-type metaphyseal chondrodysplasia is rare, we can also assume that it is unlikely that the normal individuals in each cross are all heterozygous for the allele. So it is likely that this type of dwarfism is inherited as a dominant trait with the individuals with the dwarf phenotype being heterozygous for that allele.

b. Based on your answer for part **a.**, what is the expected ratio of dwarf and normal children in the families given in the table. Use a chi-square test to determine if the total number of children for these families (52 normal, 40 dwarfs) is significantly different from the number expected.
Assuming that the trait is dominant and that one parent is heterozygous for the dwarf allele and the other parent is homozygous for the normal allele, then we would expect a ratio of 1:1 for each cross.

Phenotype	*Observed (O)*	*Expected (E)*	*$(O–E)^2/E$ or (X^2)*
dwarf	*40*	*46*	*.78*
normal	*52*	*46*	*.78*
Total	*92*	*92*	*1.56*

Because there are two phenotypic classes, the degrees of freedom (df) are 1. From the chi-square table, we can see that the calculated chi-square value falls between .455 (P of .5) and 2.706 (P of .1). The probability is relatively high that differences between what we expected and what we observed was generated by chance and that our parents are as predicted.

c. Use chi-square tests to determine if the number of children in family C (1 normal, 6 dwarf) and the number in family D (6 normal, 2 dwarf) are significantly different from the numbers expected on the basis of your proposed type of inheritance. How would you explain these deviations from the overall ratio expected?

Phenotype	*Observed (O)*	*Expected (E)*	$(O-E)^2/E$ *or* (X^2)
dwarf	6	3.5	1.79
normal	1	3.5	1.79
Total	7	7	3.58

The degrees of freedom (df) are 1. From the chi-square table, we can see that the calculated chi-square value falls between 2.706 (P *of .1) and 3.841 (* P *of .05). Essentially, the phenotypic numbers of the offspring produced have a greater than 5% probability of occurring by chance. Although the percentage is low, it is above the 0.05 probability level, which is frequently used as the cutoff value for accepting that the variation is due to chance. So the probability is sufficient that differences between what we expected and what we observed could have been generated by chance.*

Phenotype	*Observed (O)*	*Expected (E)*	$(O-E)^2/E$ *or* (X^2)
dwarf	2	4	1.0
normal	6	4	1.0
Total	8	8	2.0

The degrees of freedom (df) are 1. The calculated chi-square value falls between 0.455 (P *of .5) and 2.706 (* P *of .1). Essentially, the results of the cross have between a 10% and 50% chance of occurring by chance. So, the probability is sufficient that differences between what we expected and what we observed could have been generated by chance.*

In both cases, the differences between the expected ratio and the observed ratio may be explained by the small number of offspring. With such a small sampling size, the variation may not be statistically relevant and due strictly to chance.

37. Pink-eye and albinism are two recessive traits found in the deer mouse *Peromyscus maniculatus*. In mice with pink-eye, the eye is devoid of color and appears pink from the blood vessels within it. Albino mice are completely lacking color both in their fur and in their eyes. F. H. Clark crossed pink-eyed mice with albino mice; the resulting F_1 had normal coloration in their fur and eyes. He then crossed these F_1 mice with mice that were pink-eyed and albino and obtained the following mice. It is very hard to distinguish between mice that are albino and mice that are both pink-eyed and albino, so he combined these two phenotypes together (F. H. Clark. 1936. *Journal of Heredity* 27:259–260).

Phenotype	Number of progeny
wild-type fur, wild-type eye color	12
wild-type fur, pink-eye	62
albino	
albino, pink-eye	78
Total	152

a. Give the expected numbers of progeny with each phenotype if the genes for pink-eye and albinism assort independently.
We will define wild-type fur as "A" and albino fur as "a." For wild-type eye color, we will use "P" and for red eye we will use "p." In the first cross, Clark crossed pink-eyed-mice with albino mice. It is likely that both the pink-eyed-mice and the albino mice were homozygous wild-type for the other allele:

Pink- eye (AA pp) × *Albino* (aa PP) → F_1 *Normal eyes and fur* (Aa Pp)

He then crossed the F_1 mice with mice that were pink-eyed and albino (aa pp) which yielded wild-type, wild-type fur with pink eyes, and both albino fur with wild-type eyes and albino fur with pink eyes, which were difficult to distinguish between. If the two pairs of alleles are assorting independently, we would expect each phenotypic category to occur one-fourth of the time.

Wild-type (Aa Pp) × *Albino, pink-eye* (aa pp) → *¼ wild-type*
¼ wild-type fur and pink eyes
¼ albino fur and wild-type eyes
¼ albino fur and pink eyes

Because we cannot distinguish between albinos with wild-type eyes and albinos with pink eyes, we will combine those two phenotypic categories, and thus expect the combined ratio of albinos to be ½.

b. Use a chi-square test to determine if the observed numbers of progeny fit the number expected with independent assortment.

Phenotype	*Observed (O)*	*Expected (E)*	$(O - E)^2 / E$ *or* (X^2)
Wild-type	*12*	*38*	*17.7*
Wild-type, pink eyes	*62*	*38*	*15.2*
Albino, wild-type and albino, pink eyes	*78*	*76*	*.05*
Total	*152*	*19.5*	*32.95*

We have 3 phenotypic classes that we can observe, so the degrees of freedom = 3 – 1 = 3. The chi-square value is greater than 10.597 for a probability value less than .005, or 0.5% that random chance produced the observed ratio. The observed number of progeny do not fit the numbers expected due to independent assortment. It is likely that a phenomenon other than independent assortment accounts for the observed ratio.

*38. In the California poppy, an allele for yellow flowers (*C*) is dominant over an allele for white flowers (*c*). At an independently assorting locus, an allele for entire petals (*F*) is dominant over an allele for fringed petals (*f*). A plant that is homozygous for yellow and entire petals is crossed with a plant that is white and fringed. A resulting F_1 plant is then crossed with a plant that is white and fringed, and the following progeny are produced: 54 yellow and entire; 58 yellow and fringed, 53 white and entire, and 10 white and fringed.

a. Use a chi-square test to compare the observed numbers with those expected for the cross.
Parents: yellow, entire petals (CC FF) × *white, fringed petals* (cc ff) → F_1 (Cc Ff)
For the cross of a heterozygous F_1 individual (Cc Ff) with a homozygous recessive individual (cc ff) we would expect a phenotypic ratio of 1:1:1:1 for the different phenotypic classes.

Phenotype	*Observed (O)*	*Expected (E)*	$(O-E)^2/E$ *or* (X^2)
Yellow, entire	*54*	*43.75*	*2.40*
Yellow, fringed	*58*	*43.75*	*4.64*
White, entire	*53*	*43.75*	*1.96*
White, fringed	*10*	*43.75*	*26.0*
Total	*175*	*175*	*35*

Degrees of freedom = 4 – 1 = 3. The chi-square value is greater than 12.838 for a probability value less than .005, or 0.5% that random chance produced the observed ratio of California poppies.

b. What conclusion can you make from the results of the chi-square test?
From the chi-square value, we can see that it is unlikely that random variations produced the observed ratio. Some other phenomena must be acting.

c. Suggest an explanation for the results.
The number of plants with the cc ff *genotype is much less than expected. Possibly, the* cc ff *genotype may be sublethal. In other words, California poppies with the homozygous recessive genotypes may be less viable than the other possible genotypes.*

CHALLENGE QUESTIONS

Section 3.2

39. Dwarfism is a recessive trait in Hereford cattle. A rancher in western Texas discovers that several of the calves in his herd are dwarfs, and he wants to eliminate this undesirable trait from the herd as rapidly as possible. Suppose that the rancher hires you as a genetic consultant to advise him on how to breed the dwarfism trait out of the herd. What crosses would you advise the rancher to conduct to ensure that the allele causing dwarfism is eliminated from the herd?

    To eliminate the recessive dwarfism allele from the herd, you will need to rid the herd of any heterozygous individuals—assuming that heterozygous individuals have a phenotype similar to the homozygous normal individuals. In essence, the farmer wants to create a homozygous normal cattle population. The first step is to advise the farmer to cull from the herd any bulls and cows that when mated produced a dwarf calf. Because the dwarf calf must be homozygous for the dwarf allele, each parent had to contribute the dwarf allele to the calf; thus, each parent is heterozygous. Next, a possible way to determine if any of the remaining cows are heterozygous is to perform a series of testcrosses using a dwarf bull. In the progeny produced by such a cross, ½ are expected to be normal and ½ are expected to be dwarfs if the cow was heterozygous. Any cows that produce dwarf calves should be eliminated from the herd. If the cow is homozygous, no dwarf calves will be produced. Unfortunately, due to the limited number of offspring (typically one) produced by each cow for each mating, several matings for each cow would be necessary to determine if she is heterozygous.

    The farmer may not think it practical to purchase a dwarf bull, or even that the dwarf bull would be able to mate with the cows. A second method would be to cross the cows with a bull that is known to be heterozygous normal. For a cross between this bull and a heterozygous cow, we would expect ¾ normal offspring and ¼ dwarfs. A homozygous cow mated with a heterozygous bull would not produce any dwarf offspring. Again, several matings would be necessary to determine if the cow is heterozygous. The farmer will not want to keep the progeny of these crosses because, if the cow is heterozygous, the chance that her normal offspring are carriers of the dwarf allele is 50%.

    In either scenario, the process for building a pure-breeding herd will take several years and careful monitoring of the offspring.

*40. A geneticist discovers an obese mouse in his laboratory colony. He breeds this obese mouse with a normal mouse. All the F_1 mice from this cross are normal in size. When he interbreeds two F_1 mice, eight of the F_2 mice are normal in size and two are obese. The geneticist then intercrosses two of his obese mice, and he finds that all of the progeny from this cross are obese. These results lead the geneticist to conclude that obesity in mice results from a recessive allele.

    A second geneticist at a different university also discovers an obese mouse in her laboratory colony. She carries out the same crosses as the first geneticist and obtains

the same results. She also concludes that obesity in mice results from a recessive allele. One day, the two geneticists meet at a genetics conference, learn of each other's experiments, and decide to exchange mice. They both find that, when they cross two obese mice from the different laboratories, all the offspring are normal. However, when they cross two obese mice from the same laboratory, all the offspring are obese. Explain their results.
The first geneticist has identified an obese allele that he believes to be recessive. We will define his allele as o_1 *and the normal allele as* O_1. *The obese allele appears to be recessive based on the series of crosses he performed.*

Cross 1 with possible genotype:
Obese (o_1o_1) × *Normal* (O_1O_1) → F_1 *All normal* (O_1o_1)

Cross 2 with possible genotypes:
F_1 *normal* (Oo_1) × F_1 *normal* (O_1o_1) → F_2 *8 normal* (O_1O_1 and O_1o_1)
2 obese (o_1o_1)

Cross 3 with possible genotypes:
Obese (o_1o_1) × *Obese* (o_1o_1) → *All Obese* (o_1o_1)

A second geneticist also finds an obese mouse in her colony and performs the same types of crosses, which indicate to her that the obese allele is recessive. We will define her obese allele as o_2 *and the normal allele as* O_2.

The cross of obese mice between the two different laboratories produced only normal mice. These different alleles are both recessive. However, they are located at different gene loci. Essentially, the obese mice from the different labs have separate obesity genes that are independent of one another.
The likely genotypes of the obese mice are as follows:

Obese mouse 1 ($o_1o_1\ O_2O_2$) × *Obese mouse 2* ($O_1O_1\ o_2o_2$)
→ F_1 *All normal* ($O_1o_1\ O_2o_2$)

41. Albinism is a recessive trait in humans. A geneticist studies a series of families in which both parents are normal and at least one child has albinism. The geneticist reasons that both parents in these families must be heterozygotes and that albinism should appear in one-fourth of the children of these families. To his surprise, the geneticist finds that the frequency of albinism among the children of these families is considerably greater than ¼. Can you think of an explanation for the higher-than-expected frequency of albinism among these families?
The geneticist has indeed identified parents who are heterozygous for the albinism allele. However, by looking only at parents who have albino children, he is missing parents who are heterozygous and have no albino children. Since most parents are likely to have only a few children, the result is that the frequency of albino children produced by parents with an albino child will be higher than what would be predicted. If he were to consider the offspring of normal heterozygous parents with no albino children, along with the parents who have albino children, the expected frequency of albino offspring would most likely approach ¼.

42. Two distinct phenotypes are found in the salamander *Plethodon cinereus*: a red form and a black form. Some biologists have speculated that the red phenotype is due to an autosomal allele that is dominant over an allele for black. Unfortunately, these salamanders will not mate in captivity; so the hypothesis that red is dominant over black has never been tested.

One day a genetics student is hiking through the forest and finds 30 female salamanders, some red and some black, laying eggs. The student places each female and her eggs (about 20 – 30 eggs per female) in separate plastic bags and takes them back to the lab. There, the student successfully raises the eggs until they hatch. After the eggs have hatched, the student records the phenotypes of the juvenile salamanders, along with the phenotypes of their mothers. Thus, the student has the phenotypes for 30 females and their progeny, but no information is available about the phenotypes of the fathers.

Explain how the student can determine whether red is dominant over black with this information on the phenotypes of the females and their offspring.
To determine whether red is dominant over black the student will need to examine the colors and phenotypic ratios of the colors found in the offspring of each salamander. Certain trial assumptions will have to be made.

If black is recessive, the following assumptions about the phenotypic ratios in the offspring of the black females can be made:

Black female × black male → all black offspring
Black female × red male (heterozygous) → ½ black offspring; ½ red offspring
Black female × red male (homozygous) → all red offspring

If red is dominant, then the following assumptions about the phenotypic ratios in the offspring of the red females can be made:

Red female (homozygous) × black male → all red offspring
Red female (heterozygous) × black male → ½ red offspring; ½ black offspring
Red female (homozygous or heterozygous) × red male (homozygous) → all red
Red female (homozygous) × red male (homozygous or heterozygous) → all red
Red female (heterozygous) × red male (heterozygous) → ¾ red offspring; ¼ black offspring

The key ratio will be any female salamander that produces offspring with a 3:1 phenotypic ratio. If in the offspring of a red salamander, a ratio of 3:1 red is produced, then the red allele is dominant over the black. If, however, in the offspring of a black salamander, a 3:1 phenotypic ratio of black to red is observed, then black is the dominant allele.

Chapter Four: Sex Determination and Sex-Linked Characteristics

COMPREHENSION QUESTIONS

Section 4.1

*1. What is the most defining difference between males and females?
Males produce relatively large gametes; females produce larger gametes.

2. How do monoecious organisms differ from dioecious organisms?
Monoecious organisms exist as only one form, which has male and female reproductive structures; the same organism produces male and female gametes. Dioecious organisms exist as organisms of two distinct genders, one producing male gametes, the other producing female gametes.

3. Describe the XX-XO system of sex determination. In this system, which is the heterogametic sex and which is the homogametic sex?
In the XX-XO sex determination system, females have two copies of the sex-determining chromosome, whereas males have only one copy. Males must be considered heterogametic because they produce two different types of gametes with respect to the sex chromosome: either containing an X or not containing an X.

4. How does sex determination in the XX-XY system differ from sex determination in the ZZ-ZW system?
In the XX-XY system, males are heterogametic and produce gametes with either an X chromosome or a Y chromosome. In the ZZ-ZW system, females are heterogametic and produce gametes with either a Z or a W chromosome.

*5. What is the pseudoautosomal region? How does the inheritance of genes in this region differ from the inheritance of other Y-linked characteristics?
The pseudoautosomal region is a region of similarity between the X and Y chromosomes that is responsible for pairing the X and Y chromosomes during meiotic prophase I. Genes in this region are present in two copies in males and females and thus are inherited like autosomal genes, whereas other Y-linked genes are passed on only from father to son.

*6. How is sex determined in insects with haplodiploid sex determination?
Diploid individuals are female, whereas haploid individuals are male. Eggs that are fertilized by a sperm develop into females, and eggs that are not fertilized develop as males.

7. What is meant by genic sex determination?
In organisms that follow this system, there is no cytogenetically recognizable difference in the chromosome contents of males and females. Instead of a sex chromosome that differs between males and females, alleles at one or more loci determine the sex of the individual.

8. How does sex determination in *Drosophila* differ from sex determination in humans?
In humans, the presence of a functional Y chromosome determines maleness. People with XXY and XXXY are phenotypically male. In Drosophila, *the ratio of X chromosome material to autosomes determines the sex of the individual, regardless of the Y chromosome. Flies with XXY are female, and flies with XO are sterile males.*

9. Give the typical sex chromosomes found in the cells of people with Turner syndrome, Klinefelter syndrome, and androgen insensitivity syndrome, as well as in poly-X females.
Turner syndrome: XO
Klinefelter syndrome: XXY (rarely XXXY or XXYY)
Androgen insensitivity: XY
Poly-X females: XXX (rarely XXXX or even XXXXX)

Section 4.2

*10. What characteristics are exhibited by an X-linked trait?
Males show the phenotypes of all X-linked traits, regardless of whether the X-linked allele is normally recessive or dominant. Males inherit X-linked traits from their mothers, pass X-linked traits to their daughters, and through their daughters to their daughters' descendents, but not to their sons or their sons' descendents.

11. Explain how Bridges's study of nondisjunction in *Drosophila* helped prove the chromosome theory of inheritance.
Bridges showed that in crosses with white-eyed flies, a sex-linked trait, exceptional progeny had abnormal inheritance of sex chromosomes. In matings of white-eyed females with red-eyed males, most of the progeny followed the expected pattern of white-eyed males and red-eyed females. Exceptional red-eyed male progeny were XO and exceptional white-eyed females were XXY. These karyotypes were exactly as Bridges predicted with his hypothesis that the exceptional red-eyed males inherited their X chromosome with the red-eye allele from their red-eyed fathers and were male because they did not inherit an X chromosome from their mothers, resulting in an XO condition that is phenotypically male but sterile. Moreover, the exceptional white-eyed females inherited two X chromosomes from their white-eyed mothers, as a result of nondisjunction in meiosis I of the female, and none from their red-eyed fathers, receiving instead a Y chromosome to make them XXY females. Calvin Bridges linked exceptional inheritance of a sex-linked trait to exceptional inheritance of the X chromosome: the linked exceptions proved the rule that genes reside on chromosomes.

12. What are some of its characteristics that make *Drosophila melanogaster* a good model genetic organism?
These fruit flies have a relatively short generation time of about 10 days, and produce large numbers of progeny, with females producing 400 to 500 eggs in a 10-day period. They are easily and inexpensively cultured in the laboratory, and being small, require little laboratory space. Nevertheless, they have a complex life cycle and morphology and are large enough that males and females are easily distinguished, and many morphological mutations may be observed with just a hand lens or dissecting scope.

Drosophila melanogaster *also has very large polytene chromosomes in the salivary glands, which have facilitated cytological studies of chromosomes. At a molecular level, the relatively small genome of* Drosophila, *amounting to 180 million base pairs of DNA or only about 5% of the size of the human genome, has been completely sequenced, and techniques have been developed for facile genetic engineering of the* Drosophila *genome.*

13. Explain why tortoiseshell cats are almost always female and why they have a patchy distribution of orange and black fur.
Tortoiseshell cats have two different alleles of an X-linked gene: X^+ (non-orange, or black) and X^o (orange). The patchy distribution results from X-inactivation during early embryo development. Each cell of the early embryo randomly inactivates one of the two X chromosomes, and the inactivation is maintained in all of the daughter cells. So each patch of black fur arises from a single embryonic cell that inactivated the X^o, and each patch of orange fur arises from an embryonic cell that inactivated the X^+.

14. What are Barr bodies? How are they related to the Lyon hypothesis?
Barr bodies are darkly staining bodies in the nuclei of female mammalian cells. Mary Lyon correctly hypothesized that Barr bodies are inactivated (condensed) X chromosomes. By inactivating all X chromosomes beyond one, female cells achieve dosage compensation for X-linked genes.

*15. What characteristics are exhibited by a Y-linked trait?
Y-linked traits appear only in males and are always transmitted from fathers to sons, thus following a strict paternal lineage. Autosomal male-limited traits also appear only in males, but they can be transmitted to boys through their mothers.

APPLICATION QUESTIONS AND PROBLEMS

Section 4.1

*16. What is the sexual phenotype of fruit flies with the following chromosomes?

	Sex chromosomes	Autosomal chromosomes	Sexual phenotype
a.	XX	All normal	*Female (X:A = 1.0)*
b.	XY	All normal	*Male (X:A = 0.5)*
c.	XO	All normal	*Male, sterile (X:A = 0.5)*
d.	XXY	All normal	*Female (X:A = 1.0)*
e.	XYY	All normal	*Male (X:A = 0.5)*
f.	XXYY	All normal	*Female (X:A = 1.0)*
g.	XXX	All normal	*Metafemale (X:A > 1.0)*
h.	XX	Four haploid sets	*Male (X:A = 0.5)*
i.	XXX	Four haploid sets	*Intersex (X:A between 0.5 and 1.0)*
j.	XXX	Three haploid sets	*Female (X:A = 1.0)*
k.	X	Three haploid sets	*Metamale, sterile (X:A < 0.5)*
l.	XY	Three haploid sets	*Metamale (X:A < 0.5)*
m.	XX	Three haploid sets	*Intersex (X:A between 0.5 and 1.0)*

In fruit flies, the X to autosome ratio determines sex: males have X:A ratios of 0.5 (metamales have X:A ratios less than 0.5), and females have X:A ratios of 1.0 (metafemales have ratios greater than 1.0).

17. What will be the phenotypic sex of a human with the following gene or chromosomes or both?
    - **a.** XY with the *SRY* gene deleted—*female*
    - **b.** XY with the *SRY* gene located on an autosomal chromosome—*male*
    - **c.** XX with a copy of *SRY* gene on an autosomal chromosome—*male*
    - **d.** XO with a copy of *SRY* gene on an autosome—*male*
    - **e.** XXY with the *SRY* gene deleted—*female*
    - **f.** XXYY with one copy of the *SRY* gene deleted—*male*

    In humans, a single functional copy of the SRY gene, normally located on the Y chromosome, determines phenotypic maleness by causing gonads to differentiate into testes. In the absence of a functional SRY gene, gonads differentiate into ovaries and the individual is phenotypically female.

18. A normal female *Drosophila* produces abnormal eggs that contain all (a complete diploid set) of her chromosomes. She mates with a normal male *Drosophila* that produces normal sperm. What will be the sex of the progeny from this cross?
    In this cross, all the eggs will have two X chromosomes and two sets of autosomes. These eggs will be fertilized by two kinds of sperm produced in equal proportions: half the sperm will have one X chromosome and one set of autosomes, and the other half will have one Y chromosome and one set of autosomes. Thus, half the progeny will have 3 X chromosomes and 3 sets of autosomes and will be female because the X:autosome ratio will be 1. The other half will have 2 X chromosomes and 3 sets of autosomes, for an X:autosome ratio of 2/3, and will be intersex (see answer to problem 16m, above).

19. In certain salamanders, the sex of a genetic female can be altered, making her into a functional male; these salamanders are called sex-reversed males. When a sex-reversed male is mated with a normal female, approximately ⅔ of the offspring are female and ⅓ are male. How is sex determined in these salamanders? Explain the results of this cross.
    The 2:1 ratio of females to males could be explained if a quarter of the progeny are embryonic lethals. The sex-reversed male has the same chromosome complement as normal females. If females are homogametic (XX), then matings between sex-reversed males (XX) and normal females (XX) must result in all (XX) female progeny, with no obvious reason for embryonic lethality. However, if females are heterogametic (ZW), then ZW (sex-reversed male) crossed to ZW (normal female) results in ¼ ZZ (male), ½ ZW (female), and ¼ WW (embryonic lethal). The net result is a 2:1 ratio of females to males.

20. In some mites, males pass genes to their grandsons, but they never pass genes to male offspring. Explain.
    A system in which males are haploid and females are diploid would explain these results. Haploid males pass genes only to female progeny. The female progeny can then generate haploid male grandsons that contain the grandfather's genes.

*21. In organisms with the ZZ-ZW sex-determining system, from which of the following possibilities can a female inherit her Z chromosome?

Her mother's mother—*no*
Her mother's father—*no*
Her father's mother—*yes*
Her father's father—*yes*

Females inherit the W chromosome from the mother and the Z chromosome from the father. If we exclude the possibility that the mother and father are siblings of the same paternal line, then no Z chromosome can be inherited from the mother's parents. Males inherit one copy of the Z chromosome from each parent, so females have equal probability of inheriting the Z chromosome from the father's mother or the father's father.

Section 4.2

22. When Bridges crossed white-eyed females with red-eyed males, he obtained a few red-eyed males and white-eyed females (see Figure 4.13). What types of offspring would be produced if these red-eyed males and white-eyed females were crossed with each other?
Bridges exceptional white-eyed females were X^wX^wY and the red-eyed males were X^+Y. The results of crossing such white-eyed females with red-eyed males are shown in Figure 4.13. Meiosis in this female generates 45% X^wY, 45% X^w, 5% X^wX^w, and 5% Y gametes. Meiosis in the red-eyed male generates 50% X^+ and 50% Y gametes.

	.5 X^+	.5 Y
.45 X^wY	.225 X^wX^+Y red-eyed females	.225 X^wYY white-eyed males
.45 X^w	.225 X^wX^+ red-eyed females	.225 X^wY white-eyed males
.05 X^wX^w	.025 $X^wX^wX^+$ metafemale, dies	.025 X^wX^wY white-eyed females
.05 Y	.025 X^+Y red-eyed males	.025 YY dies

*23. Joe has classic hemophilia, an X-linked recessive disease. Could Joe have inherited the gene for this disease from the following persons?

	Yes	No
a. His mother's mother	*X*	
b. His mother's father	*X*	
c. His father's mother		*X*
d. His father's father		*X*

X-linked traits are passed on from mother to son. Therefore, Joe must have inherited the hemophilia trait from his mother. His mother could have inherited the trait from either her mother (a) or her father (b). Because Joe could not have inherited the trait from his father (Joe inherited the Y chromosome from his father), he could not have inherited hemophilia from either (c) or (d).

*24. In *Drosophila,* yellow body is due to an X-linked gene that is recessive to the gene for gray body.

a. A homozygous gray female is crossed with a yellow male. The F_1 are intercrossed to produce F_2. Give the genotypes and phenotypes, along with the expected proportions, of the F_1 and F_2 progeny.

We will use X^+ as the symbol for the dominant gray body color, and X^y for the recessive yellow body color. The homozygous gray female parent is thus X^+X^+, and the yellow male parent is X^yY.

Male progeny always inherit the Y chromosome from the male parent and either of the two X chromosomes from the female parent. Female progeny always inherit the X chromosome from the male parent and either of the two X chromosomes from the female parent.

F_1 males inherit the Y chromosome from their father, and X^+ from their mother; hence, their genotype is X^+Y and they have gray bodies.

F_1 females inherit X^y from their father and X^+ from their mother; hence, they are X^+X^y and also have gray bodies.

When the F_1 progeny are intercrossed, the F_2 males again inherit the Y from the F_1 male, and they inherit either X^+ or X^y from their mother. Therefore, we should get ½ X^+Y (gray body) and ½ X^yY (yellow body). The F_2 females will all inherit the X^+ from their father and either X^+ or X^y from their mother. Therefore, we should get ½ X^+X^+ and ½ X^+X^y (all gray body).

In summary:

P X^+X^+ *(gray female)* × X^yY *(yellow male)*
F_1 ½ X^+Y *(gray males)*
½ X^+X^y *(gray females)*
F_2 ¼ X^+Y *(gray males)*
¼ X^yY *(yellow males)*
¼ X^+X^y *(gray females)*
¼ X^+X^+ *(gray females)*

The net F_2 phenotypic ratios are ½ gray females, ¼ gray males, and ¼ yellow males.

The F_2 progeny can also be predicted using a Punnett square.

	X^+	Y
X^+	X^+X^+ *gray females*	X^+Y *gray males*
X^y	X^+X^y *gray females*	X^yY *yellow males*

b. A yellow female is crossed with a gray male. The F_1 are intercrossed to produce the F_2. Give the genotypes and phenotypes, along with the expected proportions, of the F_1 and F_2 progeny.

The yellow female must be homozygous X^yX^y because yellow is recessive, and the gray male, having only one X chromosome, must be X^+Y. The F_1 male progeny are all X^yY (yellow) and the F_1 females are all X^+X^y (heterozygous gray).

P X^yX^y *(yellow female)* × X^+Y *(gray male)*

F_1 ½ X^yY *(yellow males)*
½ X^+X^y *(gray females)*

F_2 ¼ X^+Y *(gray males)*
¼ X^yY *(yellow males)*
¼ X^+X^y *(gray females)*
¼ X^yX^y *(yellow females)*

c. A yellow female is crossed with a gray male. The F_1 females are backcrossed with gray males. Give the genotypes and phenotypes, along with the expected proportions, of the F_2 progeny.

If the F_1 X^+X^y females are backcrossed to X^+Y gray males, then

F_2 ¼ X^+Y *(gray males)*
¼ X^yY *(yellow males)*
¼ X^+X^+ *(gray females)*
¼ X^+X^y *(gray females)*

d. If the F_2 flies in part (b) mate randomly, what are the expected phenotypic proportions of flies in the F_3?

The outcome of F_2 flies from (b) mating randomly should be equivalent to random union of the male and female gametes. We need to predict the overall male and female gamete types and their frequencies.

As a result of meiosis, half of the male gametes will have the Y chromosome. Because there are equal numbers of males with either the X^+ or X^y, the X-bearing male gametes will be split equally: (½ with X)(½ with X^+) = ¼ X^+; (½ with X)(½ with X^y) = ¼ X^y

Male gametes: ½ Y, ¼ X^+, ¼ X^y

Half the F_2 females in part (b) are homozygous X^yX^y, so all their gametes will be X^y: ½ X^y. The other half are heterozygous and will produce equal proportions of X^+ and X^y gametes: ½(½X^+) = ¼ X^+; ½(½ X^y) = ¼ X^y.

Female gametes: ¼ X^+, ¾ X^y

Now using a Punnett square:

	½ Y	¼ X^+	¼ X^y
¼ X^+	1/8 X^+Y	1/16 X^+X^+	1/16 X^+X^y
¾ X^y	3/8 X^yY	3/16 X^+X^y	3/16 X^yX^y

Overall genotypic ratios are 1/8 X^+Y, 3/8 X^yY, 1/16 X^+X^+, 4/16 X^+X^y, 3/16 X^yX^y. Overall phenotypic ratios are 1/8 gray males, 3/8 yellow males, 5/16 gray females, and 3/16 yellow females.

*25. Red-green color blindness in humans is due to an X-linked recessive gene. Both John and Cathy have normal color vision. After 10 years of marriage to John, Cathy gave birth to a color-blind daughter. John filed for divorce, claiming he is not the father of the child. Is John justified in his claim of nonpaternity? Explain why. If Cathy had given birth to a color-blind son, would John be justified in claiming nonpaternity?
Since color blindness is a recessive trait, the color-blind daughter must be homozygous recessive. Because the color blindness is X-linked, then John has grounds for suspicion. Normally, their daughter would have inherited John's X chromosome. Because John is not color blind, he could not have transmitted a color-blind X chromosome to his daughter.

A remote alternative possibility is that the daughter is XO, having inherited a recessive color-blind allele from her mother and no sex chromosome from her father. In that case, the daughter would have Turner syndrome. A new X-linked color-blind mutation is also possible, albeit even less likely. We do not consider autosomal color blindness because the problem states that this is an X-linked trait.

If Cathy had a color-blind son, then John would have no grounds for suspicion. The son would have inherited John's Y chromosome and the color-blind X chromosome from Cathy.

26. Red-green color blindness in humans is due to an X-linked recessive gene. A woman whose father is color-blind possesses one eye with normal color vision and one eye with color blindness.

a. Propose an explanation for this woman's vision pattern.
The woman is heterozygous, with one X chromosome bearing the allele for normal vision and one X chromosome with the allele for color blindness. One of the two X chromosomes is inactivated at random during early embryogenesis. If one eye derived exclusively from progenitor cells that inactivated the normal X, then that eye would be color blind, whereas the other eye may be derived from progenitor cells that inactivated the color-blind X, or is a mosaic with sufficient normal retinal cells to permit color vision.

b. Would it be possible for a man to have one eye with normal color vision and one eye with color blindness?
One way would be for the man to be XXY, and the answer to part (a) would apply. Another rare possibility would be a somatic mutation in the progenitor cells for one retina, but not the other.

*27. Bob has XXY chromosomes (Klinefelter syndrome) and is color blind. His mother and father have normal color vision, but his maternal grandfather is colorblind. Assume that Bob's chromosome abnormality arose from nondisjunction in meiosis. In which parent and in which meiotic division did nondisjunction occur? Explain your answer.

Because Bob must have inherited the Y chromosome from his father, and his father has normal color vision, there is no way a nondisjunction event from the paternal lineage could account for Bob's genotype. Bob's mother must be heterozygous X^+X^c *because she has normal color vision, and she must have inherited a color-blind X chromosome from her color-blind father. For Bob to inherit two color-blind X chromosomes from his mother, the egg must have arisen from a nondisjunction in meiosis II. In meiosis I, the homologous X chromosomes separate, so one cell has the* X^+ *and the other has* X^c. *Failure of sister chromatids to separate in meiosis II would then result in an egg with two copies of* X^c.

28. Xg is an antigen found on red blood cells. This antigen is caused by an X-linked gene (X^a) that is dominant over an allele for the absence of the antigen (X^-). The inheritance of these X-linked genes was studied in children with chromosome abnormalities to determine where nondisjunction of the sex chromosomes occurred. For each type of mating below, indicate whether nondisjunction took place in the mother or father and, if possible, whether it occurred in meiosis I or meiosis II.
    a. $X^aY \times X^-X^- \rightarrow X^a$ (Turner syndrome) *Since the child received no maternal X chromosome, the nondisjunction must have taken place in the mother, in either meiosis I or meiosis II.*
    b. $X^aY \times X^aX^- \rightarrow X^-$ (Turner syndrome) *The child received no sex chromosome from the father, so the nondisjunction took place in the father, in either meiosis I or II.*
    c. $X^aY \times X^-X^- \rightarrow X^aX^-Y$ (Klinefelter syndrome) *This child received both the* X^a *and Y from the father, so the nondisjunction took place in the father, in meiosis I, where the* X^a *and Y failed to separate.*
    d. $X^aY \times X^aX^- \rightarrow X^-X^-Y$ (Klinefelter syndrome) *This child received two copies of* X^- *from the mother. The nondisjunction took place in the mother, in meiosis II, where the sister chromatids failed to separate.*

29. The Talmud, an ancient book of Jewish civil and religious laws, states that if a woman bears two sons who die of bleeding after circumcision (removal of the foreskin from the penis), any additional sons that she has should not be circumcised. (The bleeding is most likely due to the X-linked disorder hemophilia.) Furthermore, the Talmud states that the sons of her sisters must not be circumcised, while the sons of her brothers should. Is this religious law consistent with sound genetic principles? Explain your answer.
    Yes. If a woman has a son with hemophilia, then she is a carrier. Any of her sons have a 50% chance of having hemophilia. Her sisters may also be carriers. Her brothers, if they do not themselves have hemophilia (because they survived circumcision, they most likely do not have hemophilia), cannot be carriers, and therefore there is no risk of passing hemophilia on to their children.

30. Cranio frontonasal syndrome (CFNS) is a birth defect in which premature fusion of the cranial sutures leads to abnormal head shape, widely spaced eyes, nasal clefts, and various other skeletal abnormalities. George Feldman and his colleagues looked at several families in which CFNS occurred and recorded the results shown in the following table (G. J. Feldman. 1997. *Human Molecular Genetics* 6:1937–1941).
    a. On the basis of the families given, what is the most likely mode of inheritance for CFNS?
    Children with CFNS are born in families where one parent has CFNS. This is most likely a dominant trait. Moreover, we see that in families where the father has CFNS and the mother does not, CFNS is transmitted only to girls, not to boys. If the mother has CFNS, then both boys and girls get CFNS. These data are consistent with an X-linked dominant mode of inheritance for CFNS.
    b. Give the most likely genotypes of the parents in families numbered 1 and 10a.
    Family 1: normal father (X^+Y) × CFNS mother (X^+X^C)
    Family 10a: CFNS father (X^CY) × normal mother (X^+X^+)

	Parents		Offspring			
Family number	Father	Mother	Normal male	Normal female	CFNS male	CFNS female
1	normal	CFNS	1	0	2	1
5	normal	CFNS	0	2	1	2
6	normal	CFNS	0	0	1	2
8	normal	CFNS	1	1	1	0
10a	CFNS	normal	3	0	0	2
10b	normal	CFNS	1	1	2	0
12	CFNS	normal	0	0	0	1
13a	normal	CFNS	0	1	2	1
13b	CFNS	normal	0	0	0	2
7b	CFNS	normal	0	0	0	2

*31. Miniature wings (X^m) in *Drosophila* result from an X-linked allele that is recessive to the allele for long wings (X^+). Give the genotypes of the parents in the following crosses:

	Male parent	Female parent	Male offspring	Female offspring
a.	Long	Long	231 long, 250 miniature	560 long
b.	Miniature	Long	610 long	632 long
c.	Miniature	Long	410 long, 417 miniature	412 long, 415 miniature
d.	Long	Miniature	753 miniature	761 long
e.	Long	Long	625 long	630 long

The genotype of the male parent is the same as his phenotype for an X-linked trait. Because the male progeny get their X chromosomes from their mother, the phenotypes of the male progeny give us the genotypes of the female parents.

a. *Male parent is X^+Y. Because the male offspring are 1:1 long:miniature, the female parent must be X^+X^m. You can use a Punnett square to verify that all the female progeny from such a cross will have long wings (they get the dominant X^+ from the father).*

b. *Male parent is X^mY. Because the male offspring are all long, the female parent must be X^+X^+.*

c. *Male parent is X^mY; female parent is X^+X^m.*

d. *Male parent is X^+Y; female parent is X^mX^m.*

e. *Male parent is X^+Y; female parent is X^+X^+.*

*32. In chickens, congenital baldness results from a Z-linked recessive gene. A bald rooster is mated with a normal hen. The F_1 from this cross are interbred to produce the F_2. Give the genotypes and phenotypes, along with their expected proportions, among the F_1 and F_2 progeny.

For species with the ZZ-ZW sex-determination system, the females are heterogametic ZW. So a bald rooster must be Z^bZ^b (where Z^b denotes the recessive allele for baldness), and a normal hen must be Z^+W.

P $Z^bZ^b \times Z^+W$

F_1 ½ Z^bZ^+ *(normal males)*

½ Z^bW *(bald females)*

F_2 *Using a Punnett square:*

	Z^b	W
Z^+	Z^+Z^b *(normal roosters)*	Z^+W *(normal hens)*
Z^b	Z^bZ^b *(bald roosters)*	Z^bW *(bald hens)*

*33. How many Barr bodies would you expect to see in human cells containing the following chromosomes?

a. XX—*1 Barr body*

b. XY—*0*

c. XO—*0*

d. XXY—*1*

e. XXYY—*1*

f. XXXY—*2*

g. XYY—*0*

h. XXX—*2*

i. XXXX—*3*

Human cells inactivate all X chromosomes beyond one. The Y chromosome has no effect on X-inactivation.

34. A woman with normal chromosomes mates with a man who also has normal chromosomes.

  a. Suppose that during oogenesis, the woman's sex chromosomes undergo nondisjunction in meiosis I; the man's chromosomes separate normally. Give all possible combinations of sex chromosomes that might occur in this couple's children and the number of Barr bodies you would expect to see in each of the cells of each child.

  Eggs produced by nondisjunction in meiosis I: XX and O (nullo)
  Sperm produced by normal meiosis: X and Y
  Children: XXX (two Barr bodies), XO (no Barr body), XXY (1 Barr body). YO would be embryonic lethal, so would not be seen in any human child.

  b. What chromosome combinations and numbers of Barr bodies would you expect to see if the chromosomes separate normally during oogenesis, but nondisjunction of the sex chromosomes occurs in meiosis I of spermatogenesis.

  Eggs produced by normal meiosis: X
  Sperm produced by nondisjunction in meiosis I: XY and O (nullo)
  Children: XXY (1 Barr body) and XO (no Barr body)

35. Red-green color blindness is an X-linked recessive trait in humans. Polydactyly (extra fingers and toes) is an autosomal dominant trait. Martha has normal fingers and toes and normal color vision. Her mother is normal in all respects, but her father is color blind and polydactylous. Bill is color blind and polydactylous. His mother has normal color vision and normal fingers and toes. If Bill and Martha marry, what types and proportions of children can they produce?

The first step is to deduce the genotypes of Martha and Bill. Because the two traits are independent, we can deal with just one trait at a time.

Starting with the X-linked color-blind trait, Bill must be X^cY *because he is colorblind. Bill's mother must be a carrier* (X^+X^c). *Martha must be* X^+X^c, *a carrier for color blindness because her father is color blind* (X^cY).

For polydactyly, Bill must be Dd *(D denotes the dominant polydactyly allele). Because his mother has normal fingers* (dd), *he cannot be homozygous* DD. *Martha, with normal fingers, must be* dd.

If Martha (dd, X^+X^c) *marries Bill* (Dd, X^cY), *then we can predict the types and probability ratios of children they could produce.*

For polydactyly, ½ of children will be polydactylous, and ½ will have normal fingers.

For color blindness, ¼ of children will be color-blind girls, ¼ will be girls with normal vision but carrying the color blindness allele, ¼ will be color-blind boys, and ¼ will be boys with normal vision.

Combining both traits, then:
  1/8 color-blind girls with normal fingers
  1/8 color-blind girls with polydactyly
  1/8 girls with normal vision and normal fingers
  1/8 girls with normal vision and polydactyly
  1/8 color-blind boys with normal fingers
  1/8 color-blind boys with polydactyly

1/8 boys with normal vision and normal fingers
1/8 boys with normal vision and polydactyly
This analysis can also be carried out with a Punnett square.

36. A *Drosophila* mutation called *singed* (*s*) causes the bristles to be bent and misshapen. A mutation called *purple* (*p*) is another recessive mutation that causes the fly's eyes to be purple in color instead of the normal red. In the P generation, flies homozygous for *singed* and *purple* were crossed with flies that were homozygous for normal bristles and red eyes. The F1 were intercrossed to produce the F2, and the following results were obtained.

Cross 1

P male, singed bristles, purple eyes × female, normal bristles, red eyes
F_1: 420 female, normal bristles, red eyes
426 male, normal bristles, red eyes
F_2: 337 female, normal bristles, red eyes
113 female, normal bristles, purple eyes
168 male, normal bristles, red eyes
170 male, singed bristles, red eyes
56 male, normal bristles, purple eyes
58 male, singed bristles, purple eyes

Cross 2

P female, singed bristles, purple eyes × male, normal bristles, red eyes
F_1: 504 female, normal bristles, red eyes
498 male, singed bristles, red eyes
F_2: 227 female, normal bristles, red eyes
223 female, singed bristles, red eyes
225 male, normal bristles, red eyes
225 male, singed bristles, red eyes
78 female, normal bristles, purple eyes
76 female, singed bristles, purple eyes
74 male, normal bristles, purple eyes
72 male, singed bristles, purple eyes

a. What are the modes of inheritance of singed and purple? Explain your reasoning.
We can examine each trait separately.

The singed mutation is recessive because the F_1 for Cross 1 all have normal bristles. It is also X-linked because the reciprocal crosses (Cross 1 and Cross 2) give different F_1 progeny: Cross 2 yields F_1 singed males and normal females.

The purple eye color mutation appears recessive because the F_1 progeny of both crosses all have red eyes; and autosomal because there is no difference in the progeny of the reciprocal crosses, and also because there is no significant difference between male and female progeny with respect to eye color. Cross 1 F_2 are 337 red females, 338 red males, 113 purple females and 114 purple males. Cross 2 F2 are 450 red females, 450 red males, 154 purple females and 146 purple males.

b. Give genotypes for the parents and offspring in the P, F_1, and F_2 generations of Cross 1 and Cross 2.

We define X^s *as the singed allele,* X^+ *as the normal bristles allele,* p *as the purple allele, and* P *as the red-eyed allele.*

Cross 1:

The F_1 *males all have normal bristles, so the female parent is homozygous for normal bristles:* X^+X^+. *The singed male parent is* X^sY.

The purple-eyed male parent must be homozygous recessive pp. The red-eyed female parent must be homozygous PP *because all the progeny have red eyes.*

P: male, singed bristles, purple eyes × female, normal bristles, red eyes
X^sY, pp × X^+X^+, PP

F_1: *420 female, normal bristles, red eyes* X^+X^s, Pp
426 male, normal bristles, red eyes X^+Y, Pp

F_2: *337 female, normal bristles, red eyes*
X^+X^+, PP
X^+X^s, PP
X^+X^+, Pp *(2)*
X^+X^s, Pp *(2)*

113 female, normal bristles, purple eyes
X^+X^+, pp
X^+X^s, pp

168 male, normal bristles, red eyes
X^+Y, PP
X^+Y, Pp *(2)*

170 male, singed bristles, red eyes
X^sY, PP
X^sY, Pp *(2)*

56 male, normal bristles, purple eyes
X^+Y, pp

58 male, singed bristles, purple eyes
X^sY, pp

Cross 2:

The female parent with the recessive singed bristles must be homozygous X^sX^s, *and the male parent must be* X^+Y. *The purple eyed parent is pp, and the red-eyed parent must be homozygous* PP *because the* F_1 *all have red eyes.*

P: female, singed bristles, purple eyes × male, normal bristles, red eyes
X^sX^s, pp × X^+Y, PP

F_1: *504 female, normal bristles, red eyes* X^+X^s, Pp
498 male, singed bristles, red eyes X^sY, Pp

F_2: *227 female, normal bristles, red eyes*
X^+X^s, PP
X^+X^s, Pp *(2)*

223 female, singed bristles, red eyes
X^sX^s, PP
X^sX^s, Pp *(2)*
225 male, normal bristles, red eyes
X^+Y, PP
X^+Y, Pp *(2)*
225 male, singed bristles, red eyes
X^sY, PP
X^sY, Pp *(2)*
78 female, normal bristles, purple eyes
X^+Xs, pp
76 female, singed bristles, purple eyes
X^sX^s, pp
74 male, normal bristles, purple eyes
X^+Y, pp
72 male, singed bristles, purple eyes
X^sY, pp
The (2) indicates that there are twice as many of these genotypes.

37. The following two genotypes are crossed: $AaBbCcX^+X^r \times AaBBccX^+Y$, where *a*, *b*, and *c* represent autosomal genes and X^+ and X^r represent X-linked alleles in an organism with XY sex determination. What is the probability of obtaining genotype $aaBbCcX^+X^+$ in the progeny?
We have to assume that the autosomal genes a, b, and c assort independently of each other as well as of the sex chromosomes. Given independent assortment, we can calculate the probability of the genotype for each gene separately, and then multiply the probabilities to calculate the probability of the combined genotype for all four genes.
For gene a: Aa × Aa → ¼ aa
For gene b: Bb × BB → ½ Bb
For gene c: Cc × cc → ½ Cc
For the sex-linked gene r: $X^+X^r \times X^+Y \rightarrow \frac{1}{4} X^+X^+$
Combined probability of genotype aaBbCcX^+X^+ = ¼ × ½ × ½ × ¼ = *1/64*

*38. Miniature wings in *Drosophila melanogaster* result from an X-linked gene (X^m) that is recessive to an allele for long wings (X^{m+}). Sepia eyes are produced by an autosomal gene (*s*) that is recessive to an allele for red eyes (s^+).

a. A female fly that has miniature wings and sepia eyes is crossed with a male that has normal wings and is homozygous for red eyes. The F_1 are intercrossed to produce the F_2. Give the phenotypes and their proportions expected in the F_1 and F_2 flies from this cross.

The female parent (miniature wings, sepia eyes) must be X^mX^m, ss.
The male parent (normal wings, homozygous red eyes) is $X^{m+}Y$, s^+s^+.
F_1 *males are* X^mY, s^+s *(miniature wings, red eyes)*
females are $X^{m+}X^m$, s^+s *(long wings, red eyes)*

F_2 We can analyze the expected outcome of this cross with either a branch diagram or with a Punnett square.

First, the branch diagram:

½ male (Y) → ½ X^{m+} normal wings →	*¾ red (¾ s^+)*	*= ½ × ½ × ¾ = 3/16 male, normal, red*
	¼ sepia (¼ ss)	*= ½ × ½ × ¼ = 1/16 male, normal, sepia*
½ X^m miniature →	*¾ red*	*= ½ × ½ × ¾ = 3/16 male, miniature, red*
	¼ sepia	*= ½ × ½ × ¼ = 1/16 male, miniature, sepia*
½ female (X^m) → ½ X^{m+} normal →	*¾ red*	*= ½ × ½ × ¾ = 3/16 female, normal, red*
	¼ sepia	*= ½ × ½ × ¼ = 1/16 female, normal, sepia*
½ X^m miniature →	*¾ red*	*= ½ × ½ × ¾ = 3/16 female, miniature, red*
	¼ sepia	*= ½ × ½ × ¼ = 1/16 female, miniature, sepia*

Explanation: ½ of the F_2 progeny are males because they inherit Y from the F_1 male. The other ½ are females that inherit the X^m from the F_1 male. In each case, ½ of the F_2 males and females inherit X^{m+} from the F_1 female and have normal wings, whereas the other ½ inherit X^m and have miniature wings. Finally, ¾ of the progeny will have the dominant red eyes, and ¼ will have sepia eyes.

Now the Punnett square:

	¼ $X^m s^+$	¼ $X^m s$	¼ $Y s^+$	¼ $Y s$
¼ $X^{m+} s^+$	$X^{m+} X^m s^+s^+$ Long wings, red eyes	$X^{m+} X^m s^+s$ Long wings, red eyes	$X^{m+} Y s^+s^+$ Long wings, red eyes	$X^{m+} Y s^+s$ Long wings, red eyes
¼ $X^{m+} s$	$X^{m+} X^m s^+s$ Long wings, red eyes	$X^{m+} X^m ss$ Long wings, sepia eyes	$X^{m+} Y s^+s$ Long wings, red eyes	$X^{m+} Y ss$ Long wings, sepia eyes
¼ $X^m s^+$	$X^mX^m s^+s^+$ Mini wings, red eyes	$X^mX^m s^+s$ Mini wings, red eyes	$X^mY s^+s^+$ Mini wings, red eyes	$X^mY s^+s$ Mini wings, red eyes
¼ $X^m s$	$X^mX^m s^+s$ Mini wings, red eyes	$X^mX^m ss$ Mini wings, sepia eyes	$X^mY s^+s$ Mini wings, red eyes	$X^mY ss$ Mini wings, sepia eyes

b. A female fly that is homozygous for normal wings and has sepia eyes is crossed with a male that has miniature wings and is homozygous for red eyes. The F_1 are intercrossed to produce the F_2. Give the phenotypes and proportions expected in the F_1 and F_2 flies from this cross.

Parents $X^{m+}X^{m+}$, ss *and* $X^{m}Y$, $s^{+}s^{+}$

F_1 $X^{m+}X^{m}$, $s^{+}s$ *and* $X^{m+}Y$, $s^{+}s$

F_2

Sex chromosome inherited From male	*From female*	*Autosomal phenotype*	*Combined phenotype*
½ *Y male*	½ X^{m+}	*¾ red*	*3/16 long wings, red eyes*
		¼ sepia	*1/16 long wings, sepia eyes*
	½ X^{m}	*¾ red*	*3/16 mini wings, red eyes*
		¼ sepia	*1/16 mini wings, sepia eyes*
½ X^{m+} *female*	X^{m+} *or* X^{m}	*¾ red*	*3/8 long wings, red eyes*
		¼ sepia	*1/8 long wings, sepia eyes*

Note that in this case, the X chromosome the F_2 *females inherit from the mother does not affect their phenotype because they all have a dominant* X^{m+} *for long wings from their father.*

39. Suppose that a recessive gene that produces a short tail in mice is located in the pseudoautosomal region. A short-tailed male is mated with a female mouse that is homozygous for a normal tail. The F_1 from this cross are intercrossed to produce the F_2. What will the phenotypes and proportions of the F_1 and F_2 mice be from this cross?

A gene in the pseudoautosomal region must be present on both X and Y chromosomes. We will use X^{+} *and* Y^{+} *for the normal alleles, and* X^{s} *and* Y^{s} *for the short-tail alleles. The short-tailed male must be* $X^{s}Y^{s}$*, and the normal female is* $X^{+}X^{+}$*. The expected* F_1 *would then be:*

$X^{+}Y^{s}$ *males with long tails and* $X^{+}X^{s}$ *females with long tails.*

F_2 *from the intercross will be:* ¼ $X^{+}Y^{s}$ *males with long tails*
¼ $X^{s}Y^{s}$ *males with short tails*
¼ $X^{+}X^{+}$ *females with long tails*
¼ $X^{+}X^{s}$ *females with long tails*

So all the females will have long tails, and equal proportions of the males will have short and long tails.

*40. A color-blind woman and a man with normal vision have three sons and six daughters. All the sons are color blind. Five of the daughters have normal vision, but one of them is color blind. The color-blind daughter is 16 years old, is short for her age, and has never undergone puberty. Propose an explanation for how this girl inherited her color blindness.

The trivial explanation for these observations is that this form of color blindness is an autosomal recessive trait. In that case, the father would be a heterozygote, and we would expect equal proportions of color-blind and normal children, of either sex.

If, on the other hand, we assume that this is an X-linked trait, then the mother is $X^{c}X^{c}$ *and the father must be* $X^{+}Y$*. Normally, all the sons would be color blind, and all the daughters should have normal vision. The most likely way to have a daughter that is color blind would be for her not to have inherited an* X^{+} *from her father. The observation that the* *color-blind daughter is short in stature and has failed to undergo puberty is consistent with Turner syndrome (XO). The color-blind daughter would then be X^cO.*

41. Anhidrotic ectodermal dysplasia is an X-linked recessive disorder in humans characterized by small teeth, no sweat glands, and sparse body hair. The trait is usually seen in men, but women who are heterozygous carriers of the trait often have irregular patches of skin with few or no sweat glands.
    a. Explain why women who are heterozygous carriers of a recessive gene for anhidrotic ectodermal dysplasia have irregular patches of skin lacking sweat glands.
    X-inactivation occurs randomly in each of the cells of the early embryo, then is maintained in the mitotic progeny cells. The irregular patches of skin lacking sweat glands arose from skin precursor cells that inactivated the X chromosome with the normal allele.
    b. Why does the distribution of the patches of skin lacking sweat glands differ among the females depicted in the illustration, even between the two identical twins?
    The X-inactivation event occurs randomly in each of the cells of the early embryo. Even in identical twins, the different ectodermal precursor cells will inactivate different X chromosomes, resulting in different distributions of patches lacking sweat glands.

CHALLENGE QUESTIONS

Section 4.1

42. Antibiotics kill the *Wolbachia* bacteria that sometimes infect isopods and cause ZZ males to become females (see Sex Wars in Isopods at beginning of the chapter). A biologist collects some isopods from a natural population that exhibits a female-biased sex ratio. She adds antibiotics to the isopod's food to kill any bacteria. She crosses several male and female isopods and rears their offspring in the laboratory. To her surprise, the offspring of many of the crosses are all male. Can you explain her result?
    If the female bias in the population of isopods was indeed caused by Wolbachia *infection, then many of the phenotypic females in the matings were actually genotypically ZZ. Only ZZ progeny would result from matings of ZZ males with ZZ* Wolbachia*-infected phenotypic females. With antibiotics in the larval food, the* Wolbachia *would be eliminated, and all the ZZ progeny would develop as males.*

Section 4.2

43. Female humans who are heterozygous for X-linked recessive genes sometimes exhibit mild expression of the trait. However, such mild expression of X-linked traits in females who are heterozygous for X-linked alleles is not seen in *Drosophila.* What might cause this difference in the expression of X-linked genes in female humans and *Drosophila?* (Hint: In *Drosophila,* dosage compensation is accomplished by doubling the activity of genes on the X chromosome of males.)
    In humans and other mammals, X-inactivation results in females that are mosaic for X-linked heterozygous loci, with some cells expressing the dominant allele and

some cells expressing the recessive allele. In flies, however, females do not undergo X-inactivation. Therefore, all the cells in female flies express the dominant allele.

44. A geneticist discovers a male mouse in his laboratory colony with greatly enlarged testes. He suspects that this trait results from a new mutation that is either Y-linked or autosomal dominant. How could he determine if the trait is autosomal dominant or Y-linked?
Because testes are present only in males, enlarged testes could either be a sex-limited autosomal dominant trait or a Y-linked trait. Assuming the male mouse with enlarged testes is fertile, mate it with a normal female. If the trait is autosomal dominant and the parental male is heterozygous, only half the male progeny will have enlarged testes. If the trait is Y-linked, all the male progeny will have enlarged testes.

    With either outcome, however, the results from this first cross will not be conclusive. If all the male progeny do express the trait, the trait may still be autosomal dominant if the parental male was homozygous. If only some of the male progeny express the trait, the possibility still remains that the trait is Y-linked but incompletely penetrant. In either case, more conclusive evidence is needed.

    Mate the female progeny (F_1 females) with normal males. If the trait is autosomal dominant, some of the male F_2 progeny will have enlarged testes, proving that the trait can be passed through a female. If the trait is Y-linked, all the male F_2 progeny will have normal testes, like their normal male father.

45. Identical twins (also called monozygotic twins) are derived from a single egg fertilized by a single sperm, creating a zygote that later divides into two (see Chapter 6). Because two identical twins originate from a single zygote, they are genetically identical.

    Caroline Loat and her colleagues examined 9 measures of social, behavioral, and cognitive ability in 1000 pairs of identical male twins and 1000 pairs of identical female twins (C. S. Loat et al. 2004. *Twin Research* 7:54–61). They found that, for three of the measures (prosocial behavior, peer problems, and verbal ability), the two male twins of a pair tended to be more alike in their scores than were two female twins of a pair. Propose a possible explanation for this observation. What might this observation indicate about the location of genes that influence prosocial behavior, peer problems, and verbal ability?
The major cause of increased variability among female identical twins compared to male identical twins must be that only females undergo X-inactivation. As a result, human female brains are mosaic, with some neural cells expressing one allele and other cells expressing the other allele, for heterozygous X-linked loci. Since the X-inactivation is random, even identical twins will have different compositions of neural cells expressing different X-linked alleles.

46. Occasionally, a mouse X chromosome is broken into two pieces and each piece becomes attached to a different autosomal chromosome. In this event, only the genes on one of the

two pieces undergo X-inactivation. What does this observation indicate about the mechanism of X-chromosome inactivation?

One hypothesis consistent with this observation is that the X-inactivation mechanism requires or recognizes a specific region or locus on the X-chromosome and inactivates chromatin attached to this center of inactivation. When the X-chromosome breaks, the fragment containing the X-inactivation locus, or center, becomes inactivated. The other fragment escapes inactivation because it is no longer attached.

Chapter Five: Extensions and Modifications of Basic Principles

COMPREHENSION QUESTIONS

Section 5.1

*1. How do incomplete dominance and codominance differ?
Incomplete dominance means the phenotype of the heterozygote is intermediate to the phenotypes of the homozygotes. Codominance refers to situations in which both alleles are expressed and both phenotypes are manifested simultaneously.

Section 5.2

*2. What is incomplete penetrance and what causes it?
Incomplete penetrance occurs when an individual with a particular genotype does not express the expected phenotype. Environmental factors, as well as the effects of other genes, may alter the phenotypic expression of a particular genotype.

Section 5.5

3. What is gene interaction? What is the difference between an epistatic gene and a hypostatic gene?
Gene interaction is the determination of a single trait or phenotype by genes at more than one locus; the effect of one gene on a trait depends on the effects of a different gene located elsewhere in the genome. One type of gene interaction is epistasis. The alleles at the epistatic gene mask or repress the effects of alleles at another gene. The gene whose alleles are masked or repressed is called the hypostatic gene.

4. What is a recessive epistatic gene?
Recessive epistasis occurs when the epistatic gene in a homozygous recessive state masks the interacting gene or genes. In the example from the text, being homozygous recessive at the locus for deposition of color in hair shafts (ee) *completely masked the effect of the color locus regardless of whether it had the dominant black* (B-) *or recessive brown* (bb) *allele.*

*5. What is a complementation test and what is it used for?
Complementation tests are used to determine whether different recessive mutations affect the same gene or locus (are allelic) or whether they affect different genes. The two mutations are introduced into the same individual by crossing homozygotes for each of the mutants. If the progeny show a mutant phenotype, then the mutations are allelic (in the same gene). If the progeny show a wild-type (dominant) phenotype, then the mutations are in different genes and are said to complement each other because each of the mutant parents can supply a functional copy (or dominant allele) of the gene mutated in the other parent.

Section 5.6

*6. What characteristics are exhibited by a cytoplasmically inherited trait?
Cytoplasmically inherited traits are encoded by genes in the cytoplasm. Because the cytoplasm usually is inherited from a single (most often the female) parent, reciprocal crosses do not show the same results. Cytoplasmically inherited traits often show great variability because different egg cells (female gametes) may have differing proportions of cytoplasmic alleles from random sorting of mitochondria (or plastids in plants).

7. What is genomic imprinting?
Genomic imprinting refers to different expression of a gene depending on whether it was inherited from the male parent or the female parent.

8. What is the difference between genetic maternal effect and genomic imprinting?
In genetic maternal effect, the phenotypes of the progeny are determined by the genotype of the mother only. The genotype of the father and the genotype of the affected individual have no effect. In genomic imprinting, the phenotype of the progeny differs based on whether a particular allele is inherited from the mother or the father. The phenotype is therefore based on both the individual's genotype and the paternal or maternal origins of the genotype.

9. What is the difference between a sex-influenced gene and a gene that exhibits genomic imprinting?
For a sex-influenced gene, the phenotype is influenced by the sex of the individual bearing the genotype. For an imprinted gene, the phenotype is influenced by the sex of the parent *from which each allele was inherited.*

Section 5.7

10. What characteristics do you expect to see in a trait that exhibits anticipation?
Traits that exhibit anticipation become stronger or more pronounced, or are expressed earlier in development, as they are transmitted to each succeeding generation.

Section 5.8

*11. What are continuous characteristics and how do they arise?
Continuous characteristics, also called quantitative characteristics, exhibit many phenotypes with a continuous distribution. They result from the interaction of multiple genes (polygenic traits), the influence of environmental factors on the phenotype, or both.

APPLICATION QUESTIONS AND PROBLEMS

Sections 5.1 through 5.8

12. Match each term with its correct definition.

Term	Definition
d phenocopy	**a.** the percentage of individuals with a particular genotype that express the expected phenotype
h pleiotrophy	**b.** a trait determined by an autosomal gene that is more easily expressed in one sex
e polygenic trait	**c.** a trait determined by an autosomal gene that is expressed in only one sex
a penetrance	**d.** a trait that is determined by an environmental effect and has the same phenotype as a genetically determined trait
c sex-limited trait	**e.** a trait determined by genes at many loci
i genetic maternal effect	**f.** the expression of a trait is affected by the sex of the parent that transmits the gene to the offspring
f genomic imprinting	**g.** the trait appears earlier or more severely in succeeding generations
b sex-influenced trait	**h.** a gene affects more than one phenotype
g anticipation	**i.** the genotype of the maternal parent influences the phenotype of the offspring

Section 5.1

*13. Palomino horses have a golden yellow coat, chestnut horses have a brown coat, and cremello horses have a coat that is almost white. A series of crosses between the three different types of horses produced the following offspring:

Cross	**Offspring**
palomino × palomino	13 palomino, 6 chestnut, 5 cremello
chestnut × chestnut	16 chestnut
cremello × cremello	13 cremello
palomino × chestnut	8 palomino, 9 chestnut
palomino × cremello	11 palomino, 11 cremello
chestnut × cremello	23 palomino

a. Explain the inheritance of the palomino, chestnut, and cremello phenotypes in horses.
The results of the crosses indicate that cremello and chestnut are pure-breeding traits (homozygous). Palomino is a hybrid trait (heterozygous) that produces a 2:1:1 ratio when palominos are crossed with each other. The simplest hypothesis consistent with these results is incomplete dominance, with palomino as the phenotype of the heterozygotes resulting from chestnuts crossed with cremellos.

b. Assign symbols for the alleles that determine these phenotypes and list the genotypes of all parents and offspring given in the preceding table.
Let C^B = chestnut, C^W = cremello, C^BC^W = palomino.

Cross	**Offspring**
palomino × palomino	*13 palomino, 6 chestnut, 5 cremello*
$C^BC^W \times C^BC^W$	C^BC^W C^BC^B C^WC^W
chestnut × chestnut	*16 chestnut*
$C^BC^B \times C^BC^B$	C^BC^B
cremello × cremello	*13 cremello*
$C^WC^W \times C^WC^W$	C^WC^W
palomino × chestnut	*8 palomino, 9 chestnut*
$C^BC^W \times C^BC^B$	C^BC^W C^BC^B
palomino × cremello	*11 palomino, 11 cremello*
$C^BC^W \times C^WC^W$	C^BC^W C^WC^W
chestnut × cremello	*23 palomino*
$C^BC^B \times C^WC^W$	C^BC^W

*14. The L^M and L^N alleles at the MN blood group locus exhibit codominance. Give the expected genotypes and phenotypes and their ratios in progeny resulting from the following crosses:

a. $L^ML^M \times L^ML^N$
½ L^ML^M (type M), ½ L^ML^N (type MN)

b. $L^NL^N \times L^NL^N$
All L^NL^N (type N)

c. $L^ML^N \times L^ML^N$
½ L^ML^N (type MN), ¼ L^ML^M (type M), ¼ L^NL^N (type N)

d. $L^ML^N \times L^NL^N$
½ L^ML^N (type MN), ½ L^NL^N (type N)

e. $L^ML^M \times L^NL^N$
All L^ML^N (type MN)

Section 5.2

15. Assume that long ear lobes in humans are an autosomal dominant trait that exhibits 30% penetrance. A person who is heterozygous for long ear lobes mates with a person who is homozygous for normal ear lobes. What is the probability that their first child will have long ear lobes?
To have long ear lobes, the child must inherit the dominant allele and also express it. The probability of inheriting the dominant allele is 50%; the probability of expressing it is 30%. The combined probability of both is 0.5(0.3) = 0.15, or 15%

16. The eastern mosquito fish (*Gambusia affinis holbrooki*) has XX-XY sex determination. Its spotting is inherited as a Y-linked trait. The trait exhibits 100% penetrance when the fish are raised at 22°C, but the penetrance drops to 42% when the fish are raised at 26°C. A male with spots is crossed with a female without spots, and the F_1 are

intercrossed to produce the F_2. If all the offspring are raised at 22°C, what proportion of the F_1 and F_2 will have spots? If all the offspring are raised at 26°C, what proportion of the F_1 and F_2 will have spots?
Because spotting is Y-linked, the parental genotypes are: XY^s *and XX, where* Y^s *denotes the spotted allele on the Y chromosome. The* F_1 *genotypes will be:* ½ XY^s *and* ½ *XX, like the parents. The* F_2 *genotypes will also be* ½ XY^s *and* ½ *XX. Note that incomplete penetrance and expressivity do not affect genotypic ratios. At 22°C, where penetrance is 100%, the phenotypic ratios will be all spotted males and all unspotted females in both the* F_1 *and* F_2 *progeny. At 26°C, where penetrance is only 42%, then 42% of the* XY^s *males will be spotted, in the* F_1 *and* F_2.

Section 5.3

*17. When a Chinese hamster with white spots is crossed with another hamster that has no spots, approximately ½ of the offspring have white spots and ½ have no spots. When two hamsters with white spots are crossed, 2/3 of the offspring possess white spots and 1/3 have no spots.

a. What is the genetic basis of white spotting in Chinese hamsters?
The 2:1 ratio when two spotted hamsters are mated suggests lethality, and the 1:1 ratio when spotted hamsters are mated to hamsters without spots indicates that spotted is a heterozygous phenotype. Using S *and* s *to symbolize the locus responsible for white spotting, spotted hamsters are* Ss and *solid-colored hamsters are* ss. *One-quarter of the progeny expected from a mating of two spotted hamsters is* SS, *embryonic lethal, and missing from those progeny, resulting in the 2:1 ratio of spotted to solid progeny.*

b. How might you go about producing Chinese hamsters that breed true for white spotting?
Because spotting is a heterozygous phenotype, it should not be possible to obtain Chinese hamsters that breed true for spotting, unless the locus that produces spotting can somehow be separated from the lethality.

18. As discussed in the introduction to this chapter, Cuénot studied the genetic basis of yellow coat color in mice. He carried out a number of crosses between two yellow mice and obtained what he thought was a 3:1 ratio of yellow to gray mice in the progeny. The following table gives Cuénot's actual results, along with the results of a much larger series of crosses carried out by Castle and Little (Castle, W.E., and C. C. Little. 1910. *Science* 32:868–870).

Progeny resulting from crosses of yellow × yellow mice

Investigators	Yellow progeny	Non-yellow progeny	Total progeny
Cuénot	263	100	363
Castle and Little	800	435	1,235
Both combined	1,063	535	1,598

a. Using a chi-square test, determine whether Cuénot's results are significantly different from the 3:1 ratio that he thought he observed. Are they different from a 2:1 ratio?

Testing Cuénot's data for a 3:1 ratio -

	Obs	*Expected (3:1)*	*O – E*	*(O – E)²/E*
Yellow	*263*	*272.25*	*–9.25*	*0.314*
Non-yellow	*100*	*90.75*	*9.25*	*0.943*
Total	*363*	*363*		*1.257 = χ^2*

d.f. = 2 – 1 = 1
0.1 < p < .5
Cannot reject hypothesis of 3:1 ratio

Now test for 2:1 ratio -

	Obs	*Expected (2:1)*	*O – E*	*(O – E)²/E*
Yellow	*263*	*242*	*21*	*1.82*
Non-yellow	*100*	*121*	*–21*	*3.64*
Total	*363*	*363*		*5.46 = χ^2*

d.f. = 1; p < .025
The observations are inconsistent with a 2:1 ratio.

b. Determine whether Castle and Little's results are significantly different from a 3:1 ratio. Are they different from a 2:1 ratio?

	Obs	*Expected (3:1)*	*O – E*	*(O – E)²/E*
Yellow	*800*	*926.25*	*–126.25*	*17.2*
Non-yellow	*435*	*308.75*	*126.25*	*51.6*
Total	*1,235*	*1,235*		*68.8 = χ^2*

d.f. = 1; p << .005
Reject 3:1 ratio

	Obs	*Expected (2:1)*	*O – E*	*(O – E)²/E*
Yellow	*800*	*823.3*	*–23.3*	*0.66*
Non-yellow	*435*	*411.7*	*23.3*	*1.32*
Total	*1,235*	*1,235*		*1.98 = χ^2*

d.f. = 1; 0.1 < p < 0.5
Cannot reject 2:1 ratio

c. Combine the results of Castle and Cuénot and determine whether they are significantly different from a 3:1 ratio and a 2:1 ratio.

	Obs	*Expected (3:1)*	*O – E*	*(O – E)²/E*
Yellow	*1,063*	*1,198.5*	*–135.5*	*15.3*
Non-yellow	*535*	*399.5*	*135.5*	*46.0*
Total	*1,598*	*1,598*		$61.3 = \chi^2$

d.f. = 1; $p << .005$
Reject 3:1 ratio

	Obs	*Expected (2:1)*	*O – E*	*(O – E)²/E*
Yellow	*1,063*	*1,065.3*	*–2.3*	*0.005*
Non-yellow	*535*	*532.7*	*2.3*	*0.010*
Total	*1,598*	*1,598*		$0.015 = \chi^2$

d.f. = 1; $0.9 < p < 0.975$
Cannot reject 2:1 ratio

d. Offer an explanation for the different ratios that Cuénot and Castle obtained.
Cuénot had far smaller numbers of progeny, so his ratios are more susceptible to error from chance deviation. Indeed, only a slight shift in numbers of progeny would make Cuénot's data compatible with a 2:1 ratio as well as a 3:1 ratio. Investigator bias may also have played a role, based on the expectation of a 3:1 ratio.

Section 5.4

19. In the pearl millet plant, color is determined by three alleles at a single locus: Rp^1 (red), Rp^2 (purple), and *rp* (green). Red is dominant over purple and green, and purple is dominant over green ($Rp^1 > Rp^2 > rp$). Give the expected phenotypes and ratios of offspring produced by the following crosses:

a. $Rp^1/Rp^2 \times Rp^1/rp$
We expect ¼ Rp^1/Rp^1 *(red), ¼* Rp^1/rp *(red), ¼* Rp^2/Rp^1 *(red), ¼* Rp^2/rp *(purple), for overall phenotypic ratio of ¾ red, ¼ purple.*

b. $Rp^1/rp \times Rp^2/rp$
¼ Rp^1/Rp^2 *(red), ¼* Rp^1/rp *(red), ¼* Rp^2/rp *(purple), ¼* rp/rp *(green), for overall phenotypic ratio of ½ red, ¼ purple, ¼ green.*

c. $Rp^1/Rp^2 \times Rp^1/Rp^2$
This cross is equivalent to a two-allele cross of heterozygotes, so the expected phenotypic ratio is ¾ red, ¼ purple.

d. $Rp^2/rp \times rp/rp$
Another two-allele cross of a heterozygote with a homozygous recessive. Phenotypic ratio is ½ purple, ½ green.

e. $rp/rp \times Rp^1/Rp^2$
½ Rp^1/rp *(red), ½* Rp^2/rp *(purple)*

20. If there are five alleles at a locus, how many genotypes may there be at this locus? How many different kinds of homozygotes will there be? How many genotypes and homozygotes would there be with eight alleles?
Mathematically, this question is the same as asking how many different groups of two (diploid genotypes have two alleles for each locus) are possible from n *objects (alleles). Assign numbers 1, 2, 3, 4, and 5 to each of the five alleles and group the possible genotypes according to the following table:*

1,1					
1,2	*2,2*				
1,3	*2,3*	*3,3*			
1,4	*2,4*	*3,4*	*4,4*		
1,5	*2,5*	*3,5*	*4,5*	*5,5*	

Such an arrangement allows us to easily see that the number of genotypes for any n *number of alleles is simply* $\Sigma(1, 2, 3 \ldots n) = n(n+1)/2$. *Looking at the table, we see that the number of filled boxes (genotypes) is equal to half the number of boxes in a rectangle of dimensions* $n \times (n+1)$. *So, the number of genotypes* $= n(n+1)/2$. *For five alleles* ($n = 5$), *we get 15 possible genotypes and five homozygotes. For eight alleles, there are* $8(8+1)/2 = 36$ *possible genotypes.*

21. Turkeys have black, bronze, or black-bronze plumage. Examine the results of the following crosses:

Parents	**Offspring**
Cross 1: black and bronze	All black
Cross 2: black and black	¾ black, ¼ bronze
Cross 3: black-bronze and black-bronze	All black-bronze
Cross 4: black and bronze	½ black, ¼ bronze, ¼ black-bronze
Cross 5: bronze and black-bronze	½ bronze, ½ black-bronze
Cross 6: bronze and bronze	¾ bronze, ¼ black-bronze

Do you think these differences in plumage arise from incomplete dominance between two alleles at a single locus? If yes, support your conclusion by assigning symbols to each allele and providing genotypes for all turkeys in the crosses. If your answer is no, provide an alternative explanation and assign genotypes to all turkeys in the crosses.
The results of Cross 2 tell us that black is dominant to bronze. Similarly, the results of Cross 6 tell us that bronze is dominant to black-bronze. We can use B^L *for black,* B^R *for bronze, and* b *for black-bronze.*

Parents	***Offspring***
Cross 1: black (B^LB^L) × *bronze* (B^RB^R)	*All black* (B^LB^R)
Cross 2: black (B^LB^R) × *black* (B^LB^R)	¾ *black* (B^L--), ¼ *bronze* (B^RB^R)
Cross 3: black-bronze (bb) × *black-bronze* (bb)	*All black-bronze* (bb)
Cross 4: black (B^Lb) × *bronze* (B^Rb)	½ *black* (B^L--), ¼ *bronze* (B^Rb), ¼ *black-bronze* (bb)
Cross 5: bronze (B^Rb) × *black-bronze* (bb)	½ *bronze* (B^Rb), ½ *black-bronze* (bb)
Cross 6: bronze (B^Rb) × *bronze* (B^Rb)	¾ *bronze* (B^R--), ¼ *black-bronze* (bb)

22. In rabbits, an allelic series helps to determine coat color: C (full color), c^{ch} (chinchilla, gray color), c^h (Himalayan, white with black extremities), and c (albino, all white). The C allele is dominant over all others, c^{ch} is dominant over c^h and c, c^h is dominant over c, and c is recessive to all the other alleles. This dominance hierarchy can be summarized as $C > c^{ch} > c^h > c$. The rabbits in the following list are crossed and produce the progeny shown. Give the genotypes of the parents for each cross.

    a. full color × albino → ½ full color, ½ albino
    Cc × cc. *1:1 phenotypic ratios in the progeny result from a cross of a heterozygote with a homozygous recessive. Because albino is recessive to all other alleles, the full-color parent must have an albino allele, and the albino parent must be homozygous for the albino allele.*

    b. himalayan × albino → ½ himalayan, ½ albino
    c^hc × cc. *Again, the 1:1 ratio of the progeny indicate the parents must be a heterozygote and a homozygous recessive.*

    c. full color × albino → ½ full color, ½ chinchilla
    Cc^{ch} × cc. *This time, we get a 1:1 ratio, but we have chinchilla progeny instead of albino. Therefore, the heterozygous full-color parent must have a chinchilla allele as well as a dominant full-color allele. The albino parent has to be homozygous albino because albino is recessive to all other alleles.*

    d. full color × himalayan → ½ full color, ¼ himalayan, ¼ albino
    Cc × c^hc. *The 1:2:1 ratio in the progeny indicates that both parents are heterozygotes. Both must have an albino allele because the albino progeny must have inherited an albino allele from each parent.*

    e. full color × full color → ¾ full color, ¼ albino
    Cc × Cc. *The 3:1 ratio indicates that both parents are heterozygous. Both parents must have an albino allele for albino progeny to result.*

23. In this chapter, we discussed Joan Barry's paternity suit against Charlie Chaplin and how, on the basis of blood types, Chaplin could not have been the father of her child.

    a. What blood types are possible for the father of Barry's child?
    Because Barry's child inherited an I^B *allele from the father, the father could have been B or AB.*

b. If Chaplin had possessed one of these blood types, would that prove that he fathered Barry's child?
No. Many other men have these blood types. The results would have meant only that Chaplin cannot be eliminated as a possible father of the child.

*24. A woman has blood type A MM. She has a child with blood type AB MN. Which of the following blood types could *not* be that of the child's father? Explain your reasoning.

George	O	NN
Tom	AB	MN
Bill	B	MN
Claude	A	NN
Henry	AB	MM

The child's blood type has a B allele and an N allele that could not have come from the mother and must have come from the father. Therefore, the child's father must have a B and an N. George, Claude, and Henry are eliminated as possible fathers because they lack either a B or an N.

Section 5.5

*25. In chickens, comb shape is determined by alleles at two loci (*R*, *r* and *P*, *p*). A walnut comb is produced when at least one dominant allele *R* is present at one locus and at least one dominant allele *P* is present at a second locus (genotype *R_ P_*). A rose comb is produced when at least one dominant allele is present at the first locus and two recessive alleles are present at the second locus (genotype *R_ pp*). A pea comb is produced when two recessive alleles are present at the first locus and at least one dominant allele is present at the second (genotype *rr P_*). If two recessive alleles are present at the first and at the second locus (*rr pp*), a single comb is produced. Progeny with what types of combs and in what proportions will result from the following crosses?

a. *RR PP* × rr pp
All walnut (Rr Pp)
b. *Rr Pp* × *rr pp*
¼ *walnut* (Rr Pp), ¼ *rose* (Rr pp), ¼ *pea* (rr Pp), ¼ *single* (rr pp)
c. *Rr Pp* × *Rr Pp*
9/16 walnut (R_ P_), *3/16 rose* (R_ pp), *3/16 pea* (rr P_), *1/16 single* (rr pp)
d. *Rr pp* × *Rr pp*
¾ *rose* (R_ pp), ¼ *single* (rr pp)
e. *Rr pp* × *rr Pp*
¼ *walnut* (Rr Pp), ¼ *rose* (Rr pp), ¼ *pea* (rr Pp), ¼ *single* (rr pp)
f. *Rr pp* × *rr pp*
½ *rose* (Rr pp), ½ *single* (rr pp)

*26. Tatuo Aida investigated the genetic basis of color variation in the Medaka *(Aplocheilus latipes)*, a small fish that occurs naturally in Japan (T. Aida. 1921. *Genetics* 6:554–573). Aida found that genes at two loci (*B, b* and *R, r*) determine the color of the fish: fish with a dominant allele at both loci (*B_ R_*) are brown, fish with a dominant allele at the *B* locus only (*B_ rr*) are blue, fish with a dominant allele at the *R* locus only (*bb R_*) are red, and fish with recessive alleles at both loci (*bb rr*) are white. Aida crossed a homozygous brown fish with a homozygous white fish. He then backcrossed the F_1 with the homozygous white parent and obtained 228 brown fish, 230 blue fish, 237 red fish, and 222 white fish.

a. Give the genotypes of the backcross progeny.
Each of the backcross progeny received recessive alleles b *and* r. *Their phenotype is therefore determined by the alleles received from the other parent:*
Brown fish are Bb Rr; *blue fish are* Bb rr; *red fish are* bb Rr; *and white fish are* bb rr.

b. Use a chi-square test to compare the observed numbers of backcross progeny with the number expected. What conclusion can you make from your chi-square results?
We expect a 1:1:1:1 ratio of the four phenotypes.

	Observed	*Expected*	*O – E*	*$(O – E)^2/E$*
Brown	*228*	*229.25*	*–1.25*	*.007*
Blue	*230*	*229.25*	*0.75*	*.002*
Red	*237*	*229.25*	*7.75*	*.262*
White	*222*	*229.25*	*–7.25*	*.229*
Total	*917*	*917*		*$.5 = \chi^2$*

d.f. = 4 – 1 = 3; .9 < p < .975; we cannot reject the hypothesis.

c. What results would you expect for a cross between a homozygous red fish and a white fish?
The homozygous red fish would be bb RR, *crossed to* bb rr. *All progeny would be* bb Rr, *or red fish.*

d. What results would you expect if you crossed a homozygous red fish with a homozygous blue fish and then backcrossed the F_1 with a homozygous red parental fish?
Homozygous red fish bb RR × *homozygous blue fish* BB rr
F_1 will be all brown: Bb Rr *backcrossed to* bb RR *(homozygous red parent)*
Backcross progeny will be in equal proportions Bb RR *(brown);* Bb Rr *(brown);* bb RR *(red); and* bb Rr *(red). Overall, ½ brown and ½ red.*

27. A variety of opium poppy (*Papaver somniferum L.*) having lacerate leaves was crossed with a variety that has normal leaves. All the F_1 had lacerate leaves. Two F_1 plants were interbred to produce the F_2. Of the F_2, 249 had lacerate leaves and 16 had normal leaves. Give genotypes for all the plants in the P, F_1, and F_2 generations. Explain how lacerate leaves are determined in the opium poppy.
The F_1 progeny tell us that lacerate is dominant over normal leaves. In the F_2, 249:16 does not come close to a 3:1 ratio. Let's see if these numbers fit a dihybrid ratio. Dividing 265

total progeny by 16 (because dihybrid ratios are based on 16ths), we see that 1/16 of 265 is 16.56. Therefore, the F_2 progeny are very close to 15/16 lacerate, 1/16 normal, a modified dihybrid ratio. If we symbolize the two genes as A *and* B, *then:*

F_1 Aa Bb × Aa Bb *all lacerate*
F_2 *9/16* A- B- *lacerate (like F_1)*
3/16 A- bb *lacerate*
3/16 aa B- *lacerate*
1/16 aa bb *normal*

A dominant allele at either gene A *or gene* B, *or both, results in lacerate leaves. Finally, the parents must have been* AA BB *lacerate* × aa bb *normal. Note that only* AA BB *for the lacerate parent would result in F_1 that are* Aa Bb.

28. E. W. Lindstrom crossed two corn plants with green seedlings and obtained the following progeny: 3583 green seedlings, 853 virescent-white seedlings, and 260 yellow seedlings. (E. W. Lindstrom. 1921. *Genetics* 6:91–110).
    a. Give the genotypes for the green, virescent-white, and yellow progeny.
    There are 4,696 total progeny. Green appears dominant. The ratios at first glance don't fit any type of incomplete dominance for a single locus, so we hypothesize multiple loci with gene interactions. The simplest case is two loci, so we look for a fit to a ratio based on 1/16 of the total: 293.5. Quick computation with a calculator shows that these numbers are close to a 12:3:1 ratio of green:virescent-white:yellow, a modified 9:3:3:1 ratio. Let's define G *and* g *for one locus, and* Y *and* y *for the other locus.*

    9 G_ Y_ + *3* G_ yy = *12 green*
    3 gg Y_ = *3 virescent-white*
    1 gg yy = *1 yellow*

    b. Provide an explanation for how color is determined in these seedlings.
    The green arises when the G *locus is dominant, regardless of the alleles at the other* Y *locus. Yellow requires that both loci be recessive, and virescent-white arises when the* G *locus is homozygous recessive, and the* Y *locus has a dominant allele.*
    c. Does epistasis occur among the genes that determine color in the maize seedlings? If so, which gene is epistatic and which is hypostatic?
    As defined above, the G *locus is the epistatic locus. It is an example of dominant epistatis, because a dominant allele at this locus masks the effect of the* Y *locus. The* Y *locus is hypostatic, and its effect revealed only when the epistatic locus is homozygous recessive.*

*29. A dog breeder liked yellow and brown Labrador retrievers. In an attempt to produce yellow and brown puppies, he bought a yellow Labrador male and a brown Labrador female and mated them. Unfortunately, all the puppies produced in this cross were black. (See pp. 113–114 for a discussion of the genetic basis of coat color in Labrador retrievers.)

    a. Explain this result.
    Labrador retrievers vary in two loci, B *and* E. *Black dogs have dominant alleles at both loci* (B- E-), *brown dogs have* bb E-, *and yellow dogs have* B- ee *or* bb ee.

Because all the puppies were black, they must all have inherited a dominant B *allele from the yellow parent, and a dominant* E *allele from the brown parent. The brown female parent must have been* bb EE, *and the yellow male must have been* BB ee. *The black puppies were all* Bb Ee.

b. How might the breeder go about producing yellow and brown Labradors?
Simply mating yellow with yellow will produce all yellow Labrador puppies. Mating two brown Labradors will produce either all brown puppies, if at least one of the parents is homozygous EE, *or ¾ brown and ¼ yellow if both parents are heterozygous* Ee.

30. When a yellow female Labrador retriever was mated with a brown male, half of the puppies were brown and half were yellow. The same female, when mated to a different brown male, produced all brown males. Explain these results.
The first brown male was heterozygous for the E *locus, hence he was* bb Ee. *The yellow female has to be* bb ee. *The puppies from this first mating were therefore ½* bb Ee *(brown) and ½* bb ee *(yellow). The second brown male was homozygous* bb EE. *Thus, all the puppies from the second mating were* bb Ee *(brown).*

*31. A summer squash plant that produces disc-shaped fruit is crossed with a plant that produces long fruit. All the F_1 have disc-shaped fruit. When the F_1 are intercrossed, F_2 progeny are produced in the following ratio: 9/16 disc-shaped fruit: 6/16 spherical fruit: 1/16 long fruit. Give the genotypes of the F_2 progeny.
The modified dihybrid ratio in the F_2 indicates that two genes interact to determine fruit shape. Using generic gene symbols A *and* B *for the two loci, the F_1 heterozygotes are* Aa Bb.

The F_2 are:
9/16 A- B- *disc-shaped (like F_1)*
3/16 A- bb *spherical*
3/16 aa B- *spherical*
1/16 aa bb *long*

32. Some sweet-pea plants have purple flowers and other plants have white flowers. A homozygous variety of pea that has purple flowers is crossed with a homozygous variety that has white flowers. All the F_1 have purple flowers. When these F_1 are self-fertilized, the F_2 appear in a ratio of 9/16 purple to 7/16 white.

a. Give genotypes for the purple and white flowers in these crosses.
The F_2 ratio of 9:7 is a modified dihybrid ratio, indicating two genes interacting. Using A *and* B *as generic gene symbols, we can start with the F_1 heterozygotes:*

F_1 AaBb *purple self-fertilized*
F_2 *9/16* A- B- *purple (like F_1)*
3/16 A- bb *white*
3/16 aa B- *white*
1/16 aa bb *white*

Now we see that purple requires dominant alleles for both genes, so the purple parent must have been AA BB, *and the white parent must have been* aa bb *to give all purple F_1.*

b. Draw a hypothetical biochemical pathway to explain the production of purple and white flowers in sweet peas.

White precursor 1 → *white intermediate 2* → *purple pigment*

Enzyme A *Enzyme B*

33. For the following questions, refer to pages 113–114 for a discussion of how coat color and pattern are determined in dogs.

a. Why are Irish setters reddish in color?
According to the information in Table 5.3, Irish setters are BBeeSS, *and* A *or* a^t.
The B *permits expression of black pigment, but the* ee *genotype prevents black color on the body coat,resulting in a reddish color except on the nose and in the eyes. The* S *prevents spotting, resulting in a uniform coat color.*

b. Can a poodle crossed with any other breed produce spotted puppies? Why or why not?
Poodles are SS. *Because the dominant* S *allele prevents spotting, no puppies from matings with poodles will have spotting.*

c. If a St. Bernard is crossed with a Doberman, will the offspring have solid, yellow, saddle, or bicolor coats?
St. Bernards are a^ya^yBB, *and Dobermans are* a^ta^tEESS. *The offspring will be of genotype* a^ya^tB-E-S-. *Because* a^y *specifying yellow is dominant over* a^t, *and the* E *allele allows expression of the* A *genotype throughout, the offspring will have yellow coats.*

d. If a Rottweiler is crossed with a Labrador Retriever, will the offspring have solid, yellow, saddle, or bicolor coats?
Rottweilers are a^ta^tBBEESS, *and Labrador Retrievers are* A^sA^sSS. *The offspring will be* A^sa^tB-E-SS. *The combination of the dominant* A^s *and* E *alleles should create solid coats.*

Section 5.6

34. Male-limited precocious puberty results from a rare, sex-limited autosomal allele (*P*) that is dominant over the allele for normal puberty (*p*) and is expressed only in males. Bill undergoes precocious puberty, but his brother Jack and his sister Beth underwent puberty at the usual time, between the ages of 10 and 14. Although Bill's mother and father underwent normal puberty, two of his maternal uncles (his mother's brothers) underwent precocious puberty. All of Bill's grandparents underwent normal puberty. Give the most likely genotypes for all the relatives mentioned in this family.
Since precocious puberty is dominant, all the males who experienced normal puberty, such as Jack, Bill's father, and Bill's grandfathers, must be pp. *Bill and his two maternal uncles, who all experienced precocious puberty, are* Pp. *We know they are*

heterozygotes not only because P *is a rare allele but also because these individuals all had fathers that are* pp. *This means Bill inherited* P *from his mother, who must have been* Pp. *Bill's sister Beth could be either* Pp *or* pp.

35. In some goats, the presence of horns is produced by an autosomal gene that is dominant in males and recessive in females. A horned female is crossed with a hornless male. The F_1 offspring are intercrossed to produce the F_2. What proportion of the F_2 females will have horns?
Let H^+ represent the allele for the presence of horns and H^- represent the allele for hornlessness. Since H^+ is recessive in females, the horned female parent must be H^+H^+. The hornless male is H^-H^- because the absence of horns is recessive in males. Then their F_1 progeny must be all heterozygous H^+H^-. An intercross of the F_1 would produce both male and female progeny in the ratio of 1 H^+H^+, 2 H^+H^-, and 1 H^-H^-. Again, remembering that H^+ is recessive in females, we would expect a ratio of 3:1 hornless to horned females.

36. In goats, a beard is produced by an autosomal allele that is dominant in males and recessive in females. We'll use the symbol B^b for the beard allele and B^+ for the beardless allele. Another independently assorting autosomal allele that produces a black coat (*W*) is dominant over the allele for white coat (*w*). Give the phenotypes and their proportions expected for the following crosses:

a. B^+B^b *Ww* male × B^+B^b *Ww* female
Because beardedness and coat color independently assort, we can treat them independently. The difference between this cross and a dihybrid cross is that the bearded allele B^b *is dominant in males and recessive in females. So we deal with male and female progeny separately. For each sex, then, we should get a typical dihybrid ratio. In males, the dominant phenotype is bearded, so we should get ¾ bearded, ¼ beardless. In females the dominant phenotype is beardless, so we should get ¾ beardless and ¼ bearded. Each sex will have ¾ black and ¼ white coats.*

Males:		*Females:*	
	9/16 bearded, black		*9/16 beardless, black*
	3/16 bearded, white		*3/16 beardless, white*
	3/16 beardless, black		*3/16 bearded, black*
	1/16 beardless, white		*1/16 bearded, white*

b. B^+B^b *Ww* male × B^+B^b *ww* female
Here the males will again be ¾ bearded and ¼ beardless, and the females will be ¾ beardless and ¼ bearded. This time half the progeny of either sex will be black, and half will be white.

Males:		*Females:*	
	3/8 bearded, black		*3/8 beardless, black*
	3/8 bearded, white		*3/8 beardless, white*
	1/8 beardless, black		*1/8 bearded, black*
	1/8 beardless, white		*1/8 bearded, white*

c. B^+B^+ *Ww* male × B^bB^b *Ww* female
In this cross, all of the male progeny will be bearded, and all of the female progeny will be beardless. All will be ¾ black, ¼ white.

Males:	*¾ bearded, black*	*Females:*	*¾ beardless, black*
	¼ bearded, white		*¼ beardless, white*

d. $B^{+}B^{b}$ *Ww* male × $B^{b}B^{b}$ *ww* female
Males will be all bearded, and females will be ½ bearded, ½ beardless. Both males and females will be ½ black, ½ white.

Males:	*½ bearded, black*	*Females:*	*¼ beardless, black*
	½ bearded, white		*¼ beardless, white*
			¼ bearded, black
			¼ bearded, white

37. J. K. Breitenbecher (1921. *Genetics* 6:65–86) investigated the genetic basis of color variation in the four-spotted cowpea weevil *(Bruchus quadrimaculatus).* The weevils were red, black, white, or tan. Breitenbecher found that four alleles (R, R^{b}, R^{w}, and r) at a single locus determine color. The alleles exhibit a dominance hierarchy, with red (R) dominant over all other alleles, black (R^{b}) dominant over white (R^{w}) and tan (r), white dominant over tan, and tan recessive to all others ($R > R^{b} > R^{w} > r$). The following genotypes encode each of the colors:

RR, RR^{b}, RR^{w}, Rr	red
$R^{b}R^{b}$, $R^{b}R^{w}$, $R^{b}r$	black
$R^{w}R^{w}$, $R^{w}r$	white
rr	tan

Color variation in this species is sex-limited to females: males carry color genes but are always tan regardless of their genotype. For each of the following crosses carried out by Breitenbecher, give all possible genotypes of the parents.

	Parents	**Progeny**
a.	tan ♀ × tan ♂	78 red ♀, 70 white ♀, 184 tan ♂

rr × RR^{w}; *the tan female has only one possible genotype. We ignore the male progeny, and see that the female progeny are 1:1 red white, so the tan male parent must have been heterozygous with both red and white alleles.*

b. black ♀ × tan ♂ — 151 red ♀, 49 black ♀, 61 tan ♀, 249 tan ♂
The black female is $R^{b}r$, *the male is* Rr. *The tan female progeny indicates that both parents had a tan allele. The black female must then have a black allele, and the red progeny can arise only if the male has a red allele.*

c. white ♀ × tan ♂ — 32 red ♀, 31 tan ♂
The white female could be either $R^{w}R^{w}$ *or* $R^{w}r$. *The male parent must be* RR, *to produce all red female progeny.*

d. black ♀ × tan ♂ — 3586 black ♀, 1282 tan ♀, 4791 tan ♂
Black female is $R^{b}r$, *and the male is* $R^{b}r$, *to produce a 3:1 ratio of black to tan.*

e. white ♀ × tan ♂ — 594 white ♀, 189 tan ♀, 862 tan ♂
$R^{w}r$ *female and* $R^{w}r$ *male, to produce 3:1 white to tan.*

f. black ♀ × tan ♂ — 88 black ♀, 88 tan ♀, 186 tan ♂

Black female is $R^b r$, *the male is* rr, *to produce a 1:1 ratio of black to tan.*

g. tan ♀ × tan ♂ 47 white ♀, 51 tan ♀, 100 tan ♂
Tan female can be only rr; *male must be* $R^w r$.

h. red ♀ × tan ♂ 1932 red ♀, 592 tan ♀, 2587 tan ♂
Red female is Rr; *male is also* Rr.

i. white ♀ × tan ♂ 13 red ♀, 6 white ♀, 5 tan ♀, 19 tan ♂
White female is $R^w r$; *male is* Rr.

j. red ♀ × tan ♂ 190 red ♀, 196 black ♀, 311 tan ♂
RR^b *female and* $R^b R^b$ *male. The 1:1 ratio is produced by a heterozygote crossed with a homozygous recessive.*

k. black ♀ × tan ♂ 1412 black ♀, 502 white ♀, 1766 tan ♂
$R^b R^w$ *female and* $R^b R^w$ *male, to produce 3:1 black to white progeny. Additionally, either one of the parents, but not both, could be* $R^b r$.

38. Shell coiling of the snail *Limnaea peregra* results from a genetic maternal effect. An autosomal allele for a right-handed shell (s^+), called dextral, is dominant over the allele for a left-handed shell (*s*), called sinistral. A pet snail called Martha is sinistral and reproduces only as a female (the snails are hermaphroditic). Indicate which of the following statements are true and which are false. Explain your reasoning in each case.

a. Martha's genotype *must* be *ss*.
False. For maternal effect genes, the phenotype of the individual is determined solely by the genotype of the individual's mother. So we know Martha's mother must have been ss *because Martha is sinistral. If Martha was produced as a result of self-fertilization, then Martha must indeed be* ss. *But if Martha was produced by cross-fertilization, then we cannot know Martha's genotype without more information.*

b. Martha's genotype cannot be s^+s^+.
True. As explained in the answer to part (a), Martha's mother is ss, *so Martha must be either* s^+s *or* ss.

c. All the offspring produced by Martha *must* be sinistral.
False. Because we do not know Martha's genotype, we cannot yet predict the phenotype of her offspring.

d. At least some of the offspring produced by Martha *must* be sinistral.
False. If Martha is s^+s, *then all her children will be dextral. If Martha is* ss, *then all her children will be sinistral.*

e. Martha's mother *must* have been sinistral.
False. Martha's mother's phenotype is determined by the genotype of her *mother (Martha's maternal grandmother). We know Martha's mother's genotype must have been* ss, *so her mother's mother had at least one* s *allele. But we cannot know if she was a heterozygote or homozygous* ss.

f. All Martha's brothers *must* be sinistral.
True. Because Martha's mother must have been ss, *all her progeny must be sinistral.*

39. Hypospadias, a birth defect in male humans in which the urethra opens on the shaft instead of the tip of the penis, results from an autosomal dominant gene in some

families. Females who carry the gene show no effects. This is an example of: (a) an X-linked trait, (b) a Y-linked trait, (c) a sex-limited trait, (d) a sex influenced trait, or (e) genetic maternal effect? Explain your answer.
Knowing that the condition arises from an autosomal gene, we can eliminate either (a) an X-linked trait or (b) a Y-linked trait. Dominant inheritance also eliminates (e) maternal effect. If it were (d) a sex influenced trait, females would be affected to a lesser degree or differently than males. Because females who carry the gene show no effects, this condition is (c) a sex-limited trait.

40. In unicorns, two autosomal loci interact to determine the type of tail. One locus controls whether a tail is present at all; the allele for a tail (*T*) is dominant over the allele for tailless (*t*). If a unicorn has a tail, then alleles at a second locus determine whether the tail is curly or straight. Farmer Baldridge has two unicorns with curly tails. When he crosses these two unicorns, ½ of the progeny have curly tails, ¼ have straight tails, and ¼ do not have a tail. Give the genotypes of the parents and progeny in Farmer Baldridge's cross. Explain how he obtained the 2:1:1 phenotypic ratio in his cross.
We are given the symbols T *for dominant tailed and* t *for recessive tailless. We are not given any information about dominance or recessiveness for the second locus. We will use* S *and* s *for the second locus that determines whether the tail is curly or straight. Although two genes are interacting, we can analyze one locus at a time. Farmer Baldridge crossed two unicorns with tails and got a 3:1 ratio of tailed to tailless. Therefore, the two unicorns were heterozygous for the tail locus:* Tt. *The parents were both curly, and the progeny were both curly and straight, in a 2:1 ratio of curly:straight. The fact that he got straight-tailed progeny indicates that the curly tailed parents were heterozygous* Ss. *The fact that he got a 2:1 ratio instead of a 3:1 ratio indicates that this locus may not have a dominant:recessive relationship. A 2:1 ratio may be obtained if ¼ of the progeny, one of the homozygote classes, are missing (because of embryonic lethality).*

Having deduced the genotypes of the parents, we can determine the expected genotypes of the progeny:

P: curly tailed Tt Ss × *curly tailed* Tt Ss
Using a branch diagram:

¾ T- → ¼ SS = *3/16 straight-tailed*
¾ T- → ½ Ss = *6/16 curly tailed*
¾ T- → ¼ ss = *3/16 lethal*

and

¼ tt → ¼ SS = *1/16 tailless*
¼ tt → ½ Ss = *2/16 tailless*
¼ tt → ¼ ss = *1/16 lethal*

Because 4/16 of the progeny die, leaving 12/16 viable, the surviving progeny would be 6/12 curly tailed, 3/12 straight-tailed, and 3/12 tailless, which reduces to a 2:1:1 ratio of curly tailed to straight-tailed to tailless.

If S *were dominant over* s *without lethality, we would expect the following* F_1:

9/16 T- S- *curly tailed*
3/16 T- ss *straight-tailed*
3/16 tt S- *tailless*
1/16 tt ss *tailless*

However, these predictions do not fit the observed 2:1:1 ratio.

41. In 1983, a sheep farmer in Oklahoma noticed a ram in his flock that possessed increased muscle mass in his hindquarters. Many of the offspring of this ram possessed the same trait, which became known as the callipyge mutant (*callipyge* is Greek for "beautiful buttocks"). The mutation that caused the callipyge phenotype was eventually mapped to a position on the sheep chromosome 18.

When the male callipyge offspring of the original mutant ram were crossed with normal females, they produced the following progeny: ¼ male callipyge, ¼ female callipyge, ¼ male normal, and ¼ female normal. When female callipyge offspring of the original mutant ram were crossed with normal males, all of the offspring were normal. Analysis of the chromosomes of these offspring of callipyge females showed that half of them received a chromosome 18 with the callipyge gene from their mother. Propose an explanation for the inheritance of the callipyge gene. How might you test your explanation?

Here we get different results depending on the sex of the parent with the callipyge mutation. Because we know the gene is on chromosome 18, we can eliminate sex linkage as a possible cause. The phenotype is also not sex-limited because male callipyge sire equal proportions of male and female callipyge offspring. We can further eliminate maternal inheritance for the same reason. That leaves imprinting as a possible explanation. We note that half the progeny are callipyge if the father has the mutation, but none express the callipyge mutation if the mother has the mutation. We can therefore hypothesize that maternal alleles of this gene undergo imprinting and are silenced, so that the embryo expresses only the paternal allele.

We can test this hypothesis by mating the phenotypically normal male and female progeny that inherited the chromosome 18 with the callipyge gene from their mother. The hypothesis predicts that males will have normal and callipyge progeny if mated to either a normal female or a callipyge female. Conversely, the females will have all normal progeny if mated to a normal male, and both normal and callipyge progeny if mated to a callipyge male. In short, the progeny will reflect the genotype of the father, and the genotype of the mother will not be expressed.

Section 5.8

42. Which of the following statements is an example of a phenocopy? Explain your reasoning.

 a. Phenylketonuria results from a recessive mutation that causes light skin as well as mental retardation.
 Phenocopy is an environmentally induced phenotype that resembles a phenotype produced by a genotype. Since phenylketonuria has a genetic basis, this is not a phenocopy. One genotype affecting multiple traits is called pleitrophy.

 b. Human height is influenced by genes at many different loci.
 This is again not an example of phenocopy, but of a continuous characteristic or a quantitative trait.

 c. Dwarf plants and mottled leaves in tomatoes are caused by separate genes that are linked.
 Linkage of genes is not an example of phenocopy.

 d. Vestigial wings in *Drosophila* are produced by a recessive mutation. This trait is also produced by high temperature during development.
 This is indeed an example of phenocopy because an environmental factor produces a phenotype that resembles the phenotype generated by a genotype.

 e. Intelligence in humans is influenced by both genetic and environmental factors.
 As long as there is a significant effect of the underlying genotype, this is not a phenocopy. The expression of many genotypes is indeed influenced by environmental factors.

43. Long ears in some dogs are an autosomal dominant trait. Two dogs mate and produce a litter in which 75% of the puppies have long ears. Of the dogs with long ears in this litter, 1/3 are known to be phenocopies. What are the most likely genotypes of the two parents of this litter?
 Accounting for the phenocopies, we have 50% (subtracting 1/3 that are phenocopies from the 75%) of the puppies having the autosomal dominant genotype for long ears, and 50% having the recessive genotype. Therefore, one parent is homozygous recessive, and the other parent is a heterozygote.

CHALLENGE QUESTION

Section 5.1

44. Pigeons have long been the subject of genetic studies. Indeed, Charles Darwin bred pigeons in the hope of unraveling the principles of heredity but was unsuccessful. A series of genetic investigations in the early 1900s worked out the hereditary basis of color variation in these birds. W. R. Horlancher was interested in the genetic basis of kiteness, a color pattern that consists of a mixture of red and black stippling of the feathers. Horlancher knew from earlier experiments that black feather color was dominant over red. He carried out the following crosses to investigate the genetic relationship of kiteness to black and red feather color (W. R. Horlancher. 1930. *Genetics* 15:312–346).

Cross	**Offspring**
kitey × kitey	16 kitey, 5 black, 3 red
kitey × black	6 kitey, 7 black
red × kitey	18 red, 9 kitey, 6 black

a. On the basis of these results and the assumption that black is dominant over red, propose an hypothesis to explain the inheritance of kitey, black, and red feather color in pigeons.

The first cross of kitey × kitey indicates that both kitey parents must be heterozygous, to produce black and red progeny in addition to kitey progeny. The second cross, yielding a 1:1 ratio of the two parental phenotypes, is consistent with a mating between a heterozygote and a homozygous recessive. The third cross, also yielding progeny of three different phenotypes, again must be between two heterozygotes. The challenge is to assign genotypes to the parents and progeny that is consistently explains all three crosses.

Given that black is dominant over red, we have to introduce at least one additional allele to account for kitey. Starting with the simplest possible model, let's define C^B *as the allele for black,* C^R *as the allele for red, and* C^K *as the allele for kitey. Starting with cross 1, if the two kitey parents are heterozygotes, then kitey must be the dominant allele. The recessive alleles must be different in the two kitey parents, or else the progeny would show a 3:1 phenotypic ratio. So we can assign* C^KC^B *and* C^KC^R *for the two parents, and the progeny will be:*

C^KC^B *– kitey, like one parent*
C^KC^R *– kitey, like the other parent*
C^BC^R *– black, because black is dominant over red*
C^KC^K *– red, because all other genotypes are accounted for*

The problem with this scheme is that it cannot account for the third cross of red × kitey. This scheme indicates that red is either homozygous C^RC^R *or* C^KC^K*, and the red parent in the third cross must be heterozygous. In fact, there is no way three alleles can be sufficient to account for all three crosses.*

Let's try accounting for the two kitey parents in cross 1 with different kitey alleles, C^{K1} *and* C^{K2}*:*

P: $C^{K1}C^B$ *(kitey) ×* $C^{K2}C^R$ *(kitey)*
F₁: $C^{K2}C^B$ *– kitey?*
$C^{K1}C^R$ *– red?*
C^BC^R *– black*
$C^{K1}C^{K2}$ *– kitey?*

Cross 2 would then be:

$C^{K1}C^B$ *(kitey) ×* C^BC^B *(black) →* $C^{K1}C^B$ *(kitey) +* C^BC^B *(black) in 1:1 ratio*

Cross 3 would then be:

P: $C^{K1}C^R$ *(red) ×* $C^{K1}C^B$ *(black)*
F₁: $C^{K1}C^B$ *– kitey*
$C^{K1}C^R$ *– red*
$C^{K1}C^{K1}$ *– red?*

Expected progeny phenotypic ratio: 2 red, 1 kitey, 1 black

In this revised model, C^{K2} *is a kitey allele that is dominant over all other alleles.* C^{B} *is dominant over* C^{R}*, as given in the problem, and* C^{R} *is dominant over* C^{K1}*.* C^{K1}*, however, is dominant over* C^{B} *– modifies the black allele to produce a kitey phenotype. In the absence of the black allele,* C^{K1} *produces red; hence* C^{K1} *homozygotes or heterozygotes with* C^{R} *are all red.*

b. For each cross given above, test your hypothesis by using a chi-square test.

Cross 1:

	Observed	*Expected*	*O – E*	*(O – E)²/E*
Kitey	*16*	*12*	*4*	*1.33*
Black	*5*	*6*	*–1*	*0.17*
Red	*3*	*6*	*–3*	*1.50*
Total	*24*	*24*		*3.0 =* χ^2

d.f. = 2; .1 < p < .5; do not reject hypothesis.

Cross 2:

	Observed	*Expected*	*O – E*	*(O – E)²/E*
Kitey	*6*	*6.5*	*–0.5*	*.04*
Black	*7*	*6.5*	*0.5*	*.04*
Total	*13*	*13*		*.08 =* χ^2

d.f. = 1; .5 < p < .9; do not reject hypothesis

Cross 3:

	Observed	*Expected*	*O – E*	*(O – E)²/E*
Kitey	*9*	*8.25*	*0.75*	*0.07*
Black	*6*	*8.25*	*–2.25*	*0.61*
Red	*18*	*16.5*	*1.5*	*0.14*
Total	*33*	*33*		*0.82 =* χ^2

d.f. = 2; .5 < p < .9; do not reject hypothesis

Section 5.6

45. Suppose that you are tending a mouse colony at a genetics research institute and one day you discover a mouse with twisted ears. You breed this mouse with twisted ears and find that the trait is inherited. Male and female mice have twisted ears, but when you cross a twisted-eared male with normal-eared female, you obtain different results from those you obtained when you cross a twisted-eared female with normal-eared male—the reciprocal crosses give different results. Describe how you would go about determining whether this trait results from a sex-linked gene, a sex-influenced gene, a genetic maternal effect, a cytoplasmically inherited gene, or genomic imprinting. What crosses would you conduct and what results would be expected with these different types of inheritance?
Each of these is a distinct pattern of inheritance. Because male and females have twisted ears (te)*, Y-linkage is eliminated. X-linked genes are passed from mother to son and from father to daughter. A sex-influenced gene shows a different phenotype*

depending on the sex but is inherited autosomally. A genetic maternal effect depends only on the genotype of the mother; the genotype of the zygote is immaterial. A cytoplasmically inherited trait is serially perpetuated from mother to all her progeny. Genomic imprinting results in the gene of only one of the parents being expressed.

To distinguish among these possibilities, you will need pure-breeding lines of mice with twisted ears and normal ears. Perform reciprocal crosses of males with twisted ears to females with normal ears (cross A) and males with normal ears to females with twisted ears (cross B).

	A: te *male × normal female*		*B: normal male ×* te *female*	
	F_1 males	*F_1 females*	*F_1 males*	*F_1 females*
Sex-linked	*All normal*	*All dominant*	*All te*	*All dominant*
Sex-influenced	*Het male*	*Het female*	*Het male*	*Het female*
Maternal	*Normal*	*Normal*	*te*	*te*
Cytoplasmic	*Normal*	*Normal*	*te*	*te*
Imprinting pat	*Normal*	*Normal*	*te*	*te*
Imprinting mat	te	te	*Normal*	*Normal*

What we see from the table above is that if the trait is sex-linked, cross A and cross B give different phenotypes for the F_1 males, which match the phenotypes of their mothers. The F_1 females have the same dominant phenotype in either cross. If the trait is sex-influenced (heterozygous males have a different phenotype than heterozygous females), these reciprocal crosses with pure-breeding parents give the same results.

Both of these results are distinct from the results with maternal inheritance, cytoplasmic inheritance, or paternal imprinting, which all give the same results: no difference between male and female F_1 progeny, but the two crosses result in opposite phenotypes.

A further cross is needed to distinguish among maternal effect, cytoplasmic inheritance, and paternal imprinting. For these modes of inheritance, the phenotypes of the progeny depend solely on the maternal contribution, and no phenotypic differences are expected among male and female progeny. The F_1 female progeny from cross A and cross B should have the same genotype (heterozygous), but they have different phenotypes. The three remaining modes of inheritance predict different phenotypes of F_2 progeny from these females, as shown in the table below.

	Phenotypes of progeny of normal male × F_1 female from:	
Mode of inheritance	*Cross A (normal ears)*	*Cross B (twisted ears)*
Maternal	*Dominant phenotype*	*Dominant phenotype*
Cytoplasmic	*Normal ears*	*Twisted ears*
Paternal imprinting	*1:1 normal:twisted*	*1:1 normal:twisted*

In the case of maternal inheritance, the progeny depend on the genotype of the mother, and because the F_1 females from both crosses have the same heterozygous genotype, *their progeny will have the same phenotype: normal ears or twisted ears, whichever is dominant.*

For cytoplasmic inheritance, the phenotype of the progeny will be the same as the phenotype of the mother. Because the F_1 females have different phenotypes, their progeny will have different phenotypes.

For paternal imprinting, only the maternal genes are expressed in the progeny. Because the mother is heterozygous, the progeny should have 1:1 ratio of normal ears and twisted ears.

Other solutions are possible; this is just one.

Chapter Six: Pedigree Analysis, Applications, and Genetic Testing

COMPREHENSION QUESTIONS

Section 6.1

*1. What three factors complicate the task of studying the inheritance of human characteristics?

(1) Mating cannot be controlled. It is not ethical or feasible to set up controlled mating experiments.

(2) Humans have a long generation time, so it takes a long time to track inheritance of traits over more than one generation.

(3) The number of progeny per mating is limited, so phenotypic ratios are uncertain.

Section 6.2

2. Who is the proband in a pedigree? Is the proband always found in the last generation of the pedigree? Why or why not?

The proband is the person of interest for whom the pedigree chart has been drawn. The proband is not necessarily found in the last generation because the proband's children, or the children of the proband's siblings, often provide information about the genotype of the proband.

Section 6.3

*3. For each of the following modes of inheritance, describe the features that will be exhibited in a pedigree in which the trait is present: autosomal recessive, autosomal dominant, X-linked recessive, X-linked dominant, and Y-linked inheritance.

Pedigrees with autosomal recessive traits will show affected males and females arising with equal frequency from unaffected parents. The trait often appears to skip generations. Unaffected people with an affected parent will be carriers.

Pedigrees with autosomal dominant traits will show affected males and females arising with equal frequency from a single affected parent. The trait does not usually skip generations.

X-linked recessive traits will affect males predominantly and will be passed from an affected male through his unaffected daughter to his grandson. X-linked recessive traits are not passed from father to son.

X-linked dominant traits will affect males and females and will be passed from an affected male to all his daughters, but not to his sons. An affected woman (usually heterozygous for a rare dominant trait) will pass on the trait equally to half her daughters and half her sons.

Y-linked traits will show up exclusively in males, passed from father to son.

4. How does the pedigree of an autosomal recessive trait differ from the pedigree of an X-linked recessive trait?
Pedigrees of autosomal recessive traits will have equal frequencies of affected males and females, whereas pedigrees of X-linked recessive traits will show mostly affected males. Also, both parents must be carriers to have children with autosomal recessive traits, whereas a mother carrying an X-linked trait can have affected sons regardless of the genotype of the father. Finally, an X-linked trait is never passed from the father to his sons.

5. Other than the fact that a Y-linked trait appears only in males, how does the pedigree of a Y-linked trait differ from the pedigree of an autosomal dominant trait?
A Y-linked dominant trait is passed from a father to all of his sons, whereas an autosomal dominant trait would be passed to only half of his sons.

Section 6.4

*6. What are the two types of twins and how do they arise?
The two types of twins are monozygotic and dizygotic. Monozygotic twins arise when a single fertilized egg splits into two embryos in early embryonic cleavage divisions. They are genetically identical. Dizygotic twins arise from two different eggs fertilized at the same time by two different sperm. They share, on the average, 50% of the same genes.

7. Explain how a comparison of concordance in monozygotic and dizygotic twins can be used to determine the extent to which the expression of a trait is influenced by genes or environmental factors.
Monozygotic twins have 100% genetic identity, whereas dizygotic twins have 50% genetic identity. Any trait that is completely genetically determined will therefore be 100% concordant in monozygotic twins and 50% concordant in dizygotic twins. Conversely, any trait that is completely environmentally determined will have the same degree of concordance in monozygotic and dizygotic twins. To the extent that a trait has greater concordance in monozygotic twins than in dizygotic twins, the trait is genetically influenced. Environmental influences will reduce the concordance in monozygotic twins below 100%.

Section 6.5

8. How are adoption studies used to separate the effects of genes and environment in the study of human characteristics?
Studies of adoptees, their biological parents, and their adoptive parents separate environmental and genetic influences on traits. Adoptees share similar environments with their adoptive parents (because they live in the same house and eat similar foods), but they share 50% of their genes with each of their biological parents. If adoptees have greater similarity for a trait with their adoptive parents,

then the trait is environmentally influenced. If the adoptees have greater similarity for the trait with their biological parents, then the trait is genetically influenced.

Section 6.6

*9. What is genetic counseling?
Genetic counseling provides assistance to clients by interpreting results of genetic testing and diagnosis; providing information about relevant disease symptoms, treatment, and progression; assessing and calculating the various genetic risks that the person or couple faces; and helping clients and family members cope with the stress of decision-making and facing up to the drastic changes in their lives that may be precipitated by a genetic condition.

10. Give at least four different reasons that a person might seek genetic counseling.
(1) The person may be aware of a genetic disease or risk factor in the person's family.
(2) An older woman may be pregnant or contemplating pregnancy and may need information about risks and options for prenatal genetic testing.
(3) A person may have tested positive for a genetic disease or risk factor and may need help with interpretation.
(4) A person or couple may have a child with a genetic disease, or may be caregivers for a person with a genetic disease, and require counseling on treatment options and management of the disease.
(5) A married couple may be closely related (e.g., first cousins) and may need advice about pedigree analysis and genetic testing options.
(6) A couple has difficulty achieving pregnancy or carrying a pregnancy to term.
(7) A person has been exposed to mutagens or chemicals that cause a higher risk of birth defects.
(8) A couple contemplating starting a family may both be carriers of a recessive genetic condition.
(9) A couple needs advice on interpretation of results of a prenatal test.

Section 6.7

11. Briefly define newborn screening, heterozygote screening, presymptomatic testing, and prenatal diagnosis.
Newborn screening: Newborn infants are tested for various treatable genetic disorders by sampling a few drops of their blood soon after birth.

Heterozygote screening: Normal or asymptomatic individuals in a population or community are tested for recessive disease alleles to determine the frequency of the disease allele in the population and to identify carriers, particularly if there is a relatively high incidence of the disease in the population or community.

Presymptomatic testing: People known to be at higher risk for a disease that occurs later in life are tested before symptoms appear.

Prenatal diagnosis: Results from prenatal testing for any of a number of genetic conditions. Techniques, such as amniocentesis or chorionic villus sampling, are used to obtain tissue samples of the still developing fetus, or fetal protein or cells in the maternal circulation are characterized.

12. Compare the advantages and disadvantages of amniocentesis versus chorionic villus sampling for prenatal diagnosis.
Amniocentesis samples the amniotic fluid by inserting a needle into the amniotic sac, usually performed at about 16 weeks of pregnancy, and requires culturing the fetal cells. Chorionic villus sampling can be performed several weeks earlier (10th or 11th week of pregnancy) and samples a small piece of the chorion by inserting a catheter through the vagina. Amniocentesis is relatively safe, but results are not available until week 17 or 18 of pregnancy. Chorionic villus sampling has a slightly higher risk of complication, including fetal injury, but results are available several weeks earlier.

13. What is preimplantation genetic diagnosis?
Preimplantation genetic diagnosis may be performed on embryos created through in vitro fertilization. The embryos are cultured until they reach the 8–16 cell stage, and one cell is removed from each embryo for genetic testing.

14. How does heterozygote screening differ from presymptomatic genetic testing?
Both involve testing healthy individuals, but heterozygote screening refers to testing randomly selected individuals in populations to determine carrier frequency for recessive genetic disorders. Presymptomatic genetic testing refers to testing apparently healthy members of families to determine whether they have inherited a disease allele for diseases that manifest symptoms later in life.

Section 6.8

15. Briefly describe some of the recently discovered genes that contribute to human uniqueness and the importance that they may have had in human evolution.
Comparative genomic approaches aim to identify genes that are either unique to humans or have undergone changes that are unique to humans, and that may help to explain human evolution. Two questions of fundamental interest are the evolution of the large human brain and the evolution of language abilities. Mutations at six loci, named microcephalins (MCPH) 1–6, *can cause severely reduced brain size. These loci may have undergone strong selection for alleles that promote large brain size. Another gene, the* FOXP2 *gene, was identified by mutations that cause speech and language disabilities. The human variant of* FOXP2 *appears to have emerged only about 200,000 years ago, coinciding with the emergence of modern humans.*

APPLICATION QUESTIONS AND PROBLEMS

Section 6.1

16. If humans have characteristics that make them unsuitable for genetic analysis, such as long generation time, small family size, and uncontrolled crosses, why do geneticists study humans? Give several reasons why humans have been the focus of so much genetic study.
Study of human genetics is necessary to understand and overcome human genetic diseases. Because of the long life span, relatively large body size, and uniquely human lifestyle and behaviors, animal models are nonexistent or insufficient for many genetic disorders. The careful preservation of marriage, birth, death, and health records in many societies provide a wealth of data for genetic analysis. The completion of the human genome project now facilitates mapping and identifying human genes. We humans have a strong sense of identity and worth as individuals, and wish to understand how an individual's genetic profile contributes to our health, our behavior, our abilities and disabilities, and our individual future prospects.

Section 6.2

*17. Joe is color blind. His mother and father have normal vision, but his mother's father (Joe's maternal grandfather) is color blind. All Joe's other grandparents have normal color vision. Joe has three sisters—Patty, Betsy, and Lora, all with normal color vision. Joe's oldest sister, Patty, is married to a man with normal color vision; they have two children, a 9-year-old color-blind boy and a 4-year-old girl with normal color vision.

a. Using correct symbols and labels, draw a pedigree of Joe's family.

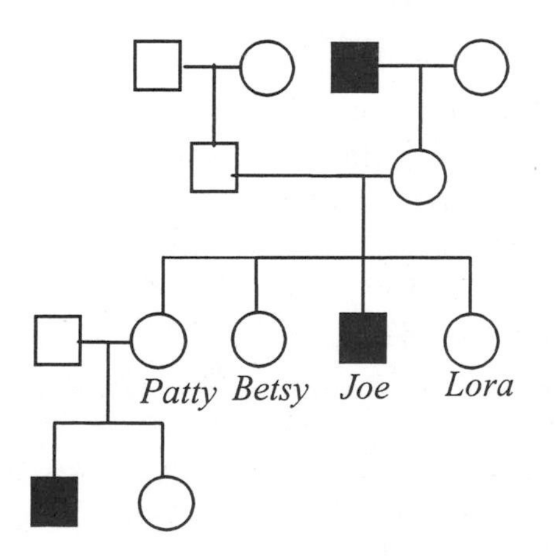

b. What is the most likely mode of inheritance for color blindness in Joe's family?

X-linked recessive. Only males have the trait, and they inherit the trait from their mothers, who are carriers. It cannot be a Y-linked trait because it is not passed from father to son. It is unlikely to be an autosomal recessive trait, because we would not expect two unrelated males marrying into the pedigree (Joe's father and Patty's husband) to be both carriers for a relatively rare trait.

c. If Joe marries a woman who has no family history of color blindness, what is the probability that their first child will be a color-blind boy?
Barring a new mutation or nondisjunction, zero. Joe cannot pass his color-blind X chromosome to his son.

d. If Joe marries a woman who is a carrier of the color-blind allele, what is the probability that their first child will be a color-blind boy?
The probability is ¼. There is ½ probability that their first child will be a boy, and there is an independent ½ probability that the first child will inherit the color-blind X chromosome from the carrier mother. ½(½) = ¼.

e. If Patty and her husband have another child, what is the probability that it will be a color-blind boy?
Again, ¼. Patty is a carrier because she had a color-blind son. The same reasoning applies as in part (d). Each child is an independent event.

Section 6.3

18. Many studies have suggested a strong genetic predisposition to migraine headaches, but the mode of inheritance is not clear. L. Russo and colleagues examined migraine headaches in several families, two of which are shown below (L. Russo et al. 2005. *American Journal of Human Genetics* 76:327–333). What is the most likely mode of inheritance for migraine headaches in these families? Explain your reasoning.

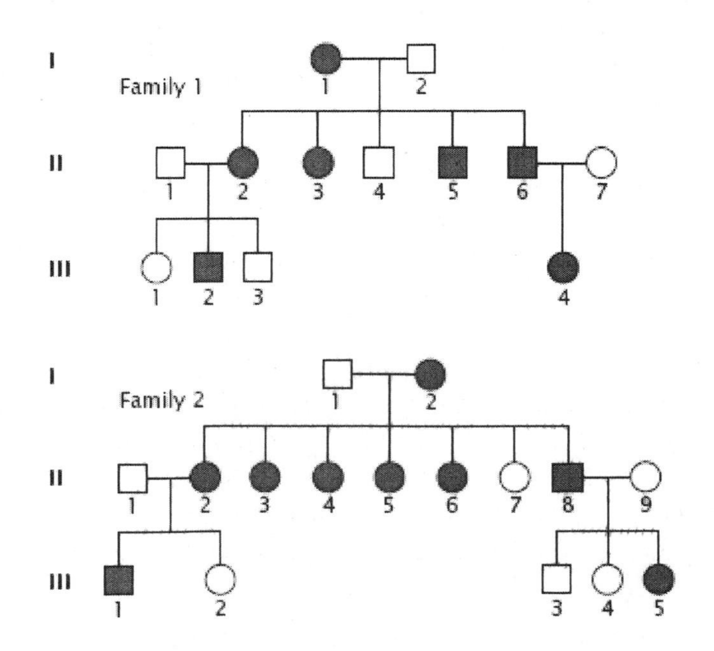

In both families, the trait is most likely dominant because it does not skip generations, and affected individuals have one affected parent. In family 2, it is not X-linked because the affected male II-8 has an unaffected daughter. For X-linked loci, an affected male would transmit the trait to all his daughters. It could be either X-linked or autosomal in family 1.

19. Dent disease is a rare disorder of the kidney in which there is impaired reabsorption of filtered solutes and progressive renal failure. R. R. Hoopes and colleagues studied mutations associated with Dent disease in the following family (R. R. Hoopes et al. 2005. *American Journal of Human Genetics* 76:260–267).

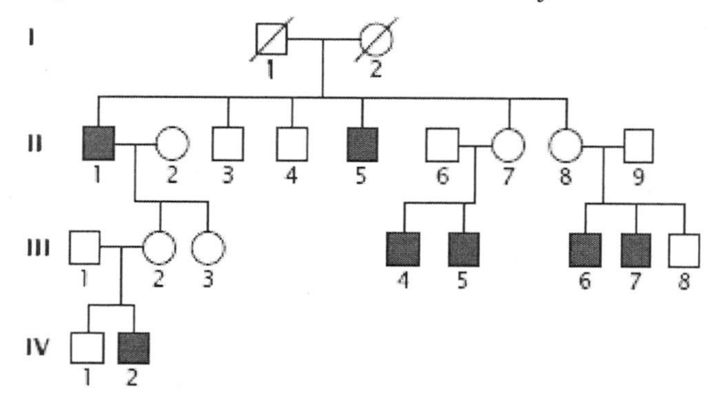

a. On the basis of this pedigree, what is the most likely mode of inheritance for the disease? Explain your reasoning.
Only males have the disease, it skips generations, and unaffected female carriers have both affected and affected sons. These observations are consistent with a recessive X-linked trait. Y-linked traits are transmitted directly from father to son, and do not skip generations.

b. From your answer to part **a.**, give the most likely genotypes for all persons in the pedigree.
We will use X^+ to denote the normal X allele and X^d to denote the Dent allele.
I: 1 – X^+Y; 2 – X^+X^d
II: 1 and 5 are X^dY, 7 and 8 are X^+X^d, the rest do not have the disease allele
III: 2 and 3 are carriers X^+X^d, 4–7 are X^dY, and the rest do not have the disease allele
IV: 2 is X^dY, 1 is X^+Y

20. A man with a specific unusual genetic trait marries an unaffected woman and they have four children. Pedigrees of this family are shown in parts (a) through (e), but the presence or absence of the trait in the children is not indicated. For each type of inheritance, indicate how many children of each sex are expected to express the trait by filling in the appropriate circles and squares. Assume that the trait is rare and fully penetrant.

a. Autosomal recessive trait—*none*

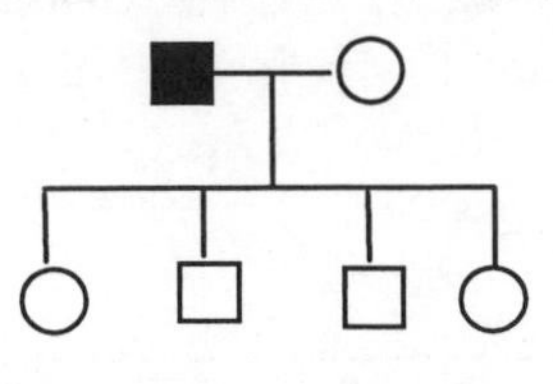

b. Autosomal dominant trait—*½ of each sex*

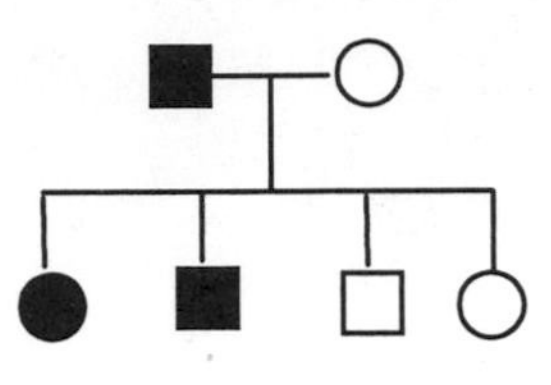

c. X-linked recessive trait—*none*

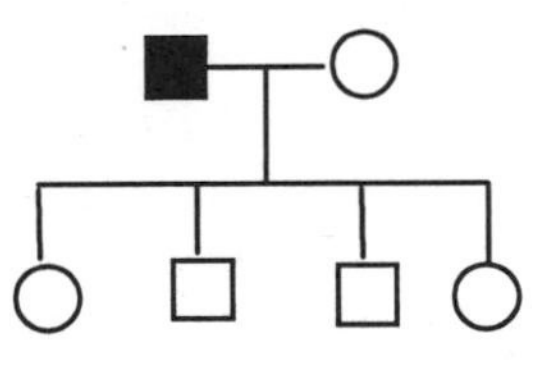

d. X-linked dominant trait—*all the female children*

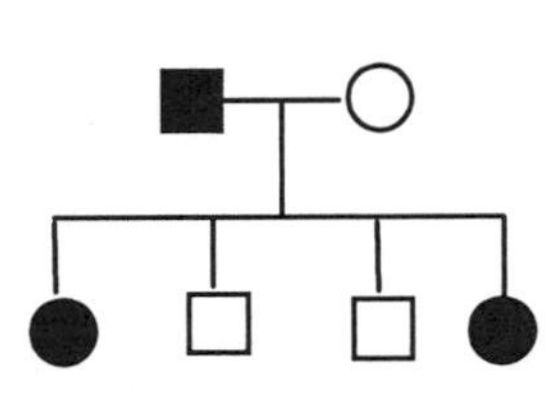

e. Y-linked trait—*all the male children*

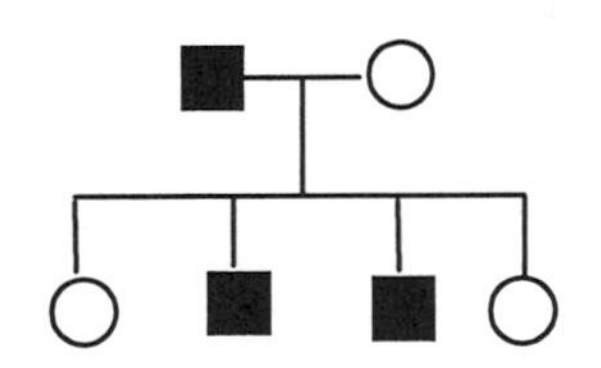

*21. For each of the following pedigrees, give the most likely mode of inheritance, assuming that the trait is rare. Carefully explain your reasoning.

a.

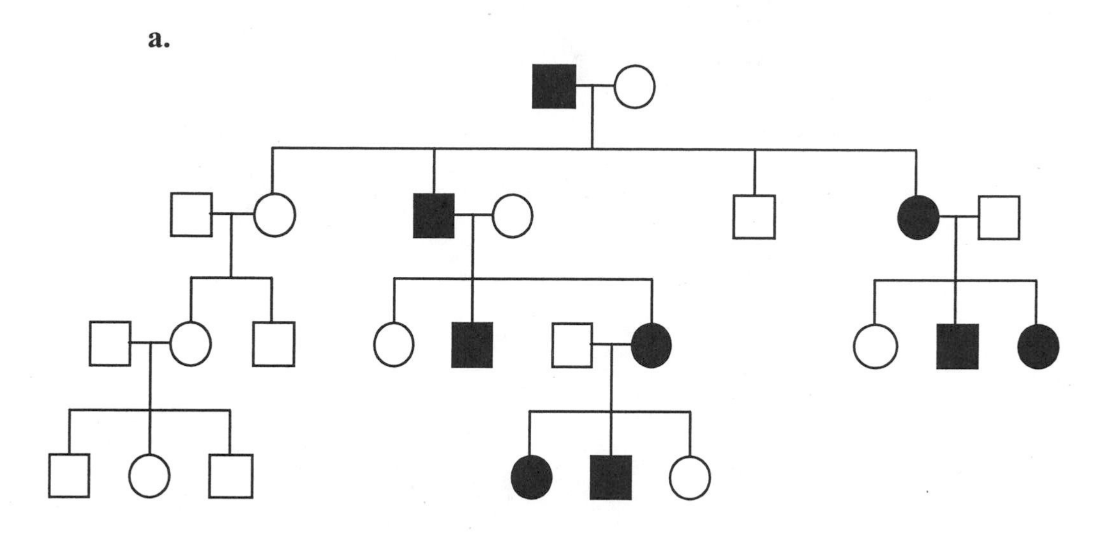

Autosomal dominant. The trait must be autosomal because affected males pass on the trait to both sons and daughters. It is dominant because it does not skip generations, all affected individuals have affected parents, and it is extremely unlikely that multiple unrelated individuals mating into the pedigree would be carriers for a rare trait.

b.

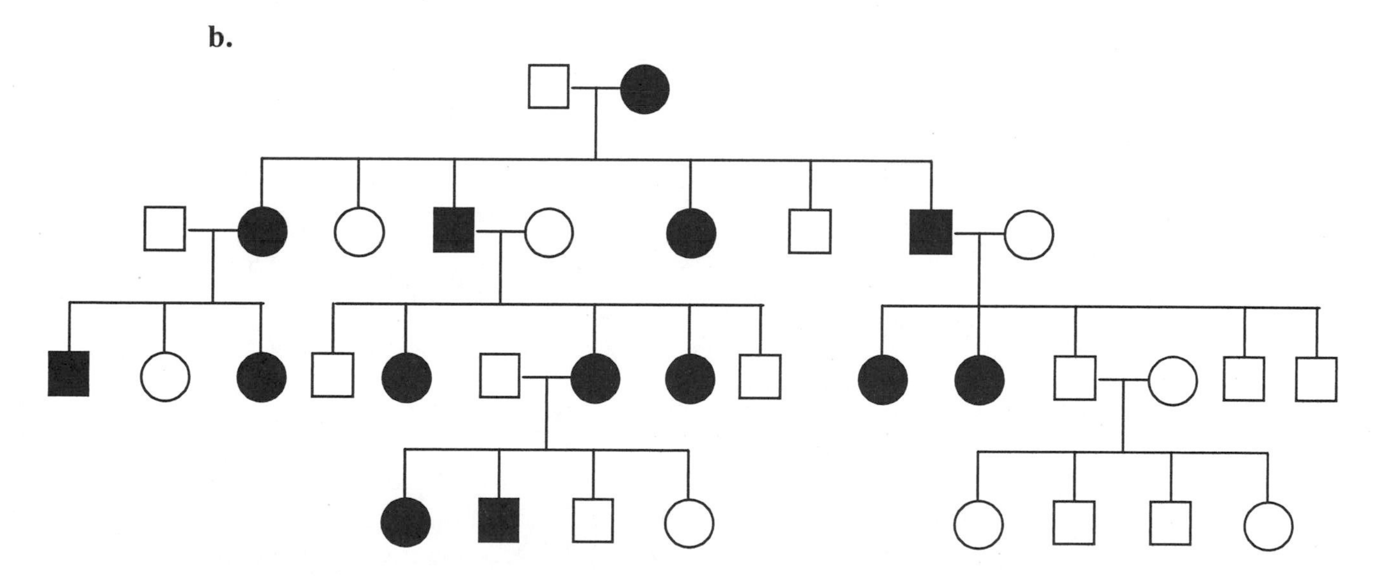

X-linked dominant. Superficially this pedigree appears similar to the pedigree in part (a) in that both males and females are affected, and it appears to be a dominant trait. However, closer inspection reveals that, whereas affected females can pass on the trait to either sons or daughters, affected males pass on the trait only to all daughters.

c.

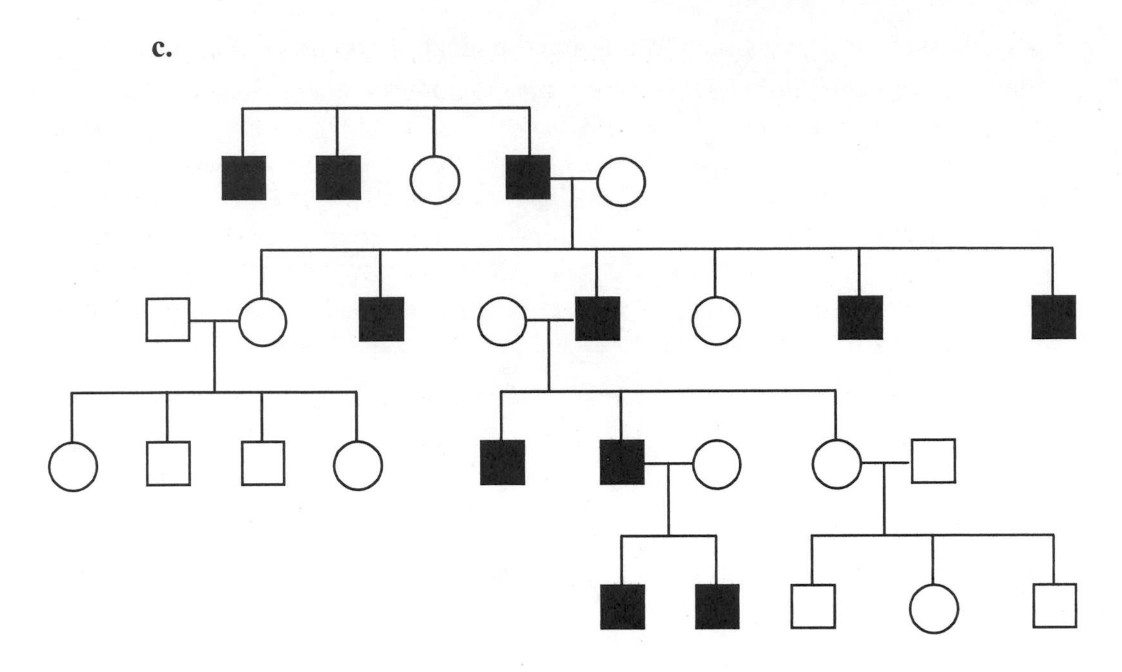

Y-linked. The trait affects only males and is passed from father to son. All sons of an affected male are affected.

d.

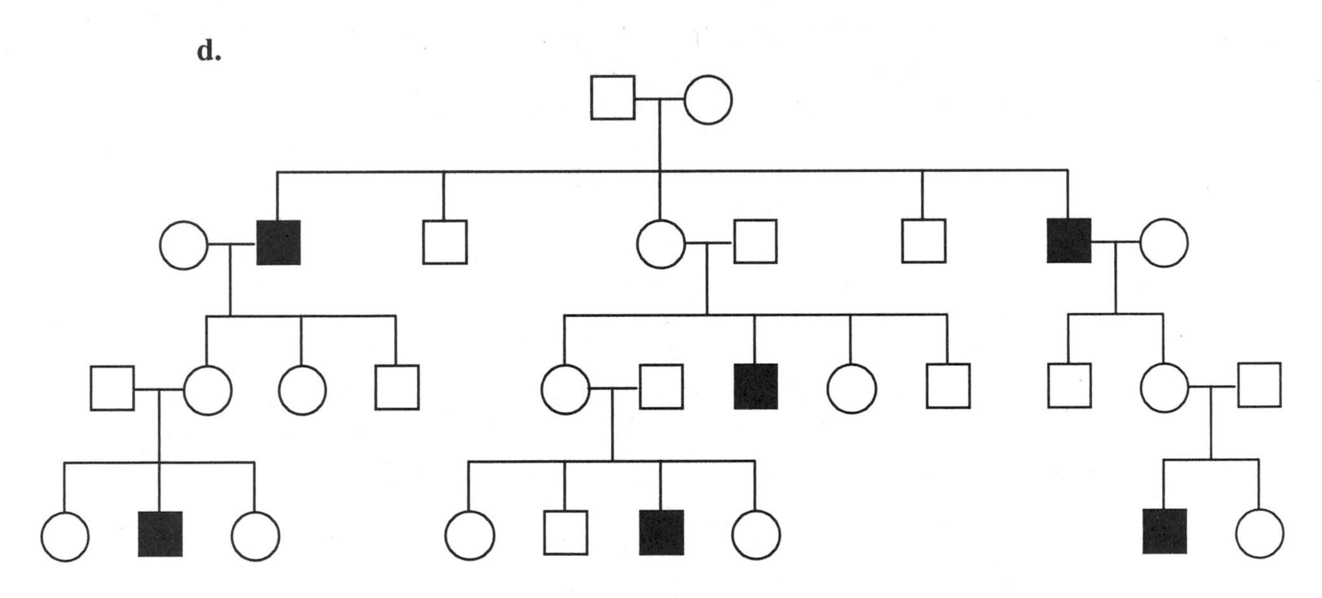

X-linked recessive or sex-limited autosomal dominant. Because only males show the trait, the trait could be X-linked recessive, Y-linked, or sex-limited. We can eliminate Y-linkage because affected males do not pass on the trait to their sons. X-linked recessive inheritance is consistent with the pattern of unaffected female carriers producing both affected and unaffected sons and affected males producing unaffected female carriers but no affected sons. Sex-limited autosomal dominant inheritance is also consistent with unaffected heterozygous females producing affected heterozygous sons, unaffected homozygous recessive sons, and unaffected heterozygous or homozygous recessive daughters. The two remaining possibilities of X-

linked recessive versus sex-limited autosomal dominant could be distinguished if we had enough data to determine whether affected males could have both affected and unaffected sons, as expected from autosomal dominant inheritance, or whether affected males can have only unaffected sons, as expected from X-linked recessive inheritance. Unfortunately, this pedigree shows only two sons from affected males. In both cases, the sons are unaffected, consistent with X-linked recessive inheritance, but two instances are not enough to conclude that affected males cannot produce affected sons.

e.

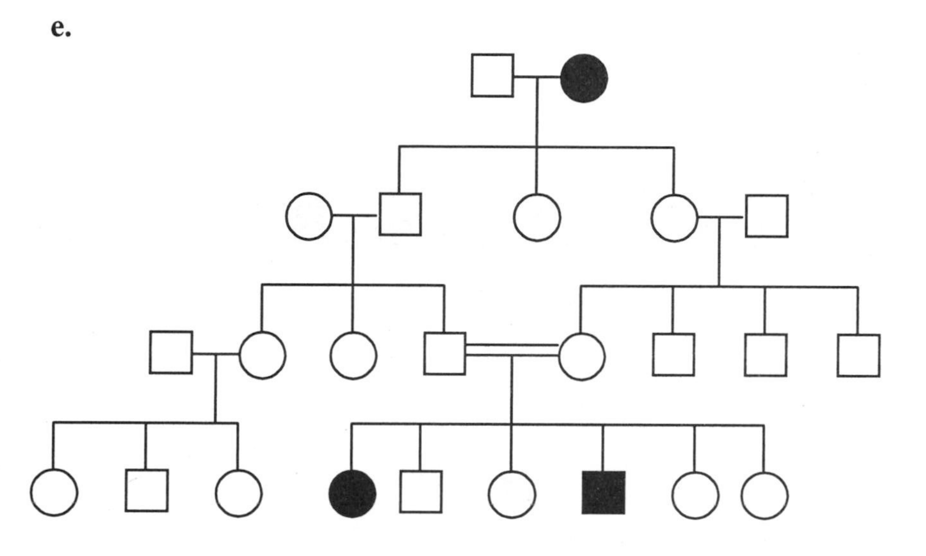

Autosomal recessive. Unaffected parents produced affected progeny, so the trait is recessive. The affected daughter must have inherited recessive alleles from both her unaffected parents, so it must be autosomal. If it were X-linked, her father would have shown the trait.

22. The trait represented in the following pedigree is expressed only in the males of the family. Is the trait Y-linked? Why or why not? If you believe the trait is not Y-linked, propose an alternate explanation for its inheritance.

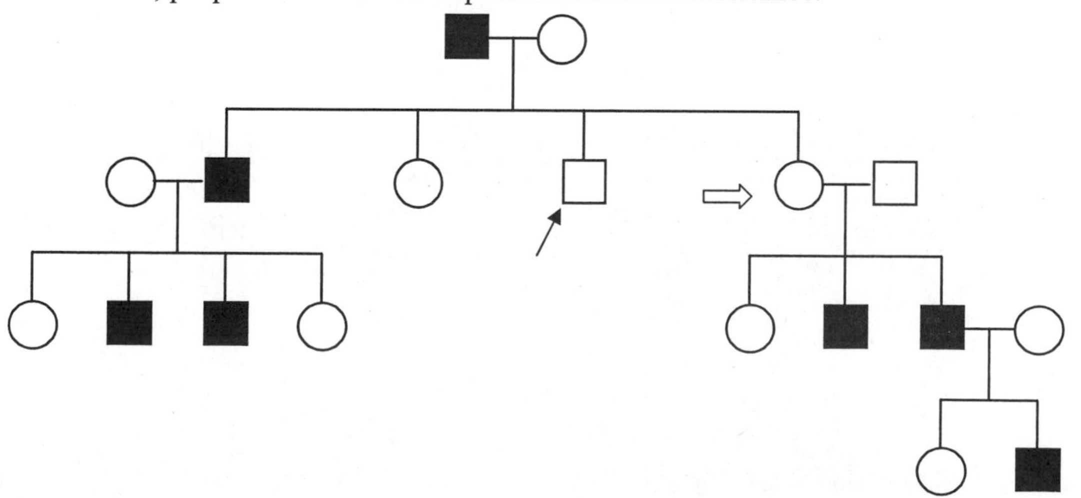

Y-linked traits are passed from father to son. This trait cannot be Y-linked because an affected father can have an unaffected son (indicated by a solid arrow) and also because we see sons inheriting the trait from their mother (indicated by an open arrow). Moreover, this trait cannot be X-linked because it is often passed from father to son, whereas X-linked traits are passed from father to daughter. The most probable mode of inheritance for this trait is sex-limited (only in males) autosomal dominant.

*23. The following pedigree illustrates the inheritance of Nance–Horan syndrome, a rare genetic condition in which affected persons have cataracts and abnormally shaped teeth.

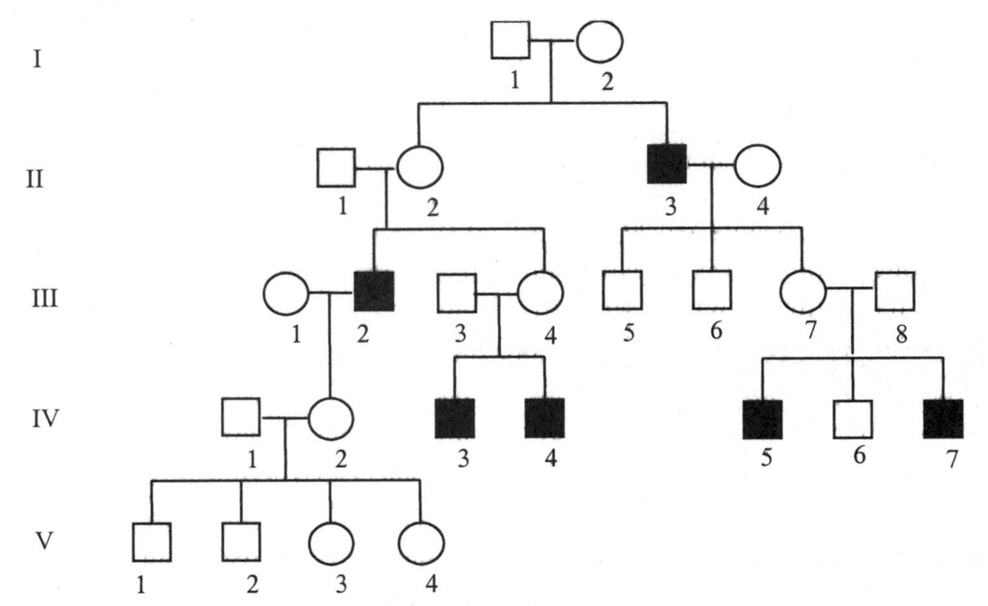

(Pedigree adapted from D. Stambolian, R. A. Lewis, K. Buetow, A. Bond, and R. Nussbaum. *American Journal of Human Genetics* 47[1990]:15.)

a. On the basis of this pedigree, what do you think is the most likely mode of inheritance for Nance–Horan syndrome?
X-linked recessive. Only males have the condition, and unaffected female carriers have affected sons.

b. If couple III-7 and III-8 have another child, what is the probability that the child will have Nance–Horan syndrome?
The probability is ¼. The female III-7 is a carrier, so there is a ½ probability that the child will inherit her X chromosome with the Nance–Horan allele and another ½ probability that the child will be a boy.

c. If III-2 and III-7 mated, what is the probability that one of their children would have Nance–Horan syndrome?
The probability is ½ because half the boys will inherit the Nance–Horan allele from the III-7 carrier female. All the girls will inherit one Nance–Horan allele from the III-2 affected male, and half of them will get a second Nance–Horan allele from the III-2 female, so half the girls will also have Nance–Horan syndrome.

24. The following pedigree illustrates the inheritance of ringed hair, a condition in which each hair is differentiated into light and dark zones. What mode or modes of inheritance are possible for the ringed hair trait in this family?

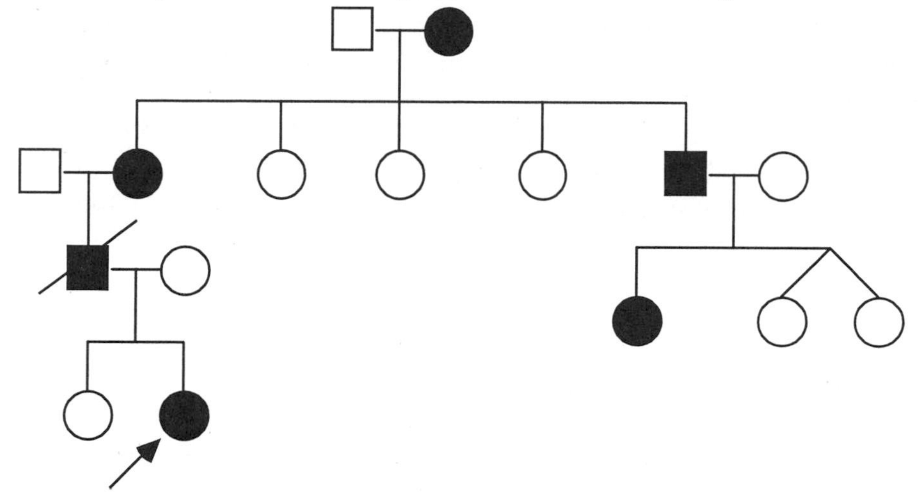

(Pedigree adapted from L. M. Ashley and R. S. Jacques, *Journal of Heredity* 41[1950]:83.)

This pedigree is consistent with autosomal dominant inheritance. Affected individuals marrying unaffected individuals have affected children, so the trait is dominant. Males do not pass the trait to all their daughters, so it cannot be X-linked dominant.

25. Ectodactyly is a rare condition in which the fingers are absent and the hand is split. This condition is usually inherited as an autosomal dominant trait. Ademar Freire-Maia reported the appearance of ectodactyly in a family in São Paulo, Brazil, whose pedigree is shown here. Is this pedigree consistent with autosomal dominant inheritance? If not, what mode of inheritance is most likely? Explain your reasoning.

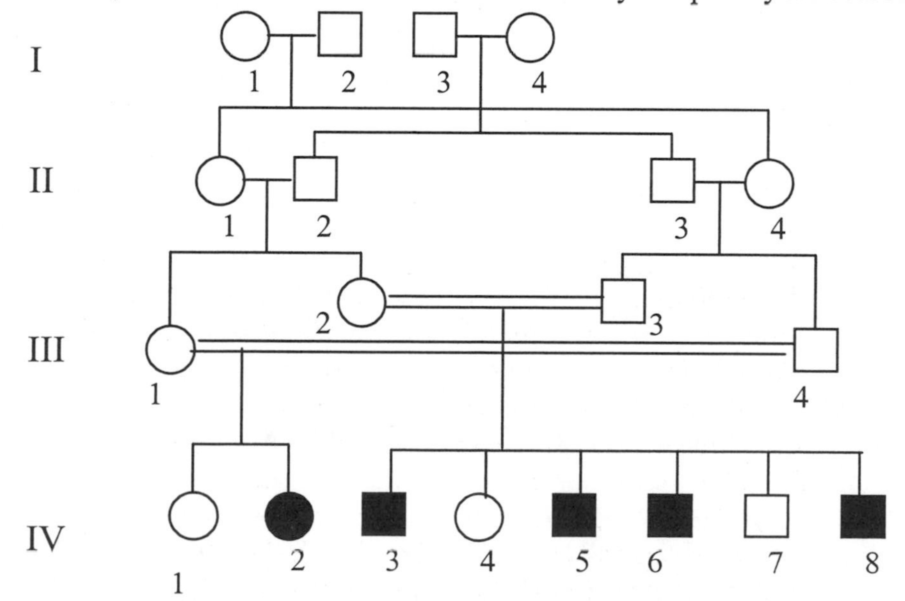

(Pedigree adapted from A.Freire-Maia, *Journal of Heredity* 62[1971]:53.)

This pedigree shows autosomal recessive inheritance, not autosomal dominant inheritance. It cannot be dominant because unaffected individuals have affected children. In generation II, two brothers married two sisters, so the members of generation III in the two families are as closely related as full siblings. A single recessive allele in one of the members of generation I was inherited by all four members of generation III. The consanguineous matings in generation III then produced children homozygous for the recessive ectodactyly allele. X-linkage is ruled out because the father of female IV-2 is unaffected; he has to be heterozygous.

26. The complete absence of one or more teeth (tooth agenesis) is a common trait in humans—indeed, more than 20% of humans lack one or more of their third molars. However, more severe absence of teeth, defined as missing six or more teeth, is less common and frequently an inherited condition. L. Lammi and colleagues examined tooth agenesis in the Finnish family shown here (L. Lammi. 2004. *American Journal of Human Genetics* 74:1043–1050).

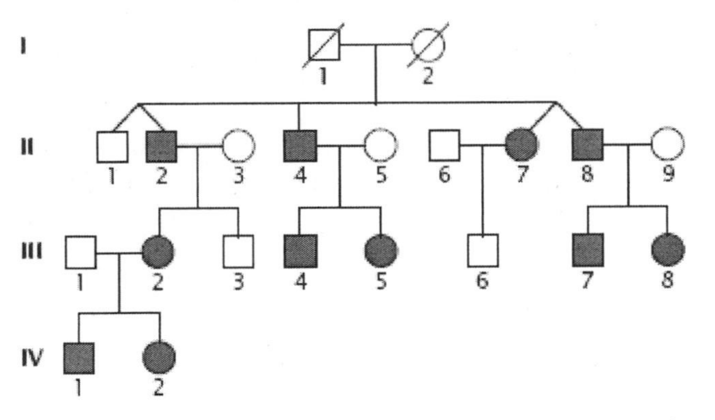

a. What is the most likely mode of inheritance for tooth agenesis in this family? Explain your reasoning.
Both males and females have the trait, and affected males and females transmit to either sons or daughters. The trait is most likely autosomal dominant.

b. Are the two sets of twins in this family monozygotic or dizygotic twins? What is the basis of your answer?
They are dizygotic, because they differ in either genotype (the first set of twin boys) or gender (the boy and girl).

c. If IV-2 married a man who has a full set of teeth, what is the probability that their child would have tooth agenesis?
People with full teeth are homozygous for the recessive allele for full teeth, and IV-2 must be heterozygous. Therefore, the probability of a child with tooth agenesis is 50%.

d. If III-2 and III-7 married and had a child, what is the probability that their child would have tooth agenesis?
They are both heterozygotes, so the probability would be ¾ for a child to have the dominant tooth agenesis phenotype, assuming that homozygotes for tooth agenesis are viable.

Section 6.4

*27. A geneticist studies a series of characteristics in monozygotic twins and dizygotic twins, obtaining the following concordances. For each characteristic, indicate whether the rates of concordance suggest genetic influences, environmental influences, or both. Explain your reasoning.

Characteristic	Monozygotic concordance (%)	Dizygotic concordance (%)
Migraine headaches	60	30
Eye color	100	40
Measles	90	90
Clubfoot	30	10
High blood pressure	70	40
Handedness	70	70
Tuberculosis	5	5

Migraine headaches appear to be influenced by genetic and environmental factors. Markedly greater concordance in monozygotic twins, who share 100% genetic identity, than in dizygotic twins, who share 50% genetic identity, is indicative of a genetic influence. However, the fact that monozygotic twins show only 60% concordance despite sharing 100% genetic identity indicates that environmental factors also play a role.

Eye color appears to be purely genetically determined because the concordance is greater in monozygotic twins than in dizygotic twins. Moreover, the monozygotic twins have 100% concordance for this trait, indicating that environment has no detectable influence.

Measles appears to have no detectable genetic influence because there is no difference in concordance between monozygotic and dizygotic twins. Some environmental influence can be detected because monozygotic twins show less than 100% concordance.

Clubfoot appears to have genetic and environmental influences, by the same reasoning as for migraine headaches. A strong environmental influence is indicated by the high discordance in monozygotic twins.

High blood pressure has genetic and environmental influences, similar to clubfoot.

Handedness, like measles, appears to have no genetic influence because the concordance is the same in monozygotic and dizygotic twins. Environmental influence is indicated by the less than 100% concordance in monozygotic twins.

Tuberculosis similarly lacks indication of genetic influence, with the same degree of concordance in monozygotic and dizygotic twins. The primacy of environmental influence is indicated by the very low concordance in monozygotic twins.

28. M. T. Tsuang and colleagues studied drug dependence in male twin pairs (M. T. Tsuang et al. 1996. *American Journal of Medical Genetics* 67:473–477). They found that 4 out of 30 monozygotic twins were concordant for dependence on opioid drugs, whereas 1 out of 34 dizygotic twins were concordant for the same trait. Calculate the concordance rates for opioid dependence in these monozygotic and dizygotic twins. On the basis of these data, what conclusion can you make concerning the roles of genetic and environmental factors in opioid dependence?
The concordance rate for monozygotic twins is 4/30 or 13%. For dizygotic twins, the concordance rate is 1/34 or 3%. Since the monozygotic twins have a higher concordance rate than dizygotic twins, these data could suggest a genetic influence. However, these rates are low and the actual numbers so few that the difference in concordance rates may not be significant. Thus, these data are somewhat suggestive of a small genetic influence, but not conclusive. The low concordance in monozygotic twins indicates a significant environmental influence.

Section 6.5

29. In a study of schizophrenia (a mental disorder including disorganization of thought and withdrawal from reality), researchers looked at the prevalence of the disorder in the biological and adoptive parents of people who were adopted as children; they found the following results:

	Prevalence of schizophrenia (%)	
Adopted persons	Biological parents	Adoptive parents
With schizophrenia	12	2
Without schizophrenia	6	4

Source: S. S. Kety et al., The biological and adoptive families of adopted individuals who become schizophrenic: prevalence of mental illness and other characteristics, *The Nature of Schizophrenia: New Approaches to Research and Treatment*, L. C. Wynne, R. L. Cromwell, and S. Matthysse, Eds. (New York: Wiley, 1978), pp. 25–37.)

What can you conclude from these results concerning the role of genetics in schizophrenia? Explain your reasoning.
These data suggest that schizophrenia has a strong genetic component. The biological parents of schizophrenic adoptees are far more likely to be schizophrenic than genetically unrelated individuals (the adoptive parents), despite the fact that the schizophrenic adoptees share the same environment as the adoptive parents. If environmental variables (such as chemicals in the water or food or power lines) were a major factor, then one would expect to see a higher frequency of schizophrenia in the adoptive parents. Another possibility is that this increased frequency of schizophrenia in the biological parents simply reflects a greater likelihood that schizophrenic parents give up their children for adoption. This latter possibility is ruled out by the data that the biological parents of nonschizophrenic adoptees do not show a similar increased frequency of schizophrenia compared to adoptive parents.

Section 6.7

30. What, if any, ethical issues might arise from the widespread use of noninvasive fetal diagnosis, which can be carried out much earlier than amniocentesis or CVS?
An issue that is common to all prenatal testing, but may be exacerbated by this new procedure, is the question of what genetic conditions should be tested. Because of the early diagnosis, mothers will have the ability to terminate the pregnancy during the first trimester, with less risk and fewer complications than at more advanced stages. To what extent should regulations determine what types of tests are offered, fearing that some mothers or couples may make decisions based on such factors as the sex of the fetus or other conditions that do not have serious health consequences?

CHALLENGE QUESTIONS

Section 6.1

31. Many genetic studies, particularly those of recessive traits, have focused on small isolated human populations, such as those on islands. Suggest one or more advantages that isolated populations might have for the study of recessive traits.
Isolated populations become inbred, so recessive phenotypes arise more frequently. Some recessive traits that are rare in other, large human populations may be more frequent in isolated populations, facilitating pedigree analysis.

Section 6.3

32. Draw a pedigree that represents an autosomal dominant trait, sex-limited to males, and that excludes the possibility that the trait is Y-linked.

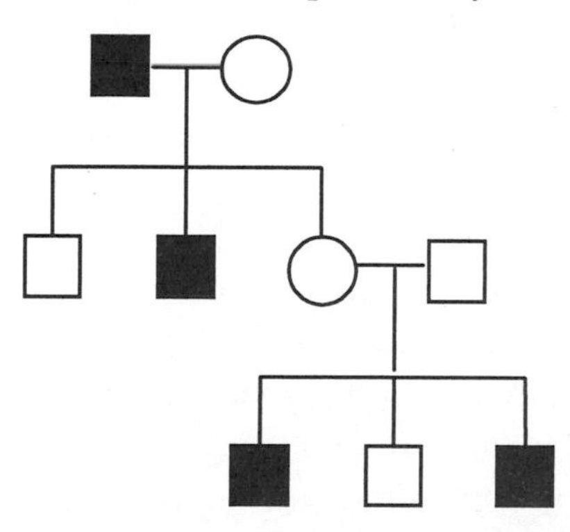

This pedigree excludes Y-linkage because not all the sons of an affected male are affected, an unaffected male has affected sons, and also because it is transmitted through an unaffected female to her sons.

33. A. C. Stevenson and E. A. Cheeseman studied deafness in a family in Northern Ireland and recorded the following pedigree (A. C. Stevenson and E. A. Cheeseman. 1956. *Annals of Human Genetics* 20:177–231).

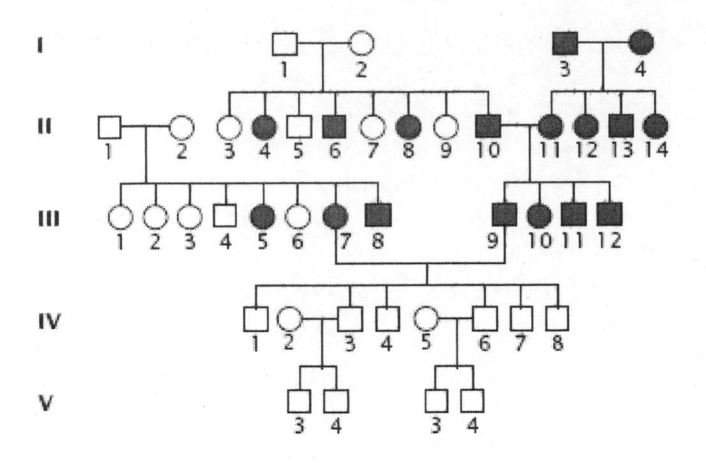

a. If you consider only generations I through III, what is the most likely mode of inheritance for this type of deafness?
Autosomal recessive. Affected children of both sexes arise from unaffected parents. Affected parents, being homozygous recessive, have all affected children.

b. Provide a possible explanation for the cross between III-7 and III-9 and the results for generations IV through V.
One possible explanation is that this deafness may be caused by recessive alleles at two different loci. If we use A *and* a *for the alleles at one locus and* B *and* b *for the alleles at the second locus, III-7 could be* aa BB *and III-9 could be* AA bb. *All their children (generation IV) would then be* Aa Bb, *and have normal hearing, having dominant alleles at both loci. Generation 5 are children of marriages with spouses from outside the pedigree, presumably being homozygous dominant for one or both loci. A second possibility is allelic complementation: two different recessive alleles at the same locus interact in such as way as to produce the dominant phenotype.*

Section 6.4

34. Dizygotic twinning often runs in families and its frequency varies among ethnic groups, whereas monozygotic twinning rarely runs in families and its frequency is quite constant among ethnic groups. These observations have been interpreted as evidence for a genetic basis for variation in dizygotic twinning but for little genetic basis for variation in monozygotic twinning. Can you suggest a possible reason for these differences in the genetic tendencies of dizygotic and monozygotic twinning?
The tendency for women to ovulate multiple eggs could be influenced by the mother's genotype. A woman's genotype would determine such factors as hormonal levels important for ovulation. In contast, the tendency for a fertilized zygote to split into two embryos may be an entirely random event not dependent on the genetic composition of either the embryo or the mother.

Chapter Seven: Linkage, Recombination, and Eukaryotic Gene Mapping

COMPREHENSION QUESTIONS

Section 7.1

*1. What does the term recombination mean? What are two causes of recombination?
Recombination means that meiosis generates gametes with different allelic combinations than the original gametes the organism inherited. If the organism was created by the fusion of an egg bearing AB *and a sperm bearing* ab, *recombination generates gametes that are* Ab *and* aB. *Recombination may be caused by loci on different chromosomes that sort independently or by a physical crossing over between two loci on the same chromosome, with breakage and exchange of strands of homologous chromosomes paired in meiotic prophase I.*

Section 7.2

*2. In a testcross for two genes, what types of gametes are produced with (a) complete linkage, (b) independent assortment, and (c) incomplete linkage?
(a) Complete linkage of two genes means that only nonrecombinant gametes will be produced; the recombination frequency is zero.
(b) Independent assortment of two genes will result in 50% of the gametes being recombinant and 50% being nonrecombinant, as would be observed for genes on two different chromosomes. Independent assortment may also be observed for genes on the same chromosome if they are far enough apart that one or more crossovers occur between them in every meiosis.
(c) Incomplete linkage means that greater than 50% of the gametes produced are nonrecombinant and less than 50% of the gametes are recombinant; the recombination frequency is greater than 0 and less than 50%.

3. What effect does crossing over have on linkage?
Crossing over generates recombination between genes located on the same chromosome, and thus renders linkage incomplete.

4. Why is the frequency of recombinant gametes always half the frequency of crossing over?
Crossing over occurs at the four-strand stage, when two homologous chromosomes, each consisting of a pair of sister chromatids, are paired. Each crossover involves just two of the four strands and generates two recombinant strands. The remaining two strands that were not involved in the crossover generate two nonrecombinant strands. Therefore, the frequency of recombinant gametes is always half the frequency of crossovers.

*5. What is the difference between genes in coupling configuration and genes in repulsion? What effect does the arrangement of linked genes (whether they are in coupling configuration or in repulsion) have on the results of a cross?
Genes in coupling configuration have two wild-type alleles on the same chromosome and the two mutant alleles on the homologous chromosome. Genes in repulsion have a wild-type allele of one gene together with the mutant allele of the second gene on the same chromosome, and vice versa on the homologous chromosome. The two arrangements have opposite effects on the results of a cross. For genes in coupling configuration, most of the progeny will be either wild type for both genes, or mutant for both genes, with relatively few that are wild type for one gene and mutant for the other. For genes in repulsion, most of the progeny will be mutant for only one gene and wild-type for the other, with relatively few recombinants that are wild-type for both or mutant for both.

6. How does one test to see if two genes are linked?
One first obtains individuals that are heterozygous for both genes. This may be achieved by crossing an individual homozygous dominant for both genes to one homozygous recessive for both genes, resulting in a heterozygote with genes in coupling configuration. Alternatively, an individual that is homozygous recessive for one gene may be crossed to an individual homozygous recessive for the second gene, resulting in a heterozygote with genes in repulsion. Then, the heterozygote is mated to a homozygous recessive tester and the progeny of each phenotypic class are tallied. If the proportion of recombinant progeny is far less than 50%, the genes are linked. If the results are not so clear-cut, then they may be tested by chi-square, first for equal segregation at each locus, then for independent assortment of the two loci. Significant deviation from results expected for independent assortment indicates linkage of the two genes.

7. What is the difference between a genetic map and a physical map?
A genetic map gives the order of genes and relative distance between them based on recombination frequencies observed in genetic crosses. A physical map locates genes on the actual chromosome or DNA sequence, and thus represents the physical distance between genes.

*8. Why do calculated recombination frequencies between pairs of loci that are located relatively far apart underestimate the true genetic distances between loci?
The further apart two loci are, the more likely it is to get double crossovers between them. Unless there are marker genes between the loci, such double crossovers will be undetected because double crossovers give the same phenotypes as nonrecombinants. The calculated recombination frequency will underestimate the true crossover frequency because the double crossover progeny are not counted as recombinants.

Section 7.3

9. Explain how to determine which of three linked loci is the middle locus from the progeny of a three-point testcross.
Double crossovers always result in switching the middle gene with respect to the two nonrecombinant chromosomes. Hence, one can compare the two double crossover phenotypes with the two nonrecombinant phenotypes and see which gene is reversed. In the diagram below, we see that the coupling relationship of the middle gene is flipped in the double crossovers with respect to the genes on either side. Therefore, whichever gene on the double crossover can be altered to make the double crossover resemble a nonrecombinant chromosome is the middle gene. If we take either of the double crossover products l M r *or* L m R, *changing the* M *gene will make it resemble a nonrecombinant.*

*10. What does the interference tell us about the effect of one crossover on another?
A positive interference value results when the actual number of double crossovers observed is less than the number of double crossovers expected from the single crossover frequencies. Thus, positive interference indicates that a crossover inhibits or interferes with the occurrence of a second crossover nearby.

Conversely, a negative interference value, where more double crossovers occur than expected, suggests that a crossover event can stimulate additional crossover events in the same region of the chromosome.

Section 7.4

11. What is a lod score and how is it calculated?
The term lod means logarithm of odds. It is used to determine whether genes are linked, usually in the context of pedigree analysis. One first determines the probability of obtaining the observed progeny given a specified degree of linkage. That probability is divided by the probability of obtaining the observed progeny if the genes are not linked and sort independently. The log of the ratio of these probabilities is the lod score. A lod score of 3 or greater, indicating that the specified degree of linkage results in at least a thousand fold greater likelihood of yielding the observed outcome than if the genes are unlinked, indicates linkage.

12. List some of the methods for physically mapping genes and explain how they are used to position genes on chromosomes.
Deletion mapping: Recessive mutations are mapped by crossing mutants with strains containing various overlapping deletions that map to the same region as the recessive mutation. If the heterozygote with the mutation on one chromosome and the deletion on the homologous chromosome has a mutant phenotype, then the

mutation must be located on the same physical portion of the chromosome that is deleted. If, on the other hand, the heterozygote has a wild-type phenotype (the mutation and the deletion complement), then the mutation lies outside the deleted region of the chromosome.

Somatic-cell hybridization: Human and mouse cells are fused. The resulting hybrid cell randomly loses human chromosomes and retains only a few. A panel of hybrids that retain different combinations of human chromosomes is tested for expression of a human gene. A correlation between the expression of the gene and the retention of a unique human chromosome in those cell lines indicates that the human gene must be located on that chromosome.

In situ hybridization: DNA probes that are labeled with either a radioactive or fluorescent tag are hybridized to chromosome spreads. Detection of the labeled hybridized probe by autoradiography or fluorescence imaging reveals which chromosome and where along that chromosome the homologous gene is located.

DNA sequencing: Overlapping DNA sequences are joined using computer programs to ultimately form chromosome-length sequence assemblies, or contigs. The locations of genes along the DNA sequence can be determined by searching for matches to known gene or protein amino acid sequences.

APPLICATION QUESTIONS AND PROBLEMS

Section 7.2

*13. In the snail *Cepaea nemoralis,* an autosomal allele causing a banded shell (B^B) is recessive to the allele for unbanded shell (B^O). Genes at a different locus determine the background color of the shell; here, yellow (C^Y) is recessive to brown (C^{Bw}). A banded, yellow snail is crossed with a homozygous brown, unbanded snail. The F_1 are then crossed with banded, yellow snails (a testcross).

a. What will be the results of the testcross if the loci that control banding and color are linked with no crossing over?
With absolute linkage, there will be no recombinant progeny. The F_1 inherited banded and yellow alleles (B^BC^Y) together on one chromosome from the banded yellow parent and unbanded and brown alleles (B^OC^{Bw}) together on the homologous chromosome from the unbanded brown parent. Without recombination, all the F_1 gametes will contain only these two allelic combinations, in equal proportions. Therefore, the F_2 testcross progeny will be ½ banded, yellow and ½ unbanded, brown.

b. What will be the results of the testcross if the loci assort independently?
With independent assortment, the progeny will be:
¼ banded, yellow
¼ banded, brown
¼ unbanded, yellow
¼ unbanded, brown

c. What will be the results of the testcross if the loci are linked and 20 map units apart?
The recombination frequency is 20%, so each of the two classes of recombinant progeny must be 10%. The recombinants are banded, brown and unbanded, yellow. The two classes of nonrecombinants are 80% of the progeny, so each must be 40%. The nonrecombinants are banded, yellow and unbanded, brown.
In summary:
40% banded, yellow
40% unbanded, brown
10% banded, brown
10% unbanded, yellow

14. In silkmoths (*Bombyx mori*), red eyes (*re*) and white-banded wing (*wb*) are encoded by two mutant alleles that are recessive to those that produce wild-type traits (re^+ and wb^+); these two genes are on the same chromosome. A moth homozygous for red eyes and white-banded wings is crossed with a moth homozygous for the wild-type traits. The F_1 have normal eyes and normal wings. The F_1 are crossed with moths that have red eyes and white-banded wings in a testcross. The progeny of this testcross are:

wild-type eyes, wild-type wings	418
red eyes, wild-type wings	19
wild-type eyes, white-banded wings	16
red eyes, white-banded wings	426

a. What phenotypic proportions would be expected if the genes for red eyes and white-banded wings were located on different chromosomes?
¼ wild-type eyes, wild-type wings
¼ red eyes, wild-type wings
¼ wild-type eyes, white-banded wings
¼ red eyes, white-banded wings

b. What is the genetic distance between the genes for red eyes and white-banded wings?
The F_1 *heterozygote inherited a chromosome with alleles for red eyes and white-banded wings* (re wb) *from one parent and a chromosome with alleles for wild-type eyes and wild-type wings* (re^+ wb^+) *from the other parent. These are therefore the phenotypes of the nonrecombinant progeny, present in the highest numbers. The recombinants are the 19 with red eyes, wild-type wings and 16 with wild-type eyes, white-banded wings.*
$RF = \text{recombinants/total progeny} \times 100\% = (19 + 16)/879 \times 100\% = 4.0\%$
The distance between the genes is 4 map units.

*15. A geneticist discovers a new mutation in *Drosophila melanogaster* that causes the flies to shake and quiver. She calls this mutation spastic (*sps*) and determines that spastic is due to an autosomal recessive gene. She wants to determine if the spastic gene is linked to the recessive gene for vestigial wings (*vg*). She crosses a fly homozygous for spastic and vestigial traits with a fly homozygous for the wild-type traits and then uses the resulting F_1 females in a testcross. She obtains the following flies from this testcross:

$vg^+ \; sps^+$	230
$vg \; sps$	224
$vg \; sps^+$	97
$vg^+ sps$	99
total	650

Are the genes that cause vestigial wings and the spastic mutation linked? Do a series of chi-square tests to determine if the genes have assorted independently.

To test for independent assortment, we first test for equal segregation at each locus, then test whether the two loci sort independently.

Test for vg:

Observed vg *= 224 + 97 = 321*

Observed vg^+ *= 230 + 99 = 329*

Expected vg *or* vg^+ *= ½ × 650 = 325*

$$\chi^2 = \Sigma\frac{(observed - expected)^2}{expected} = (321 - 325)^2/325 + (329 - 325)^2/325 = 16/325 + 16/325$$

$$= 0.098$$

We have n *– 1 degrees of freedom, where* n *is the number of phenotypic classes = 2, so just 1 degree of freedom. From Table 3.4, we see that the P value is between 0.7 and 0.8. So these results do not deviate significantly from the expected 1:1 segregation.*

Similarly, testing for sps, *we observe 327* sps^+ *and 323* sps *and expect ½ × 650 = 325 of each:*

$\chi^2 = 4/325 + 4/325 = .025$*, again with 1 degree of freedom. The P value is between 0.8 and 0.9, so these results do not deviate significantly from the expected 1:1 ratio.*

Finally, we test for independent assortment, where we expect 1:1:1:1 phenotypic ratios, or 162.5 of each.

Observed	*Expected*	*o – e*	*(o – e)²*	*(o – e)²/e*
230	*162.5*	*67.5*	*4556.25*	*28.0*
224	*162.5*	*61.5*	*3782.25*	*23.3*
97	*162.5*	*–65.5*	*4290.25*	*26.4*
99	*162.5*	*–63.5*	*4032.25*	*24.8*

We have four phenotypic classes, giving us three degrees of freedom. The chi-square value of 102.5 is off the chart, so we reject independent assortment.

Instead, the genes are linked, and the RF = (97 + 99)/650 × 100% = 30%, giving us 30 map units between them.

16. In cucumbers, heart-shaped leaves (*hl*) are recessive to normal leaves (*Hl*), and having numerous fruit spines (*ns*) is recessive to having few fruit spines (*Ns*). The genes for leaf shape and number of spines are located on the same chromosome; mapping experiments indicate that they are 32.6 map units apart. A cucumber plant

having heart-shaped leaves and numerous spines is crossed with a plant that is homozygous for normal leaves and few spines. The F_1 are crossed with plants that have heart-shaped leaves and numerous spines. What phenotypes and proportions are expected in the progeny of this cross?

The recombinants should total 32.6%, so each recombinant phenotype will be 16.3% of the progeny. Because the F_1 inherited a chromosome with heart-shaped leaves and numerous spines (hl ns) *from one parent and a chromosome with normal leaves and few spines* (Hl Ns) *from the other parent, these are the nonrecombinant phenotypes, and together they total 67.4%, or 33.7% each. The two recombinant phenotypes are heart-shaped leaves with few spines* (hl Ns) *and normal-shaped leaves with numerous spines* (Hl ns).

Heart-shaped, numerous spines	*33.7%*
Normal-shaped, few spines	*33.7%*
Heart-shaped, few spines	*16.3%*
Normal-shaped, numerous spines	*16.3%*

*17. In tomatoes, tall (*D*) is dominant over dwarf (*d*), and smooth fruit (*P*) is dominant over pubescent (*p*) fruit, which is covered with fine hairs. A farmer has two tall and smooth tomato plants, which we will call plant A and plant B. The farmer crosses plants A and B with the same dwarf and pubescent plant and obtains the following numbers of progeny:

	Progeny of	
	Plant A	Plant B
Dd Pp	122	2
Dd pp	6	82
dd Pp	4	82
dd pp	124	4

a. What are the genotypes of plant A and plant B?
The genotypes of both plants are DdPp.

b. Are the loci that determine height of the plant and pubescence linked? If so, what is the map distance between them?
Yes. From the cross of plant A, the map distance is 10/256 = 3.9%, or 3.9 m.u. The cross of plant B gives 6/170 = 3.5%, or 3.5 m.u. If we pool the data from the two crosses, we get 16/426 = 3.8%, or 3.8 m.u.

c. Explain why different proportions of progeny are produced when plant A and plant B are crossed with the same dwarf pubescent plant.
The two plants have different coupling configurations. In plant A, the dominant alleles D *and* P *are coupled; one chromosome is* D P *and the other is* d p. *In plant B, they are in repulsion; its chromosomes have* D p *and* d P.

18. Alleles *A* and *a* occur at a locus that is located on the same chromosome as a locus with alleles *B* and *b*. *Aa Bb* is crossed with *aa bb,* and the following progeny are produced:

Aa Bb	5
Aa bb	45
aa Bb	45
aa bb	5

What conclusion can be made about the arrangement of the genes on the chromosome in the *Aa Bb* parent?

The results of this testcross reveal that Aa bb *and* aa Bb, *with far greater numbers, are the progeny that received nonrecombinant chromatids from the* Aa Bb *parent. Given that all the progeny received* ab *from the* aa bb *parent, the nonrecombinant progeny received either an* Ab *or an* aB *chromatid from the* Aa Bb *parent. Therefore, the* A *and* B *loci are in repulsion in the* Aa Bb *parent.* Aa Bb *and* aa bb *are the recombinant classes, and their frequencies indicate that the genes* A *and* B *are 10 m.u. apart.*

19. Daniel McDonald and Nancy Peer determined that eyespot (a clear spot in the center of the eye) in flour beetles is caused by an X-linked gene (*es*) that is recessive to the allele for the absence of eyespot (es^+). They conducted a series of crosses to determine the distance between the gene for eyespot and a dominant X-linked gene for stripped (*St*), which acted as a recessive lethal (is lethal when homozygous in females or hemizygous in males). The following cross was carried out (D. J. McDonald and N. J. Peer. 1961. *Journal of Heredity* 52:261–264).

$$♀\ \frac{es^+\ St}{es\ St^+} \times \frac{es\ St^+}{Y}\ ♂$$

↓

$\frac{es^+\ St}{es\ St^+}$	1630	*Nonrecombinant*
$\frac{es\ St^+}{es\ St^+}$	1665	*Nonrecombinant*
$\frac{es\ St}{es\ St^+}$	935	*Recombinant*
$\frac{es^+\ St^+}{es\ St^+}$	1005	*Recombinant*
$\frac{es\ St^+}{Y}$	1661	*Nonrecombinant*
$\frac{es^+\ St^+}{Y}$	1024	*Recombinant*

a. Which progeny are the recombinants and which progeny are the nonrecombinants?

The female progeny all received es St$^+$ *from the male parent. Male progeny received a Y chromosome from the male parent. From the female parent, nonrecombinants received either* es$^+$ St *or* es St$^+$. *Recombinants received either* es St *or* es$^+$ St$^+$. *Males who received* es$^+$ St *(nonrecombinant) or* es St *(recombinant) did not survive. The recombinant and nonrecombinant progeny classes are identified above.*

b. Calculate the recombination frequency between *es* and *St.*

Using just the male progeny, RF = 1024/(1024 + 1661) = 0.38 = 38%. Using the female progeny, RF = (935 + 1005)/(935 + 1005 + 1630 + 1665) = 0.37 = 37%. Using all the progeny, RF = (935 + 1005 + 1024)/(935 + 1005 + 1024 + 1630 + 1665 + 1661) = 0.37 = 37%

c. Are some potential genotypes missing among the progeny of the cross? If so, which ones and why?

Males who received es$^+$ St *(nonrecombinant) or* es St *(recombinant) did not survive because* St *acts lethal in hemizygous males.*

20. In tomatoes, dwarf (*d*) is recessive to tall (*D*) and opaque (light green) leaves (*op*) is recessive to green leaves (*Op*). The loci that determine the height and leaf color are linked and separated by a distance of 7 m.u. For each of the following crosses, determine the phenotypes and proportions of progeny produced.

a. $\frac{D \quad Op}{d \quad op} \times \frac{d \quad op}{d \quad op}$

The recombinants in this cross would be D op *and* d Op, *and each would be 3.5% of the progeny, to total 7% recombinants. Each of the nonrecombinants would be 46.5%, to total the remaining 93%.*

Tall, green	*46.5%*
Dwarf, opaque	*46.5%*
Tall, opaque	*3.5%*
Dwarf, green	*3.5%*

b. $\frac{D \quad op}{d \quad Op} \times \frac{d \quad op}{d \quad op}$

Here with the genes in repulsion, the recombinants are D Op *and* d op.

Tall, green	*3.5%*
Dwarf, opaque	*3.5%*
Tall, opaque	*46.5%*
Dwarf, green	*46.5%*

c. $\frac{D \quad Op}{d \quad op} \times \frac{D \quad Op}{d \quad op}$

This is not a testcross, so we have to account for recombination in both parents. The most straightforward way is to do a Punnett square, including the types and proportions of gametes produced by meiosis in each parent. Because the genes are in coupling configuration in both parents, we can use the figures from part (a).

	D Op *0.465*	D op *0.035*	d Op *0.035*	d op *0.465*
D Op *0.465*	*Tall, green .216*	*Tall, green .016*	*Tall, green .016*	*Tall, green .216*
D op *0.035*	*Tall, green .016*	*Tall, opaque .001*	*Tall, green .001*	*Tall, opaque .016*
d Op *0.035*	*Tall, green .016*	*Tall, green .001*	*Dwarf, green .001*	*Dwarf, green .016*
d op *0.465*	*Tall, green .216*	*Tall, opaque .016*	*Dwarf, green .016*	*Dwarf, opaque .216*

In summary, we get

Tall, green	*3(.216) + 4(.016) + 2(.001) = .714*
Dwarf, opaque	*.216*
Tall, opaque	*2(.016) + .001 = .033*
Dwarf, green	*2(.016) + .001 = .033*

d. $\frac{D \quad op}{d \quad Op} \times \frac{D \quad op}{d \quad Op}$

Again, this is not a testcross, and recombination in both parents must be taken into account. Both are in repulsion, so we use the proportions from part (b).

	D Op *0.035*	D op *0.465*	d Op *0.465*	d op *0.035*
D Op *0.035*	*Tall, green .001*	*Tall, green .016*	*Tall, green .016*	*Tall, green .001*
D op *0.465*	*Tall, green .016*	*Tall, opaque .216*	*Tall, green .216*	*Tall, opaque .016*
d Op *0.465*	*Tall, green .016*	*Tall, green .216*	*Dwarf, green .216*	*Dwarf, green .016*
d op *0.035*	*Tall, green .001*	*Tall, opaque .016*	*Dwarf, green .016*	*Dwarf, opaque .001*

In summary, we get

Tall, green	*3(.001) + 4(.016) + 2 (.216) = .499*
Dwarf, opaque	*.001*
Tall, opaque	*2(.016) + .216 = .248*
Dwarf, green	*2(.016) + .216 = .248*

21. In German cockroaches, bulging eyes (*bu*) are recessive to normal eyes (bu^+) and curved wings (*cv*) are recessive to straight wings (cv^+). Both traits are encoded by autosomal genes that are linked. A cockroach has genotype $bu^+bu\ cv^+cv$ and the genes are in repulsion. Which of the following sets of genes will be found in the most common gametes produced by this cockroach?

a. bu^+cv^+
b. *bu cv*
c. bu^+bu
d. cv^+cv
e. $bu\ cv^+$

Explain your answer.

The most common gametes will have (e) $bu\ cv^+$. *Equally common will be gametes that have* bu^+cv, *not given among the choices. Since these genes are linked, in repulsion, the wild-types alleles are on different chromosomes. Thus the cockroach has one chromosome with* bu^+cv *and the homologous chromosome with* $bu\ cv^+$. *Meiosis always produces nonrecombinant gametes at higher frequencies than recombinants, so gametes bearing* $bu\ cv^+$ *will be produced at higher frequencies than (a) or (b), which are the products of recombination. The choices (c) and (d) have two copies of one locus and no copy of the other locus. They violate Mendelian segregation: each gamete must contain one allele of each locus.*

*22. In *Drosophila melanogaster,* ebony body (*e*) and rough eyes (*ro*) are encoded by autosomal recessive genes found on chromosome 3; they are separated by 20 map units. The gene that encodes forked bristles (*f*) is X-linked recessive and assorts independently of *e* and *ro*. Give the phenotypes of progeny and their expected proportions when each of the following genotypes is test-crossed.

a. $\dfrac{e^+ \quad ro^+}{e \quad ro} \ \dfrac{f^+}{f}$

We can calculate the four phenotypic classes and their proportions for e *and* ro, *and then each of those classes will be split 1:1 for* f *because* f *sorts independently. The recombination frequency between* e *and* ro *is 20%, so each of the recombinants* (e^+ ro *and* e ro^+) *will be 10%, and each of the nonrecombinants* (e^+ ro^+ *and* e ro) *will be 40%. Each of these will then be split equally among* f^+ *and* f.

$e^+ ro^+ f^+$	*20%*
$e^+ ro^+ f$	*20%*
$e\ ro\ f^+$	*20%*
$e\ ro\ f$	*20%*
$e^+ ro\ f^+$	*5%*
$e^+ ro\ f$	*5%*
$e\ ro^+ f^+$	*5%*
$e\ ro^+ f$	*5%*

b. $\dfrac{e^+ \quad ro \quad f}{e \quad ro^+ \quad f}$

We can do the same calculations as in part (a), except the nonrecombinants are e^+ ro *and* e ro^+ *and the recombinants are* e^+ ro^+ *and* e ro.

$e^+ ro^+ f^+$	*5%*
$e^+ ro^+ f$	*5%*
$e\ ro\ f^+$	*5%*
$e\ ro\ f$	*5%*
$e^+ ro\ f^+$	*20%*
$e^+ ro\ f$	*20%*
$e\ ro^+ f^+$	*20%*
$e\ ro^+ f$	*20%*

23. Honeybees have haplodiploid sex determination: females are diploid, developing from fertilized eggs, whereas males are haploid, developing from unfertilized eggs (see Chapter 4). Otto Mackensen studied linkage relations among eight mutations in honeybees (O. Mackensen. 1958. *Journal of Heredity* 49:99–102). The following table gives the results of two of MacKensen's crosses including three recessive mutations: *cd* (cordovan body color), *h* (hairless), and *ch* (chartreuse eye color).

Queen genotype	**Phenotypes of drone (male) progeny**
$\dfrac{cd \quad h^+}{cd^+ \quad h}$	294 cordovan, 236 hairless, 262 cordovan and hairless, 289 wild-type
$\dfrac{h \quad ch^+}{h^+ \quad ch}$	3131 hairless, 3064 chartreuse, 96 chartreuse and hairless, 132 wild-type

a. Only the genotype of the queen is given. Why is the genotype of the male parent not needed for mapping these genes? Would the genotype of the male parent be required if we examined female progeny instead of male progeny?
Since haploid males develop from unfertilized eggs, they receive no genes from the male parent. Therefore, the genotype of the male parent does not contribute to the genotype or phenotype of the progeny. Moreover, since we are given the genotypes and the coupling relationships for the female parents, we can deduce

the genotypes of the male parents from the female progeny, so again the genotype of the male parent is not required if we examine female progeny instead of male progeny.

b. Determine the nonrecombinant and recombinant progeny for each cross and calculate the map distances between *cd, h,* and *ch.* Draw a linkage map illustrating the linkage arrangements among these three genes.
For cd *and* h*: nonrecombinants are 294 cordovan and 236 hairless; recombinants are 262 cordovan and hairless, and 289 wild-type. RF = (262 + 289)/(262 + 289 + 294 + 236) = .51 = 51%. The two loci are unlinked.*
For h *and* ch*: 3131 hairless + 3064 chartreuse = 6195 nonrecombinants; 96 chartreuse and hairless + 132 wild-type = 228 recombinants. RF = 228/(6195+228) = .035 = 3.5%*

h *ch* *cd*

3.5

*24. A series of two-point crosses were carried out among seven loci (*a, b, c, d, e, f,* and *g*), producing the following recombination frequencies. Map the seven loci, showing their linkage groups, the order of the loci in each linkage group, and distances between the loci of each linkage group.

Loci	% Recombination	Loci	% Recombination
a - b	50	*c - d*	50
a - c	50	*c - e*	26
a - d	12	*c - f*	50
a - e	50	*c - g*	50
a - f	50	*d - e*	50
a - g	4	*d - f*	50
b - c	10	*d - g*	8
b - d	50	*e - f*	50
b - e	18	*e - g*	50
b - f	50	*f - g*	50
b - g	50		

50% recombination indicates that the genes assort independently. Less than 50% recombination indicates linkage. Starting with the most tightly linked genes a *and* g, *we look for other genes linked to these and find only gene* d *has less than 50% recombination with* a *and* g. *So one linkage group consists of* a, g, *and* d. *We know that gene* g *is between* a *and* d *because the* a *to* d *distance is 12.*

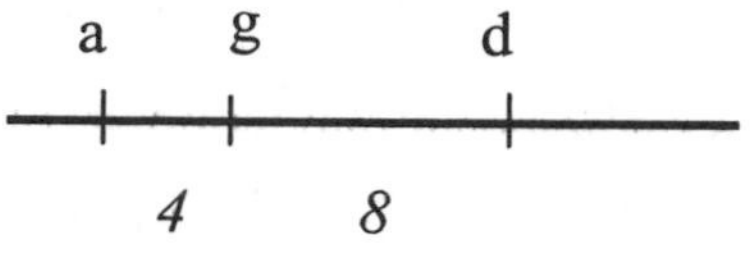

a g d

4 *8*

Similarly, we find a second linkage group of b, c, *and* e, *with* b *in the middle.*

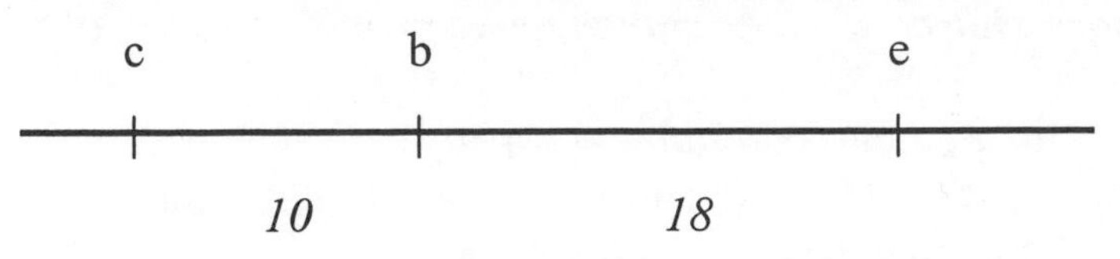

Gene f *is unlinked to either of these groups, on a third linkage group.*

25. R. W. Allard and W. M. Clement determined recombination rates for a series of genes in lima beans (R. W. Allard and W. M. Clement. 1959. *Journal of Heredity* 50:63–67). The following table lists paired recombination rates for eight of the loci (*D, Wl, R, S,* L_1, *Ms, C,* and *G*) that they mapped. On the basis of these data, draw a series of genetic maps for the different linkage groups of the genes, indicating the distances between the genes. Keep in mind that these rates are estimates of the true recombination rates and that some error is associated with each estimate. An asterisk beside a recombination frequency indicates that the recombination frequency is significantly different from 50%.

Recombination Rates (%) among Seven Loci in Lima Beans

	Wl	*R*	*S*	L_1	*Ms*	*C*	*G*
D	2.1*	39.3*	52.4	48.1	53.1	51.4	49.8
Wl		38.0*	47.3	47.7	48.8	50.3	50.4
R			51.9	52.7	54.6	49.3	52.6
S				26.9*	54.9	52.0	48.0
L_1					48.2	45.3	50.4
Ms						14.7*	43.1
C							52.0

*Significantly different from 50%.

Genes that show significantly less than 50% recombination frequency are linked. From this table we see that D, Wl, *and* R *are linked,* S *and* L_1 *are linked,* Ms *and* C *are linked, and* G *appears to be unlinked to the other genes. Of the three linked loci,* D *and* R *appear to be the farthest apart, so that* Wl *is between* D *and* R. *We then use the recombination frequency as a measure of approximate distance in drawing the linkage maps.*

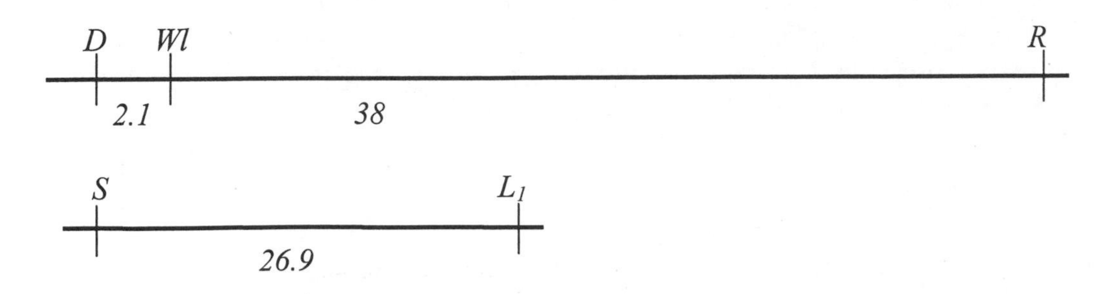

Section 7.3

26. Raymond Popp studied linkage among genes for pink eye (*p*), shaker-1 (*sh*-1), and hemoglobin (*Hb*) in mice (R. A. Popp. 1962. *Journal of Heredity* 53:73–80). He performed a series of test crosses, in which mice heterozygous for pink-eye, shaker-1, and hemoglobin 1 and 2 were crossed with mice that were homozygous for pink-eye, shaker-1, and hemoglobin 2.

$$\frac{P\ Sh\text{-}1\ Hb^1}{p\ sh\text{-}1\ Hb^2} \times \frac{p\ sh\text{-}1\ Hb^2}{p\ sh\text{-}1\ Hb^2}$$

The following progeny were produced.

Progeny genotype	Number	
$\frac{p\ sh\text{-}1\ Hb^2}{p\ sh\text{-}1\ Hb^2}$	274	nonrecombinants
$\frac{P\ Sh\text{-}1\ Hb^1}{p\ sh\text{-}1\ Hb^2}$	320	nonrecombinants
$\frac{P\ sh\text{-}1\ Hb^2}{p\ sh\text{-}1\ Hb^2}$	57	recombinants between *P* and *sh*-1
$\frac{p\ Sh\text{-}1\ Hb^1}{p\ sh\text{-}1\ Hb^2}$	45	recombinants between *p* and *Sh*-1
$\frac{P\ Sh\text{-}1\ Hb^2}{p\ sh\text{-}1\ Hb^2}$	6	recombinants between *Sh*-1 and Hb^2
$\frac{p\ sh\text{-}1\ Hb^1}{p\ sh\text{-}1\ Hb^2}$	5	recombinants between *sh-1* and Hb^1
$\frac{p\ Sh\text{-}1\ Hb^2}{p\ sh\text{-}1\ Hb^2}$	0	double crossovers
$\frac{P\ sh\text{-}1\ Hb^1}{p\ sh\text{-}1\ Hb^2}$	1	double crossovers
Total	708	

a. Determine the order of these genes on the chromosome.

All the progeny receive p sh-*1* Hb^2 *from the male parent, shown as the lower chromosome. The upper chromosomes are the products of meiosis in the heterozygous parent, and are identified above. The nonrecombinants have* p sh-*1* Hb^2 *or* P Sh-*1* Hb^1*; the double crossovers have* p Sh-*1* Hb^2 *or* P sh-*1* Hb^1. *The two classes differ in the* Sh-*1 locus; therefore,* Sh-*1 is the middle locus.*

b. Calculate the map distances between the genes.

P *and* Sh-*1: recombinants have* P sh-*1 or* p Sh-*1*. *RF = (57 + 45 + 1)/708 = 0.145 = 14.5 m.u.*

Sh-*1 and* Hb*: recombinants have* Sh-*1* Hb^2 *or* sh-*1* Hb^1. *RF = (6 + 5 + 1)/708 = 0.017 = 1.7 m.u.*

c. Determine the coefficient of coincidence and the interference among these genes.

Expected dcos = RF1 × RF2 × total progeny = 0.145(0.017)(708) = 1.7
C.o.C. = observed dcos/expected dcos = 1/1.7 = 0.59
Interference = 1 – C.o.C = 0.41

*27. Waxy endosperm (*wx*), shrunken endosperm (*sh*), and yellow seedling (*v*) are encoded by three recessive genes in corn that are linked on chromosome 5. A corn plant homozygous for all three recessive alleles is crossed with a plant homozygous for all the dominant alleles. The resulting F_1 are then crossed with a plant homozygous for the recessive genes in a three-point testcross. Following are the progeny of the testcross:

wx	*sh*	*V*	87
Wx	*Sh*	*v*	94
Wx	*Sh*	*V*	3479
wx	*sh*	*v*	3478
Wx	*sh*	*V*	1515
wx	*Sh*	*v*	1531
wx	*Sh*	*V*	292
Wx	*sh*	*v*	280
total			10,756

a. Determine order of these genes on the chromosome.

The nonrecombinants are Wx Sh V *and* wx sh v.
The double crossovers are wx sh V *and* Wx Sh v.
Comparing the two, we see that they differ only at the v *locus, so* v *must be the middle gene.*

b. Calculate the map distances between the genes.

Wx-V *distance—recombinants are* wx V *and* Wx v:
RF = (292 + 280 + 87 + 94)/10,756 = 753/10,756 = .07 = 7%, or 7 m.u.
Sh-V *distance—recombinants are* sh V *and* Sh v:
RF = (1515 + 1531 + 87 + 94)/10,756 = 3227/10,756 = 30 = 30% or 30 m.u.
The Wx-Sh *distance is the sum of these two distances: 7 + 30 = 37 m.u.l.*

c. Determine the coefficient of coincidence and the interference among these genes.

Expected dcos = RF1 × RF2 × total progeny =.07(.30)(10,756) = 226
C.o.C. = observed dcos/expected dcos = (87 + 94)/226 = 0.80
Interference = 1 – C.o.C = 0.20

28. Priscilla Lane and Margaret Green studied the linkage relations of three genes affecting coat color in mice: mahogany (*mg*), agouti (*a*), and ragged (*Ra*). They carried out a series of three-point crosses, mating mice that were heterozygous at all three loci with mice that were homozygous for the recessive alleles at these loci (P. W. Lane and M. C. Green. 1960. *Journal of Heredity* 51:228–230). The following table lists the results of the testcrosses.

Progeny of Test Crosses

Phenotype	Number
a *Ra* +	1
+ + *mg*	1
a + +	15
+ *Ra* *mg*	9
+ + +	16
a *Ra* *mg*	36
a + *mg*	76
+ *Ra* +	69
Total	213

a. Determine the order of the loci that encode mahogany, agouti, and ragged on the chromosome, the map distances between them, and the interference and coefficient of coincidence for these genes.
Based on the numbers of progeny, the nonrecombinants are a + mg *and* + Ra +. *The double crossovers are* a Ra + *and* + + mg. *These differ in the a locus; therefore, a (agouti) is the middle locus.*

The recombinants between a and mg are a *and* + *or* + *and* mg *(ignoring the* Ra *locus):* $RF = (1 + 1 + 15 + 9)/213 = 0.12 = 12\%$, *or 12 m.u.*
The recombinants between a *and* Ra *are* + + *or* a Ra *(ignoring the* mg *locus).*
$RF = (1 + 1 + 16 + 36)/213 = 0.25 = 25\%$, *or 25 m.u.*

mg a Ra

12 25

The C.o.C = observed dcos/expected dcos $= 2/(0.12)(0.25)(213) = 2/6.4 = 0.31$.
Interference $= 1 - \text{C.o.C.} = 1 - 0.31 = 0.69$

b. Draw a picture of the two chromosomes in the triply heterozygous mice used in the testcrosses, indicating which of the alleles are present on each chromosome.
The parental chromosomes have the nonrecombinant coupling relationships: a + mg *and* + Ra +. *Shown with* a *as the middle locus:*

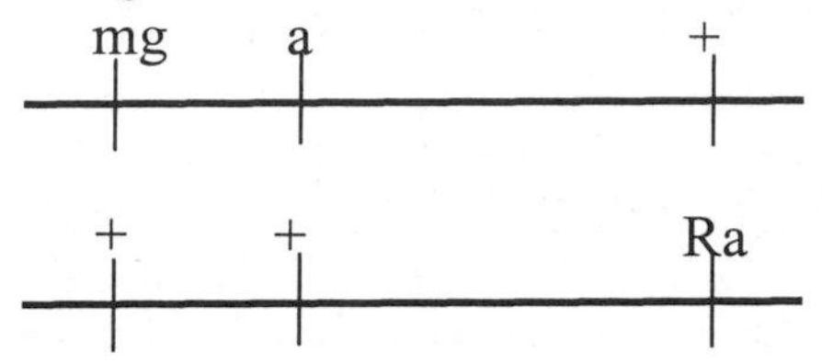

29. Fine spines (*s*), smooth fruit (*tu*), and uniform fruit color (*u*) are three recessive traits in cucumbers whose genes are linked on the same chromosome. A cucumber plant heterozygous for all three traits is used in a testcross and the progeny at the top of the following page are produced from this testcross.

S	*U*	*Tu*	2
s	*u*	*Tu*	70
S	*u*	*Tu*	21
s	*u*	*tu*	4
S	*U*	*tu*	82
s	*U*	*tu*	21
s	*U*	*Tu*	13
S	*u*	*tu*	17
Total			230

a. Determine the order of these genes on the chromosome.
Nonrecombinants are s u Tu *and* S U tu.
Double crossovers are s u tu *and* S U Tu.
Because Tu *differs between the nonrecombinants and the double crossovers,* Tu *is the middle gene.*

b. Calculate the map distances between the genes.
S-Tu *distance: recombinants are* S Tu *and* s tu.
RF = (2 + 4 + 21 + 21)/230 = 48/230 = 21%, or 21 m.u.
U-Tu *distance: recombinants are* u tu *and* U Tu.
RF = (2 + 4 + 13 + 17)/230 = 36/230 = 16%, or 16 m.u.

c. Determine the coefficient of coincidence and the interference among these genes.
Expected dcos = (48/230)(36/230)(230) = 7.5
C.o.C. = observed dcos/expected dcos = 6/7.5 = 0.8
I = 1 – C.o.C. = 0.2

d. List the genes found on each chromosome in the parents used in the testcross.
In the correct gene order for the heterozygous parent: s Tu u *and* S tu U
For the testcross parent: s tu u *and* s tu u

*30. In *Drosophila melanogaster,* black body (*b*) is recessive to gray body (b^+), purple eyes (*pr*) are recessive to red eyes (pr^+), and vestigial wings (*vg*) are recessive to normal wings (vg^+). The loci coding for these traits are linked, with the map distances:

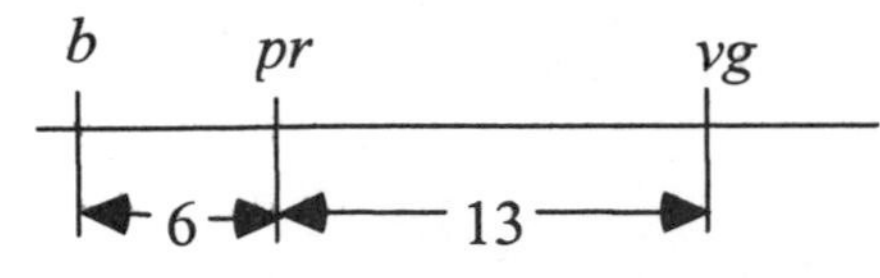

The interference among these genes is 0.5. A fly with black body, purple eyes, and vestigial wings is crossed with a fly homozygous for gray body, red eyes, and normal wings. The female progeny are then crossed with males that have black

body, purple eyes, and vestigial wings. If 1000 progeny are produced from this testcross, what will be the phenotypes and proportions of the progeny?
Although we know what the recombination frequencies are between the pairs of genes, these recombination frequencies result from both single crossover (sco) and double crossover (dco) progenies. So we must first calculate how many double crossover progeny we should get.

Working backward, given that interference = 0.5, the coefficient of coincidence = 1 – interference = 0.5.

We now use the C.o.C. to calculate the actual dco progeny:
C.o.C. = 0.5 = actual dcos/theoretical dcos = actual dcos/(.06)(.13)(1000)
The denominator calculates to 7.8, so actual dcos = 0.5(7.8) = 3.9.
We round 3.9 to 4 double crossover progeny.

Because the parents were either homozygous recessive for all three loci or homozygous dominant (wild-type) for all three loci, the F_1 heterozygote fly has chromosomes with b pr vg *and* b^+ pr^+ vg^+. *These are therefore the nonrecombinant progeny phenotypes. The double crossover progeny will be* b pr^+ vg *and* b^+ pr vg^+. *We calculated above that there will be four double crossover progeny, so we should expect two progeny flies of each double crossover phenotype.*

Next, we know that the recombination frequency between b *and* pr *is 6% or 0.06. This recombination frequency arises from the sum of the single crossovers between* b *and* pr *and the double crossover progeny:*
scos(b-pr) + *dcos = .06(1,000) = 60. But we already calculated that dcos = 4, so substituting in the above equation, we get: scos*(b-pr) + *4 = 60; scos = 56. The single crossover phenotypes between* b *and* pr *are* b pr^+ vg^+ *and* b^+ pr vg. *These total 56, or 28 each.*

Similarly, scos(pr-vg) + *dcos = .13(1000) = 130; scos*(pr-vg) *= 130 – dcos = 130 – 4 = 126. The single crossover phenotypes between* pr *and* vg *are* b pr vg^+ *and* b^+ pr^+ vg. *These total 126, or 63 each.*

The two remaining phenotypic classes are the nonrecombinants.
 # nonrecombinants = 1000 – scos(b-pr) *– scos*(pr-vg) *– dcos = 1000 – 56 – 126 – 4.*
So, # nonrecombinants = 1000 – 186 = 814. The nonrecombinant phenotypes are b^+ pr^+ vg^+ *and* b pr vg; *we expect 407 of each, to total 814.*

In summary, the expected numbers of all eight phenotypic classes are:

Phenotype	Number
b^+ pr^+ vg^+	*407*
b pr vg	*407*
b^+ pr^+ vg	*63*
b pr vg^+	*63*
b^+ pr vg	*28*
b pr^+ vg^+	*28*
b^+ pr vg^+	*2*
b pr^+ vg	*2*

31. A group of geneticists are interested in identifying genes that may play a role in susceptibility to asthma. They study the inheritance of genetic markers in a series of families that have two or more children affected with asthma. They find an association between the presence or absence of asthma and a genetic marker on the short arm of chromosome 20 and calculate a lod score of 2 for this association. What does this lod score indicate about genes that may influence asthma?
The lod score of 2 indicates that the probability of observing this degree of association if the marker is linked to asthma is 100 times higher than if the marker has no linkage to asthma. An lod score of 3, for an odds ratio of 1000-fold, is generally considered convincing evidence of linkage.

Section 7.4

*32. The locations of six deletions have been mapped to the *Drosophila* chromosome shown on the following page. Recessive mutations *a, b, c, d, e,* and *f* are known to be located in the same region as the deletions, but the order of the mutations on the chromosome is not known. When flies homozygous for the recessive mutations are crossed with flies homozygous for the deletions, the following results are obtained, where "m" represents a mutant phenotype and a plus sign (+) represents the wild type. On the basis of these data, determine the relative order of the seven mutant genes on the chromosome.

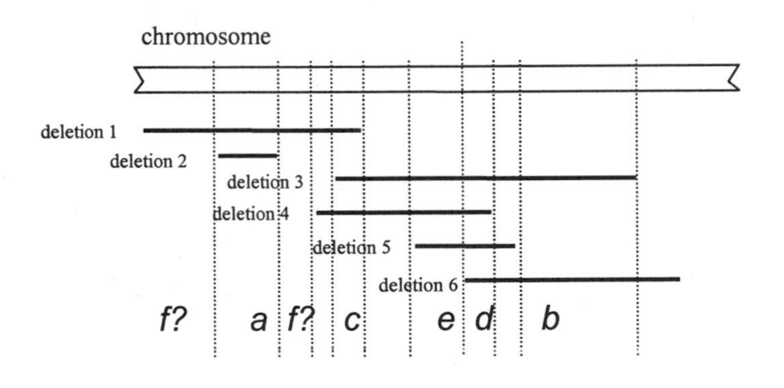

Deletion	a	b	c	d	e	f
			Mutations			
1	m	+	m	+	+	m
2	m	+	+	+	+	+
3	+	m	m	m	m	+
4	+	+	m	m	m	+
5	+	+	+	m	m	+
6	+	m	+	m	+	+

The mutations are mapped to the intervals indicated on the figure above the table. The location of f *is ambiguous; it could be in either location shown above.*

33. A panel of cell lines was created from mouse-human somatic-cell fusions. Each line was examined for the presence of human chromosomes and for the production of an enzyme. The following results were obtained:

		Human chromosomes											
Cell line	Enzyme	1	2	3	4	5	6	7	8	9	10	17	22
A	-	+	-	-	-	+	-	-	-	-	-	+	-
B	+	+	+	-	-	-	-	-	+	-	-	+	+
C	-	+	-	-	-	+	-	-	-	-	-	-	+
D	-	-	-	-	+	-	-	-	-	-	-	-	-
E	+	+	-	-	-	-	-	-	+	-	+	+	-

On the basis of these results, which chromosome has the gene that codes for the enzyme?

The enzyme is produced only in cell lines B and E. Of all the chromosomes, only chromosome 8 is present in just these two cell lines and absent in all the other cell lines that do not produce the enzyme. Therefore, the gene for the enzyme is most likely on chromosome 8.

*34. A panel of cell lines was created from mouse-human somatic-cell fusions. Each line was examined for the presence of human chromosomes and for the production of three enzymes. The following results were obtained:

	Enzyme			Human chromosomes								
Cell line	1	2	3	4	8	9	12	15	16	17	22	X
A	+	-	+	-	-	+	-	+	+	-	-	+
B	+	-	-	-	-	+	-	-	+	+	-	-
C	-	+	+	+	-	-	-	-	-	+	-	+
D	-	+	+	+	+	-	-	-	+	-	-	+

On the basis of these results, give the chromosome location of enzyme 1, enzyme 2, and enzyme 3.

Enzyme 1 is located on chromosome 9. Chromosome 9 is the only chromosome that is present in the cell lines that produce enzyme 1 and absent in the cell lines that do not produce enzyme 1.

Enzyme 2 is located on chromosome 4. Chromosome 4 is the only chromosome that is present in cell lines that produce enzyme 2 (C and D) and absent in cell lines that do not produce enzyme 2 (A and B).

Enzyme 3 is located on the X chromosome. The X chromosome is the only chromosome present in the three cell lines that produce enzyme 3 and absent in the cell line that does not produce enzyme 3.

CHALLENGE QUESTION

Section 7.5

35. Transferrin is a blood protein that is encoded by the transferrin locus (*Trf*). In house mice the two alleles at this locus (Trf^a and Trf^b) are codominant and encode three types of transferrin:

Genotype	Phenotype
Trf^a/Trf^a	Trf-a
Trf^a/Trf^b	Trf-ab
Trf^b/Trf^b	Trf-b

The dilution locus, found on the same chromosome, determines whether the color of a mouse is diluted or full; an allele for dilution (*d*) is recessive to an allele for full color (d^+):

Genotype	Phenotype
d^+d^+	d+ (full color)
d^+d	d+ (full color)
dd	d (dilution)

Donald Shreffler conducted a series of crosses to determine the map distance between the tranferrin locus and the dilution locus (D. C. Shreffler. 1963 *Journal of Heredity* 54:127 –129). The following table presents a series of crosses carried out by Shreffler and the progeny resulting from these crosses.

			Progeny phenotypes			
Cross	♂ × ♀	d^+ Trf-ab	d^+ Trf-b	d Trf-ab	d Trf-b	Total
1	$\frac{d^+\ Trf^a}{d\ Trf^b} \times \frac{d\ Trf^b}{d\ Trf^b}$	32	3	6	21	62
2	$\frac{d\ Trf^b}{d\ Trf^b} \times \frac{d^+\ Trf^a}{d\ Trf^b}$	16	0	2	20	38
3	$\frac{d^+\ Trf^a}{d\ Trf^b} \times \frac{d\ Trf^b}{d\ Trf^b}$	35	9	4	30	78
4	$\frac{d\ Trf^b}{d\ Trf^b} \times \frac{d^+\ Trf^a}{d\ Trf^b}$	21	3	2	19	45
5	$\frac{d^+\ Trf^b}{d\ Trf^a} \times \frac{d\ Trf^b}{d\ Trf^b}$	8	29	22	5	64
6	$\frac{d\ Trf^b}{d\ Trf^b} \times \frac{d^+\ Trf^b}{d\ Trf^a}$	4	14	11	0	29

a. Calculate the recombinant frequency between the *Trf* and the *d* loci by using the pooled data from all the crosses.

We sum all the recombinant progeny and divide by the total progeny of all six crosses:

RF = (3 + 6 + 0 + 2 + 9 + 4 + 3 + 2 + 8 + 5 + 4 + 0)/316 = 46/316 = 0.146 = 15%

b. Which crosses represent recombination in male gamete formation and which crosses represent recombination in female gamete formation?

Crosses representing recombination in males are those where the female parent is homozygous and the male parent is heterozygous: crosses 1, 3, and 5.

Crosses 2, 4, and 6 represent female recombination because the female parent is heterozygous and the male parent is homozygous.

c. On the basis of your answer to part **b.**, calculate the frequency of recombination among male parents and female parents separately.

RF for males = (3 + 6 + 9 + 4 + 8 + 5)/(62 + 78 + 64) = 35/204 = 0.17 = 17%

RF for females = (0 + 2 + 3 + 2 + 4 + 0)/(38 + 45 + 29) = 11/112 = 0.098 = 9.8%

d. Are the rates of recombination in males and females the same? If not, what might produce the difference?

The rate of recombination between these two genes appears to be higher in males than in females, although a statistical test should be performed to determine the significance of the apparent difference. Such gender differences in recombination rates could arise from differences in gametogenesis (meiosis is arrested for prolonged periods in mammalian oogenesis), imprinting of nearby loci affecting chromatin structure and crossing over in the region, or differences in synapsis of chromosomes in male and female meioses (Lynn et al., Am. J. Hum. Genet. *77:670–675, 2005).*

Chapter Eight: Bacterial and Viral Genetic Systems

COMPREHENSION QUESTIONS

Section 8.1

1. Explain how auxotrophic bacteria are isolated.
   Unlike prototrophic bacteria (wild-type) that can grow on minimal media, auxotrophic bacteria are mutant strains of bacteria that are unable to grow on minimal media. In other words, auxotrophs are not nutritionally self-sufficient. To isolate an auxotrophic bacterium from a culture of wild-type bacteria, first spread the bacterial culture out on a petri dish containing nutritionally complete growth medium allowing prototrophic and auxotrophic colonies to grow. Using the replica plating technique, transfer a few cells from each colony to replica plates. One of the replica plates should contain a selective medium that lacks a nutrient required by the auxotroph for growth, and the other replica plate should contain a nutritionally complete medium. The auxotrophic colonies should grow only on the nutritionally complete medium and not the selective medium. Prototrophic colonies should grow on both types of media.

2. Briefly explain the differences between F^+, F^-, Hfr, and F′ cells.
   An F^+ cell will contain the F factor as a circular plasmid separate from the chromosome. The Hfr cell has the F factor integrated into its chromosome. In F′ strains, the F factor exists as a separate circular plasmid, but the plasmid carries bacterial genes that were originally part of the bacterial chromosome. The F^- strain does not contain the F Factor and can receive DNA from cells that contain the F Factor (F^+, Hfr, and F′ cells).

*3. What types of matings are possible between F^+, F^-, Hfr, and F′ cells? What outcomes do these matings produce? What is the role of F factor in conjugation?

Types of matings	*Outcomes*
$F^+ \times F^-$	*Two F^+ cells*
Hfr $\times$ F^-	*One F^+ cell and one F^- cell*
$F' \times F^-$	*Two F′ cells*

The F factor contains a number of genes involved in the conjugation process, including genes necessary for the synthesis of the sex pilus. The F factor also has an origin of replication that allows for the factor to be replicated during the conjugation process, and genes for opening the plasmid and initiating the chromosome transfer.

*4. Explain how interrupted conjugation, transformation, and transduction can be used to map bacterial genes. How are these methods similar and how are they different?
   To map genes by conjugation, an interrupted mating procedure is used. During the conjugation process, an Hfr strain is mixed with an F^- strain. The two strains must

have different genotypes and must remain in physical contact for the transfer to occur. At regular intervals, the conjugation process is interrupted. The chromosomal transfer from the Hfr strain always begins with a portion of the integrated F factor and proceeds in a linear fashion. To transfer the entire chromosome would require approximately 100 minutes. The time required for individual genes to be transferred is relative to their position on the chromosome and the direction of transfer initiated by the F factor. Gene distances are typically mapped in minutes of conjugation. The genes that are transferred by conjugation to the recipient must be incorporated into the recipient's chromosome by recombination to be expressed.

In transformation, the relative frequency at which pairs of genes are transferred or cotransformed indicates the distance between the two genes. Closer gene pairs are cotransformed more frequently. As was the case with conjugation, the donor DNA must recombine into the recipient cell's chromosome. Physical contact of the donor and recipient cells is not needed. The recipient cell uptakes the DNA directly from the environment. Therefore, the DNA from the donor strain has to be isolated and broken up before transformation can take place.

A viral vector is needed for the transfer of DNA by transduction. DNA from the donor cell is packaged into a viral protein coat. The viral particle containing the bacterial donor DNA then infects another bacterial cell or the recipient. The donor bacterial DNA is incorporated into the recipient cell's chromosome by recombination. Only genes that are close together on the bacterial chromosome can be cotransduced. Therefore, the rate of cotransduction, like the rate of cotransformation, gives an indication of the physical distances between genes on the chromosome.

These three processes are similar in that all involve the uptake by the recipient cell of a piece of the donor chromosome and the incorporation of some of that piece into the recipient chromosome by recombination. They also all calculate the mapping distance by measuring the frequency with which recipient cells are transformed. The process use different methods to get donor DNA incorporated into the recipient cell.

5. What is horizontal gene transfer and how might it occur?
Horizontal gene transfer occurs when a bacterial cell acquires genes from another species. Genome analysis experiments have shown that bacterial species have even acquired DNA from eukaryotic organisms. Three mechanisms that could lead to horizontal gene transfer are transduction, transformation, and conjugation.

Section 8.2

*6. List some of the characteristics that make bacteria and viruses ideal organisms for many types of genetic studies.
(1) Reproduction is rapid, asexual, and produces lots of progeny.
(2) Their genomes are small.
(3) They are easy to grow in the laboratory.

(4) Techniques are available for isolating and manipulating their genes.
(5) Mutant phenotypes, especially auxotrophic phenotypes, are easy to measure.

7. What types of genomes do viruses have?
Viral genomes can consist of either DNA or RNA molecules. The viral nucleic acids can be either double-stranded or single-stranded, depending on the type of virus.

8. Briefly describe the differences between the lytic cycle of virulent phages and the lysogenic cycle of temperate phages.
Virulent phages reproduce strictly by the lytic cycle and ultimately result in the death of the host bacterial cell. During the lytic cycle, a virus injects its genome into the host cell. The genome directs production and assembly of new viral particles. A viral enzyme is produced and breaks open the cell, releasing new viral particles into the environment.

Temperate phages can utilize either the lytic or lysogenic cycle. The infection cycle begins when a viral particle injects its genome into the host cell. In the lysogenic cycle, the viral genome integrates into the host chromosome as a prophage. The inactive prophage can remain part of the bacterial chromosome for an extended period and is replicated along with the bacterial chromosome prior to cell division. Certain environmental stimuli can trigger the prophage to exit the lysogenic cycle and enter the lytic cycle.

9. Briefly explain how genes in phages are mapped.
To map genes in phages, bacterial cells are doubly infected with phage particles that differ in two or more genes. During the production of new phage progeny, the phage DNAs can undergo recombination, thus resulting in the formation of recombinant plaques. The rate of recombination is used to determine the linear order and relative distances between genes. The farther apart two genes are on the chromosome, the more frequently they will recombine.

*10. How does specialized transduction differ from generalized transduction?
In generalized transduction, randomly selected bacterial genes are transferred from one bacterial cell to another by a virus. In specialized transduction, only genes from a particular region of the bacterial chromosome are transferred to another bacterium. The process of specialized transduction requires lysogenic phages that integrate into specific locations on the host cell's chromosome. When the phage DNA excises from the host chromosome and the excision process is imprecise, the phage DNA will contain a small part of the bacterial DNA that was adjacent to the viral insertion site. The hybrid DNA must be injected by the phage into another bacterial cell during another round of infection.

Transfer of DNA by generalized transduction requires that the host DNA be broken down into smaller pieces and that a piece of the host DNA is packaged into a phage coat instead of phage DNA. The defective phage cannot produce new phage particles upon a subsequent infection, but it can inject the bacterial DNA into another bacterium. Through a double crossover event, the donor DNA can become incorporated into the bacterial recipient's chromosome.

*11. Briefly explain the method used by Benzer to determine whether two different mutations occurred at the same locus.
Benzer conducted complementation tests by infecting cells of E. coli *K with large numbers of the two mutant phage types. For successful infection to occur on the* E. coli *K strains, each mutant phage needed to supply the gene product or protein missing in the other. Complementation will happen only if the mutations are at separate loci. If the two mutations are at the same locus, then complementation of gene products will not occur and no plaques will be produced on the* E. coli *K lawns.*

*12. Explain how a retrovirus, which has an RNA genome, is able to integrate its genetic material into that of a host having a DNA genome.
Retroviruses are able to integrate their genomes into the host cell's DNA genome through the action of the enzyme reverse transcriptase. Reverse transcriptase can synthesize complementary DNA from either a RNA or DNA template. The retrovirus enzyme synthesizes a double-stranded copy of DNA using the retroviral single-stranded RNA as the template. The newly synthesized DNA molecule can then integrate into the host chromosome to form a provirus.

13. Briefly describe the genetic structure of a typical retrovirus.
Retroviral genomes all have three genes in common: gag, pol, *and* env. *Proteins that make up the viral capsid are encoded by the* gag *gene. Reverse transcriptase and an enzyme called integrase are encoded by the* pol *gene. While reverse transcriptase synthesizes double-stranded viral DNA from an RNA template, integrase results in the insertion of the viral DNA into the host chromosome. Finally, the* env *gene encodes for proteins found on the viral envelope.*

14. What are the evolutionary origins of HIV-1 and HIV-2?
Both HIV-1 and HIV-2 are related to simian immunodeficiency viruses (SIV). Analysis of DNA sequences indicates that HIV-1 is related to simian immunodeficiency virus found in chimpanzees (SIVcpz). DNA sequence analysis also reveals that SIVcpz is a hybrid virus created by recombination between a retrovirus in red-capped mangabey and a retrovirus found in the great spot monkey. HIV-2 sequence analysis shows that it evolved from a simian immunodeficiency virus found in sooty mangabeys (SIVsm).

APPLICATION QUESTIONS AND PROBLEMS

Section 8.1

*15. John Smith is a pig farmer. For the past five years, Smith has been adding vitamins and low doses of antibiotics to his pig food; he says that these supplements enhance the growth of the pigs. Within the past year, however, several of his pigs have died from infections of common bacteria, which failed to respond to large doses of antibiotics. Can you offer an explanation for the increased rate of mortality due to infection in Smith's pigs? What advice might you offer Smith to prevent this problem in the future?

Over the past five years, Farmer Smith, by treating his pigs with low doses of antibiotics, has been selecting for bacteria that are resistant to the antibiotics. The doses used killed sensitive bacteria, but not those bacteria that were moderately sensitive or slightly resistant. Over time, only resistant bacteria will be present in his pigs because any sensitive bacteria will have been eliminated by the low doses of antibiotics. The pigs that died this past year were infected by bacteria that had become so resistant they were not killed by even high doses of antibiotics.

In the future, Farmer Smith can continue to use the vitamins, but he should use the antibiotics only when a sick pig requires them. In this manner, he will not be selecting for antibiotic-resistant bacteria, and the chances of the antibiotic therapy successfully treating his sick pigs will be greater.

16. Rarely, conjugation of Hfr and F^- cells produces two Hfr cells. Explain how this occurs.

Hfr strains contain an F factor integrated into the bacterial chromosome. The F factor mediates transfer of the bacterial chromosome. During conjugation of an Hfr strain with an F^- strain, the transfer process begins within the F factor. So literally, part of the F factor is the first to arrive in F^- cell. However, the remaining part of the F factor is transferred last. Because nearly 100 minutes are required to completely transfer the donor chromosome, the two cells must remain in contact for the entire 100 minutes. So, if the donor and recipient cell are not disturbed and the transfer process is not interrupted, then the entire Hfr strain's chromosome, including the F factor, can be donated to the F^- cell. Following recombination between the donor and recipient chromosomes, this results in the production of a second Hfr strain.

17. Austin Taylor and Edward Adelberg isolated some new strains of Hfr cells that they then used to map several genes in *E. coli* by using interrupted conjugation (A. L. Taylor and E. A. Adelberg. 1960. *Genetics* 45:1233–1243). In one experiment, they mixed cells of Hfr strain AB-312, which were xyl^+ mtl^+ mal^+ met^+ and sensitive to phage T6, with F^- strain AB-531, which was xyl^- mtl^- mal^- met^- and resistant to phage T6. The cells were allowed to undergo conjugation. At regular intervals, the researchers removed a sample of cells and interrupted conjugation by killing the Hfr cells with phage T6. The F^- cells, which were resistant to phage T6, survived and

were then tested for the presence of genes transferred from the Hfr strain. The results of this experiment are shown in the accompanying graph. On the basis of these data, give the order of the *xyl, mtl, mal,* and *met* genes on the bacterial chromosome and indicate the minimum distances between them.

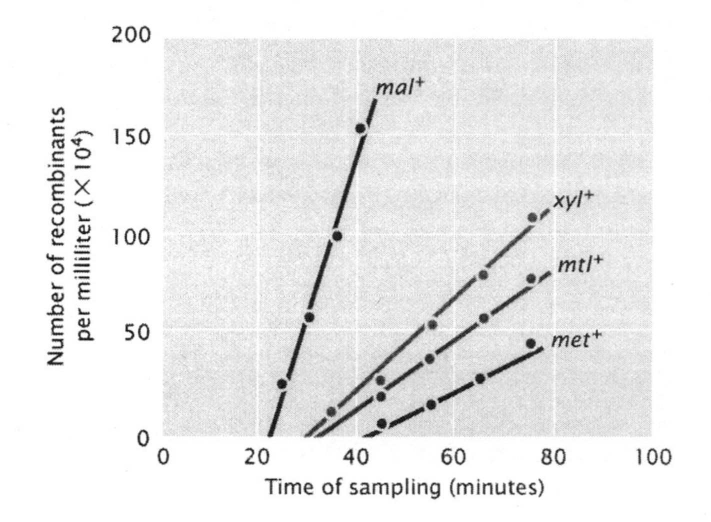

The closer genes are to the F factor, the more quickly they will be transferred and more recombinants will be produced. The transfer process will occur in a linear fashion. By interrupting the mating process, the transfer will stop and the F^- *strain will have received only genes carried on the piece of the Hfr strain's chromosome that entered the* F^- *cell prior to the disruption. From the graph, we can determine when the first recombinants for each marker were first identified and subsequently approximate the minutes that separate the different genetic markers.*

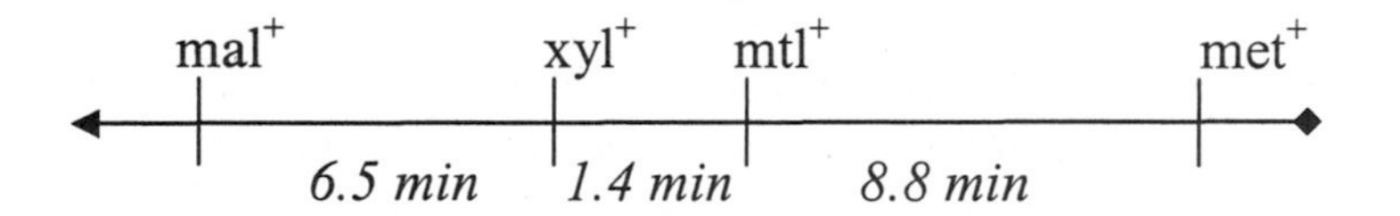

*18. A series of Hfr strains that have genotype $m^+ n^+ o^+ p^+ q^+ r^+$ are mixed with an F^- strain that has genotype $m^- n^- o^- p^- q^- r^-$. Conjugation is interrupted at regular intervals and the order of appearance of genes from the Hfr strain is determined in the recipient cells. The order of gene transfer for each Hfr strain is:

Hfr 5 $m^+ q^+ p^+ n^+ r^+ o^+$
Hfr 4 $n^+ r^+ o^+ m^+ q^+ p^+$
Hfr 1 $o^+ m^+ q^+ p^+ n^+ r^+$
Hfr 9 $q^+ m^+ o^+ r^+ n^+ p^+$

What is the order of genes on the circular bacterial chromosome? For each Hfr strain, give the location of the F factor in the chromosome and its polarity.

In each of the Hfr strains, the F factor has been inserted at a different location in the chromosome. The orientation of the F factor in the strains varies as well.

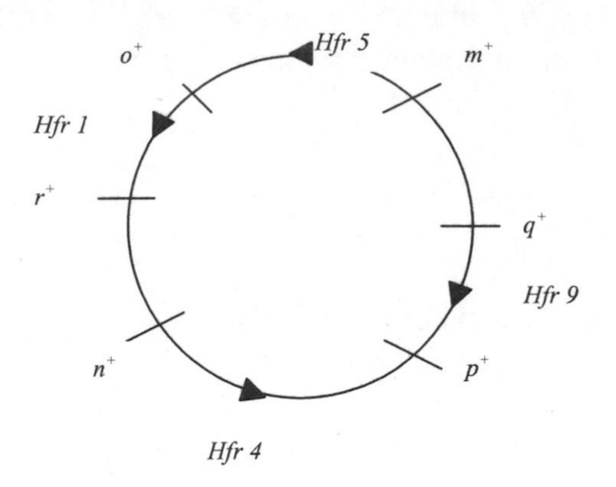

*19. Crosses of three different Hfr strains with separate samples of an F⁻ strain are carried out, and the following mapping data are provided from studies of interrupted conjugation:

Appearance of genes in F⁻ cells

Hfr1:	Genes	b^+	d^+	c^+	f^+	g^+
	Time	3	5	16	27	59
Hfr2:	Genes	e^+	f^+	c^+	d^+	b^+
	Time	6	24	35	46	48
Hfr3	Genes	d^+	c^+	f^+	e^+	g^+
	Time	4	15	26	44	58

Construct a genetic map for these genes, indicating their order on the bacterial chromosome and the distances between them.

The F factor for each Hfr strain has been inserted into a different location on the chromosome, and the orientation of the F factor varies in the different strains. Although most of the selective markers transferred from each Hfr strain to the F⁻ strain are the same, some of the markers for a given Hfr strain are not transferred due to the mating being disrupted prior to the transfer of that selective marker. The relative position of the genes to each other in minutes does not vary. So, for the different Hfr strains, the distance in minutes between each gene remains constant. The genes and their relative positions are shown below. Times are in minutes of conjugation.

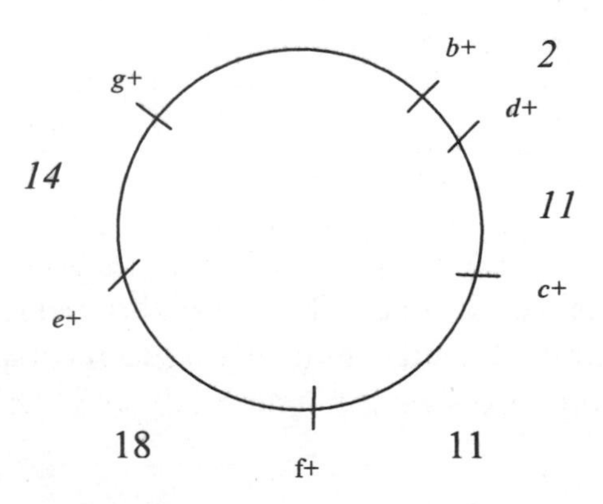

20. DNA from a strain of *Bacillus subtilis* with the genotype trp^+ tyr^+ is used to transform a recipient strain with the genotype trp^- tyr^-. The following numbers of transformed cells were recovered:

Genotype	Number of transformed cells
trp^+ tyr^-	154
trp^- tyr^+	312
trp^+ tyr^+	354

What do these results suggest about the linkage of the *trp* and *tyr* genes?
During transformation, only genes that are closely linked or located near each other on the donor's chromosome will be transformed together. In other words, a higher cotransformation frequency indicates a shorter distance between the two genes on the donor's chromosome.

To calculate the cotransformation frequency of the trp^+ *and* tyr^+ *genes from* Bacillus subtilis, *divide the number of transformed cells with the genotype* trp^+ tyr^+ *by the total number of transformed cells (354/820). The frequency of cotransformation is 0.43, or 43%. The high level of cotransformation indicates that these two genes are closely linked.*

21. DNA from a strain of *Bacillus subtilis* with genotype a^+ b^+ c^+ d^+ e^+ is used to transform a strain with genotype a^- b^- c^- d^- e^-. Pairs of genes are checked for cotransformation and the following results are obtained:

Pair of genes	Cotransformation
a^+ and b^+	no
a^+ and c^+	no
a^+ and d^+	yes
a^+ and e^+	yes
b^+ and c^+	yes
b^+ and d^+	no
b^+ and e^+	yes
c^+ and d^+	no
c^+ and e^+	yes
d^+ and e^+	no

On the basis of these results, what is the order of the genes on the bacterial chromosome?
Only genes located near each other on the bacterial chromosome will be cotransformed together. However, by performing transformation experiments and screening for different pairs of cotransforming genes, a map of the gene order can be determined. Gene pairs that never result in cotransformation must be farther apart on the chromosome, while gene pairs that result in cotransformation are more closely linked. From the data, we see that gene a^+ *cotransforms with both* e^+ *and* d^+. *However, genes* d^+ *and* e^+ *do not exhibit cotransformation, indicating that* a^+ *and* e^+ *are more closely linked than* d^+ *and* e^+. *Gene* a^+ *does not exhibit*

cotransformation with either gene b^+ *or* c^+, *yet gene* e^+ *does. This indicates that gene* e^+ *is more closely linked to genes* b^+ *and* c^+ *than is gene* a^+. *The orientation of genes* b^+ *and* c^+ *relative to* e^+ *cannot be determined from the data provided.*

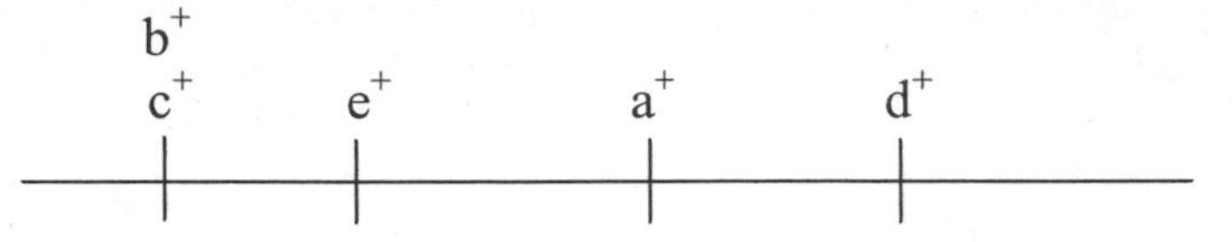

22. DNA from a bacterial strain that is *his*$^+$ *leu*$^+$ *lac*$^+$ is used to transform a strain that is *his*$^-$ *leu*$^-$ *lac*$^-$. The following percentages of cells were transformed:

Donor strain	Recipient strain	Genotype of transformed cells	Percent
his$^+$ *leu*$^+$ *lac*$^+$	*his*$^-$ *leu*$^-$ *lac*$^-$	*his*$^+$ *leu*$^+$ *lac*$^+$	0.02%
		his$^+$ *leu*$^+$ *lac*$^-$	0.00%
		his$^+$ *leu*$^-$ *lac*$^+$	2.00%
		his$^+$ *leu*$^-$ *lac*$^-$	4.00%
		his$^-$ *leu*$^+$ *lac*$^+$	0.10%
		his$^-$ *leu*$^-$ *lac*$^+$	3.00%
		his$^-$ *leu*$^+$ *lac*$^-$	1.50%

a. What conclusions can you make about the order of these three genes on the chromosome?
The percentages of cotransformation between his^+, leu^+, *and* lac^+ *loci must be examined. Genes that cotransform more frequently will be closer together on the donor chromosome. Cotransformation between* lac^+ *and* his^+ *occurs in 2.02% of the transformed cells. By comparing this value with the cotransformation of* lac^+ *and* leu^+ *at 0.12% and the cotransformation of* his^+ *and* leu^+ *at .02%, we can see that* lac^+ *and* his^+ *cotransform more frequently. Therefore,* lac^+ *and* his^+ *must be more closely linked than the other gene pairs combinations. Because* lac^+ *and* leu^+ *cotransform more frequently than* his^+ *and* leu^+, leu^+ *must be located closer to* lac^+ *than it is to* his^+.

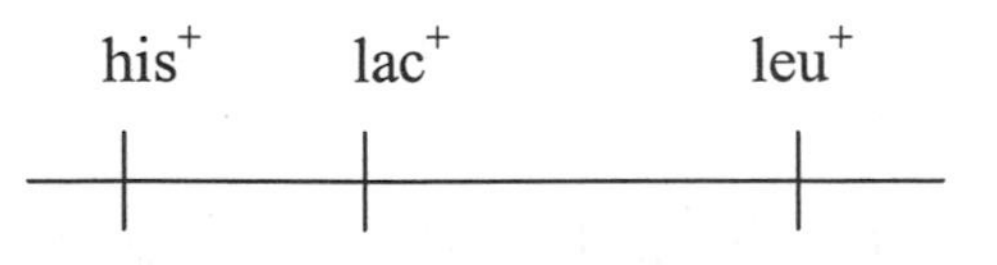

b. Which two genes are closest?
From the cotransformation frequencies, we can predict that lac^+ *and* his^+ *are the closest two genes.*

23. Rollin Hotchkiss and Julius Marmur studied transformation in the bacterium *Streptococcus pneumoniae* (R. D. Hotchkiss and J. Marmur. 1954. *Proceedings of the National Academy of Sciences* 40:55–60). They examined four mutations in this bacterium: penicillin resistance (*P*), streptomycin resistance (*S*), sulfanilamide resistance (*F*), and the ability to

utilize mannitol (*M*). They extracted DNA from strains of bacteria with different combinations of different mutations and used this DNA to transform wild-type bacterial cells ($P^+ S^+ F^+ M^+$). The results from one of their transformation experiments are shown here.

Donor DNA	**Recipient DNA**	**Transformants**	**Percentage of all cells**
$M\,S\,F$	$M^+ S^+ F^+$	$M^+ S\,F^+$	4.0
		$M^+ S^+ F$	4.0
		$M\,S^+ F^+$	2.6
		$M\,S\,F^+$	0.41
		$M^+ S\,F$	0.22
		$M\,S^+ F$	0.0058
		$M\,S\,F$	0.0071

a. Hotchkiss and Marmur noted that the percentage of cotransformation was higher than would be expected on a random basis. For example, the results show that the 2.6% of the cells were transformed into *M* and 4% were transformed into *S*. If the *M* and *S* traits were inherited independently, the expected probability of cotransformation of *M* and *S* (*M S*) would be $0.026 \times 0.04 = 0.001$, or 0.1%. However, they observed 0.41% *M S* cotransformants, four times more than they expected. What accounts for the relatively high frequency of cotransformation of the traits they observed?
It is likely that the M *and* S *traits are linked or in other words they are located very close to each other on the* Steptococcus pneunomiae *chromosome. By being located close together on the chromosome, these markers are more likely to be cotransformed on a single fragment of DNA.*

b. On the basis of the results, what conclusion can you make about the order of the *M, S,* and *F* genes on the bacterial chromosome?
Because M *and* S *cotransform quite frequently, they are likely close to each other on the chromosome as indicated above.* S *and* F *cotransform more frequently (.22) than do* M *and* F *(0.0058). The transformation data suggests that* S *and* F *are located closer together than are* M *and* F.

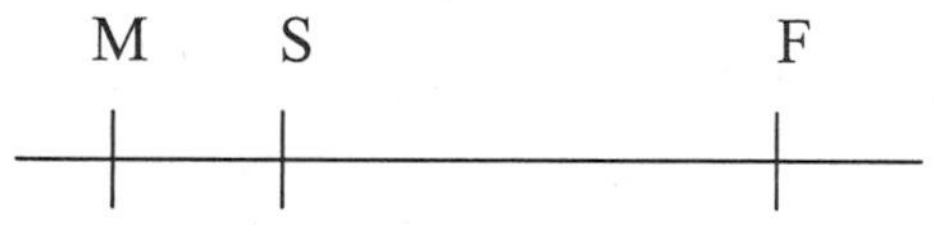

c. Why is the rate of cotransformation for all three genes (*M S F*) almost the same as the cotransformation of *M F* alone?
Genes M *and* F *cotransform infrequently, which is likely due to the physical distance between them. Genes* M S *are more closely linked on chromosome and the relative positions of* M, S, *and* F *make it likely that, if* M *and* F *are cotransformed on the same DNA molecule, then* S *will be cotransformed as well.*

24. In the course of a study on the effects of the mechanical shearing of DNA, Eugene Nester, A. T. Ganesan, and Joshua Lederberg studied the transfer, by transformation, of sheared DNA from a wild type strain of *Bacillus subtilis* (his_2^+ aro_3^+ try_2^+ aro_1^+ tyr_1^+ aro_2^+) to strains of bacteria carrying a series of mutations (E. W. Nester, A. T. Ganesan, and J. Lederberg. 1963. *Proceedings of the National Academy of Sciences* 49:61–68). They reported the following rates of cotransformation between his_2^+ and the other genes (expressed as cotransfer rate), shown here.

Genes	Rate of cotransfer
his_2^+ and aro_3^+	0.015
his_2^+ and try_2^+	0.10
his_2^+ and aro_1^+	0.12
his_2^+ and tyr_1^+	0.23
his_2^+ and aro_2^+	0.05

On the basis of these data, which gene is farthest from his_2^+? Which gene is closest?

Genes that are more closely linked or located physically nearer each other on the chromosome will cotransform more frequently than those genes that are not. Genes that are farther apart will cotransform less frequently with each other. From the above data, we can see that his_2^+ *and* tyr_1^+ *cotransform the most frequently together (0.23), which indicates* tyr_1^+ *is the closest gene to the* his_2^+ *marker. Also from the above data, we can see that* his_2^+ *and* aro_3^+ *cotransform less frequently than the other pairs (0.015), which indicates that* aro_3^+ *is the farthest gene from* his_2^+.

*25. Anagnostopoulos and I. P. Crawford isolated and studied a series of mutations that affected several steps in the biochemical pathway leading to tryptophan in the bacterium *Bacillus subtilis* (C. Anagnostopoulos and I. P. Crawford. 1961. *Proceedings of the National Academy of Sciences* 47:378–390). Seven of the strains that they used in their study are listed here, along with the mutation found in that strain.

Strain	Mutation
T3	T^-
168	I^-
168PT	I^-
TI	I^-
TII	I^-
T8	A^-
H25	H^-

To map the genes for tryptophan synthesis, they carried out a series of transformation experiments on strains having different mutations and determined the percentage of recombinants among the transformed bacteria. Their results were as follows:

Recipient	Donor	Percent recombinants
T3	168PT	12.7
T3	T11	11.8
T3	T8	43.5
T3	H25	28.6
168	H25	44.9
TII	H25	41.4
TI	H25	31.3
T8	H25	67.4
H25	T3	19.0
H25	TII	26.3
H25	TI	13.4
H25	T8	45.0

On the basis of these two-point recombination frequencies, determine the order of the genes and the distances between them. Where more than one cross was completed for a pair of genes, average the recombination rates from the different crosses. Draw a map of the genes on the chromosome.

Although this likely is not the actual case for the tryptophan operon, we will assume that each mutation listed (T^-, I^-, A^-, and H^-) represents a single gene and not just a selectable phenotype. The percent recombination can be used to determine the map distances. Some gene pairs have multiple crosses so we will have to average the percentages of recombination to determine the map distances.

T---I: *12.7% (T3 × 168PT) and 11.8% (T3 × TII)*
T---A: *43.5% (T3 × T8)*
T---H: *28.6% (T3 × H25) and 19.0 (H25 × T3)*
I---H: *44.9% (168 × H25), 41.4% (TII × H25), 31.3% (TI × H25), 26.3% (H25 × TII), and 13.4% (H25 × TI)*
A---H: *67.4% (T8 × H25) and 45.0% (H25 × T8)*

From the above crosses, we can determine the map distances by taking the averages of the crosses for each gene.

I – T = 12.3
T – A = 43.5
T – H = 23.8
I – H = 31.5
H – A= 56.2

From these results, we can see that A and H are the furthest apart and that T is between I and H. This gives a map of:

From these results, we can see that A and H are the furthest apart and that T is between I and H. This gives a map of:

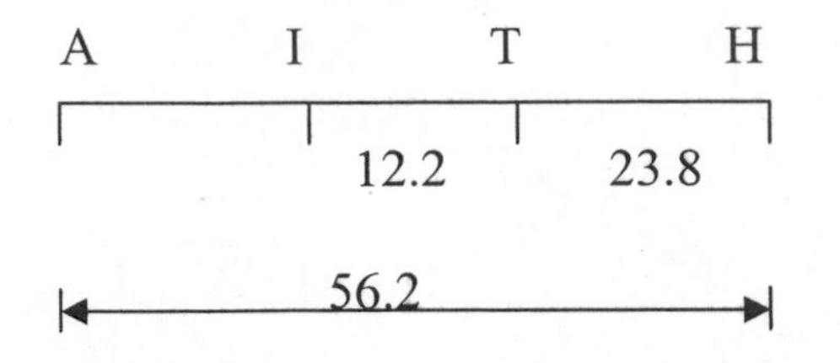

We have no direct experimental measure of the distance between A and I. However, since map distances are additive, we can calculate that this distance by subtraction. If we use the A to H distance, this is: 56.2 – (12.2 + 23.8) = 20.2. If we use the A to T distance, this is: 43.5 – 12.2 = 31.3. The true distance between A and I is probably close to these estimates. Using the average (20.2 + 31.3) / 2 = 25.8, we can complete the map.

A I T H

25.8 12.2 23.8

Section 8.2

26. Two mutations that affect plaque morphology in phages (a^- and b^-) have been isolated. Phages carrying both mutations (a^- b^-) are mixed with wild-type phages (a^+ b^+) and added to a culture of bacterial cells. Subsequent to infection and lysis, samples of the phage lysate are collected and cultured on bacterial cells. The following numbers of plaques are observed:

Plaque phenotype	Number
a^+ b^+	2043
a^+ b^-	320
a^- b^+	357
a^- b^-	2134

What is the frequency of recombination between the *a* and *b* genes?

First, we must identify the progeny phage whose plaque phenotype is different from either of the infecting phage. The original infecting phages were wild-type (a^+ b^+) *and doubly mutant* (a b). *Any phages that give rise to the* a^+ b^- *plaque phenotype or the* a^- b^+ *plaque phenotype were produced by recombination between the two types of infecting phage particles.*

Plaque phenotype	*Number*
a^+ b^+	*2043*
a^+ b	*320 (recombinant)*
a^- b^+	*357 (recombinant)*
a^- b^-	*2134*
Total plaques	*4854*

The frequency of recombination is calculated by dividing the total number of recombinant plaques by the total number of plaques (677/4854), which gives a frequency of 0.14, or 14%.

27. T. Miyake and M. Demerec examined proline-requiring mutations in the bacterium *Salmonella typhimurium* (T. Miyake and M. Demerec. 1960. *Genetics* 45:755–762). On the basis of complementation studies, they found four proline auxotrophs: *proA, proB, proC,* and *proD.* To determine if *proA, proB, proC,* and *proD* loci were located close together on the bacterial chromosome, they conducted a transduction experiment. Bacterial strains that were $proC^+$ and had mutations at *proA, proB,* or *proD* were used as donors. The donors were infected with bacteriophages, and progeny phages were allowed to infect recipient bacteria with genotype $proC^- \; proA^+ \; proB^+ \; proD^+$. The bacteria were then plated on a selective medium that allowed only $proC^+$ bacteria to grow. The following results were obtained:

Donor genotype	Transductant genotype	Number
$proC^+ \; proA^- \; proB^+ \; proD^+$	$proC^+ \; proA^+ \; proB^+ \; proD^+$	2765
	$proC^+ \; proA^- \; proB^+ \; proD^+$	3
$proC^+ \; proA^+ \; proB^- \; proD^+$	$proC^+ \; proA^+ \; proB^+ \; proD^+$	1838
	$proC^+ \; proA^+ \; proB^- \; proD^+$	2
$proC^+ \; proA^+ \; proB^+ \; proD^-$	$proC^+ \; proA^+ \; proB^+ \; proD^+$	1166
	$proC^+ \; proA^+ \; proB^+ \; proD^-$	0

a. Why are there no $proC^-$ genotypes among the transductants?
Transductants were initially screened for the presence of proC^+. *Thus, only* proC^+ *transductants were identified.*

b. Which genotypes represent single transductants and which represent cotransductants?
The wild-type genotypes ($\mathrm{proC}^+ \; \mathrm{proA}^+ \; \mathrm{proB}^+ \; \mathrm{proD}^+$) *represent single transductants of* proC^+. *Both the* $\mathrm{proC}^+ \; \mathrm{proA}^- \; \mathrm{proB}^+ \; \mathrm{proD}^+$ *and* $\mathrm{proC}^+ \; \mathrm{proA}^+ \; \mathrm{proB}^- \; \mathrm{proD}^+$ *genotypes represent cotransductants of* proC^+, proA^- *and* proC^+, proB^-.

c. Is there evidence that *proA, proB,* and *proD* are located close to *proC?* Explain your answer.
From the data, it appears that both proA *and* proB *are located close to* proC. *Both are capable of being cotransduced along with* proC. *The* proD *marker may be located at distance from* proC *so that it cannot contransduce with* proC. *However, the data is not conclusive.*

*28. A geneticist isolates two mutations in bacteriophage. One mutation causes the clear plaques (*c*) and the other produces minute plaques (*m*). Previous mapping experiments have established that the genes responsible for these two mutations are 8 map units apart. The geneticist mixes phages with genotype $c^+ \; m^+$ and genotype $c^- \; m^-$ and uses the mixture to infect bacterial cells. She collects the progeny phages and cultures a sample of them on plated bacteria. A total of 1000 plaques are

observed. What numbers of the different types of plaques ($c^+ m^+$, $c^- m^-$, $c^+ m^-$, $c^- m^+$) should she expect to see?
We know that the two genes are 8 map units apart. These 8 map units correspond to a percent recombination between the two genes of 8%. When the geneticist mixes the two phages ($m^+ c^+ \times m^- c^-$), *creating a double infection of the bacterial cell, she should expect the two types of recombinant plaque phenotypes,* $m^+ c^-$ *and* $m^- c^+$, *to comprise 8% of the progeny phage. The remaining 92% will be a combination of the wild-type phage and the doubly mutant phage.*

Plaque phenotype	*Expected number*
$c^+ m^+$	*460*
$c^- m^-$	*460*
$c^+ m^-$	*40 (recombinant)*
$c^+ m^-$	*40 (recombinant)*
Total plaques	*1000*

29. The geneticist carries out the same experiment described in Problem 23, but this time she mixes phages with genotypes $c^+ m^-$ and $c^- m^+$. What results are expected with *this* cross?
We know that the two genes are 8 map units apart, corresponding to 8% recombination. The phage used by the geneticist in this experiment will produce recombinants with different phenotypes from her previous experiment, but the number of recombinants will remain the same.

Plaque phenotype	*Expected number*
$c^+ m^+$	*40 (recombinant)*
$c^- m^-$	*40 (recombinant)*
$c^- m^+$	*460*
$c^+ m^-$	*460*
Total plaques	*1000*

*30. A geneticist isolates two r mutants (r_{13} and r_2) that cause rapid lysis. He carries out the following crosses and counts the number of plaques listed here:

Genotype of parental phage	Progeny	Number of plaques
$h^+ r_{13}^- \times h^- r_{13}^+$	$h^+ r_{13}^+$	1
	$h^- r_{13}^+$	104
	$h^+ r_{13}^-$	110
	$h^- r_{13}^-$	2
	Total	216
$h^+ r_2^- \times h^- r_2^+$	$h^+ r_2^+$	6
	$h^- r_2^+$	86
	$h^+ r_2^-$	81
	$h^- r_2^-$	7
	Total	180

a. Calculate the recombination frequencies between r_2 and h and between r_{13} and h.

To determine the recombination frequencies, the recombinant offspring must be identified. The recombination frequency is calculated by dividing the total number of recombinant plaques by the total number of plaques.

Genotype of parents	*Progeny*	*Number of plaques*
$h^+r_{13}^- \times h^-r_{13}^+$	$h^+r_{13}^+$ *(recombinant)*	*1*
	$h^-r_{13}^-$	*104*
	$h^+r_{13}^-$	*110*
	$h^-r_{13}^-$ *(recombinant)*	*2*
	total	*216*
$h^+r_2^- \times h^-r_2^+$	$h^+r_2^+$ *(recombinant)*	*6*
	$h^-r_2^+$	*86*
	$h^+r_2^-$	*81*
	$h^-r_2^-$ *(recombinant)*	*7*
	total	*180*

The recombination frequency between r_2 and h is 13/180 = .072, or 7.2%. The RF between r_{13} and h is 3/216 = 0.014, or 1.4%.

b. Draw all possible linkage maps for these three genes.

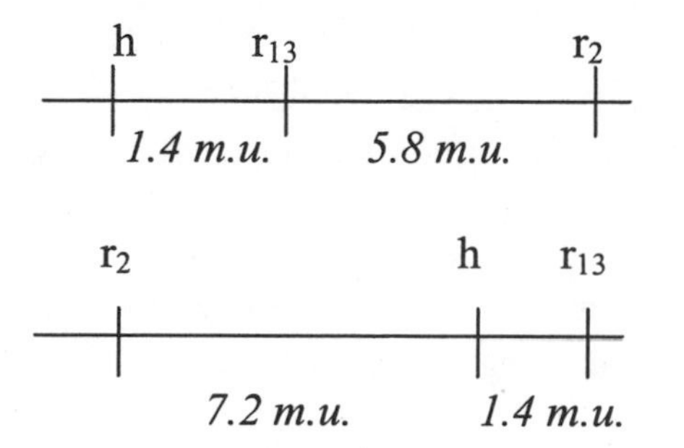

*31. *E. coli* cells are simultaneously infected with two strains of phage λ. One strain has a mutant host range, is temperature sensitive, and produces clear plaques (genotype = *h st c*); another strain carries the wild-type alleles (genotype = h^+ st^+ c^+). Progeny phages are collected from the lysed cells and plated on bacteria. These genotypes of the progeny phage are:

Progeny phage genotype	Number of plaques
h^+ c^+ st^+	321
h c st	338
h^+ c st	26
h c^+ st^+	30
h^+ c st^+	106
h c^+ st	110
h^+ c^+ st	5
h c st^+	6

a. Determine the order of the three genes on the phage chromosome.
First, we need to identify the progeny phages that have genotypes similar to the parents and the progeny phages that have genotypes that differ from the parents. The parental genotypes are h^+c^+ st^+ *and* h c st. *Any genotype that differs from these two genotypes had to be generated by recombination. By comparing the genotype of the double-recombinant phage progeny with the nonrecombinants, we can predict the gene order.*

Phage genotype	*Number of progeny*	*Type*
h^+c^+ st^+	*321*	*Parental*
h c st	*338*	*Parental*
h^+c st	*26*	*Recombinant*
h c^+ st^+	*30*	*Recombinant*
h^+ c st^+	*106*	*Recombinant*
h c^+st	*110*	*Recombinant*
h^+ c^+ st	*5*	*Double-recombinant*
h c st^+	*6*	*Double-recombinant*
Total	*942*	

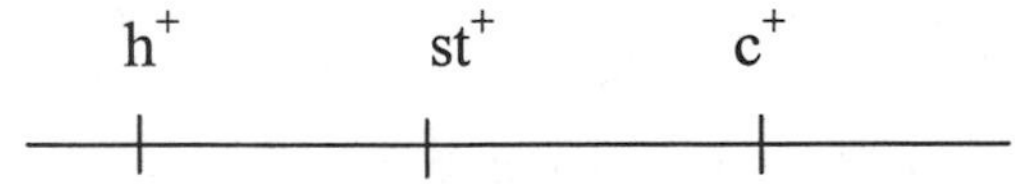

b. Determine the map distances between the genes.
The map distances can be calculated by determining the percent recombination between each gene pair. The double-recombinant progeny, h^+ c^+ *and* h c, *appear to be parentals. However, this genotype was generated by a double-crossover event. To consider the double-crossover events, multiply the number of double-recombinant progeny by two.*

$h^+ st^+$: *[(26+30+5+6)/942] × 100 % = 7.1%, or 7.1 m.u.*
$h^+ c^+$: *[(26+30+106+110+10+12)/942] × 100% = 31.2%, or 31.2 m.u.*
$st^+ c^+$: *[(106+110+5+6)/942] × 100 % = 24.1% or 24.1 m.u.*

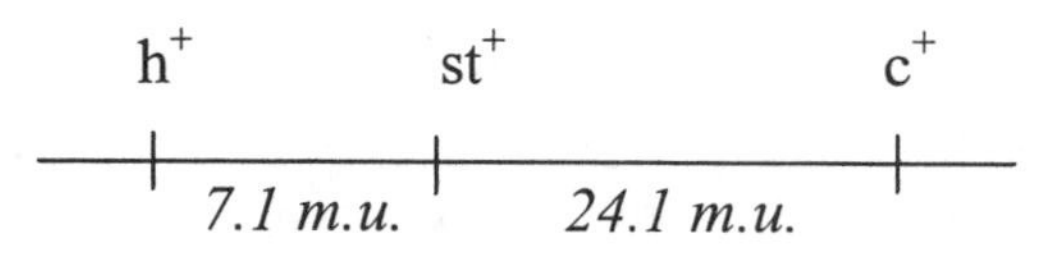

c. Determine the coefficient of coincidence and the interference.

$$COC = \frac{\text{(observed number of double recombinants)}}{\text{(expected number of double recombinants)}}$$

$$COC = (6 + 5)/(.071 \times .241 \times 942) = 0.68$$

$$\text{Interference} = 1 - COC = 1 - 0.68 = 0.32$$

32. A donor strain of bacteria with genes $a^+\ b^+\ c^+$ is infected with phages to map the donor chromosome with generalized transduction. The phage lysate from the bacterial cells is collected and used to infect a second strain of bacteria that are $a^-\ b^-\ c^-$. Bacteria with the a^+ gene are selected and the percentage of cells with cotransduced b^+ and c^+ genes are recorded.

Donor	Recipient	Selected gene	Cells with cotransduced gene (%)
$a^+\ b^+\ c^+$	$a^-\ b^-\ c^-$	a^+	25 b^+
		a^+	3 c^+

Is the *b* or *c* gene closer to *a*? Explain your reasoning.
The gene b$^+$ *cotransduces more frequently with* a$^+$*, the selective marker, than does* c$^+$*. Because genes that are closer together on the donor bacterial chromosome cotransduce more frequently, we can see that* b$^+$ *is closer to* a$^+$*.*

33. A donor strain of bacteria with genotype $leu^+\ gal^-\ pro^+$ is infected with phages. The phage lysate from the bacterial cells is collected and used to infect a second strain of bacteria that are $leu^-\ gal^+\ pro^-$. The second strain is selected for leu^+, and the following cotransduction data obtained:

Donor	Recipient	Selected gene	Cells with cotransduced gene (%)
$leu^+\ gal^-\ pro^+$	$leu^-\ gal^+\ pro^-$	leu^+	47 pro^+
		leu^+	26 gal^-

Which genes are closest, *leu* and *gal* or *leu* and *pro*?
Because leu *and* pro *cotransduce together more frequently, they must be the closest.*

34. A geneticist isolates two new mutations from the *rII* region of bacteriophage T4, called rII_X and rII_Y. *E. coli* B cells are simultaneously infected with phages carrying the rII_X mutation *and* with phages carrying the rII_Y mutation. After the cells have lysed, samples of the phage lysate are collected. One sample is grown on *E. coli* K cells and a second sample on *E. coli* B cells. There are 8322 plaques on the *E coli* B and 3 plaques on *E. coli* K. What is the recombination frequency between these two mutations?

The recombination frequency $= \dfrac{(2 \times \text{plaques on K})}{(\text{Total number of plaques})}$

Recombination frequency $= (2 \times 3)/8322 = 7.2 \times 10^{-4}$, *or 0.072%*

35. A geneticist is working with a new bacteriophage called phage Y3 that infects *E. coli*. He has isolated eight mutant phages that fail to produce plaques when grown on *E. coli* strain K. To determine whether these mutations occur at the same functional gene, he simultaneously infects *E. coli* K cells with paired combinations of the mutants and looks to see whether plaques formed. He obtains the results at the top of the following page. (A plus sign means that plaques were formed on *E. coli* K; a minus sign means no plaques were formed on *E. coli* K).

Mutant	1	2	3	4	5	6	7	8
1								
2	+							
3	+	+						
4	+	–	+					
5	–	+	+	+				
6	–	+	+	+	–			
7	+	–	+	–	+	+		
8	–	+	+	+	–	–	+	

a. To how many functional genes (cistrons) do these mutations belong?
The geneticist is essentially conducting the "trans" portion of the cis–trans test. If complementation occurs between the different phage mutants that are infecting the E. coli *K cells, then plaques will form on the lawn of* E. coli *K. Complementation can occur only when the mutations of the different phages are located on different cistrons or functional genes. Phage mutants that do not complement each other have mutations that lie on the same cistrons.*

From the formation of plaques on E. coli *K, we can see three groups of phages that failed to complement with other phages within their group but did complement the phages in the other groups. Because there are three groups, we can infer the presence of three cistrons or functional genes.*

b. Which mutations belong to the same functional gene?
We will identify the groups as group 1, group 2 and group 3.
Group 1: *Mutants 1, 5, 6, and 8*
Group 2: *Mutants 2, 4, and 7*
Group 3: *Mutant 3*

CHALLENGE QUESTIONS

Section 8.1

36. As a summer project, a microbiology student independently isolates two mutations in *E. coli* that are auxotrophic for glycine (gly^-). The student wants to know

whether these two mutants occur at the same cistron. Outline a procedure that the student could use to determine whether these two *gly*$^-$ mutations occur within the same cistron.
To determine if the two gly$^-$ *are with the same cistrons, a strain of bacteria will have to be constructed that contains both genes, but on different DNA molecules within the strain. Only by both mutations being present in the same cell can complementation of the two mutations be tested. The student will need to create a merodiploid or partial diploid strain. A method for doing this is to create an F′ that contains one of the* gly$^-$ *markers. Because* gly$^-$ *results in an auxotrophic mutant, it would be difficult to use as an initial screen for an F′ that contains it. Therefore, the student will need to use Hfr × F*$^-$ *matings to map the location of* gly$^-$ *marker and identify other protrophic markers that are nearby. By screening for F′ strains that have the nearby protrophic markers, an F′ strain that contains the* gly$^-$ *marker should be identified.*

Next, the F′ strain having the gly$^-$ *marker should be mated to a F*$^-$ *that contains the other* gly$^-$ *mutation. The identified exconjugants should contain both* gly$^-$ *mutations on separate DNA molecules within the cell. If the exconjugant can grow on minimal media not supplemented with glycine, then complementation has occurred and the two* gly$^-$ *mutations are located on different cistrons.*

37. A group of genetics students mixes two auxotrophic strains of bacteria: one is *leu*$^+$ *trp*$^+$ *his*$^-$ *met*$^-$ and the other is *leu*$^-$ *trp- his*$^+$ *met*$^+$. After mixing the two strains, they plate the bacteria on minimal medium and observe a few prototrophic colonies (*leu*$^+$ *trp*$^+$ *his*$^+$ *met*$^+$). They assume that some gene transfer has occurred between the two strains. How can they determine whether the transfer of genes is due to conjugation, transduction, or transformation?
Conjugation requires the direct contact of the donor bacterial strain and the recipient. If the transfer does not occur when the bacteria are kept physically separate, then conjugation is not the likely pathway. Another test would be to conduct interrupted mating experiments (assuming that one of the bacterial strains is an Hfr strain) to see if the transfer of the different markers is time dependent, which is also indicative of conjugation.

If the transfer occurs by transformation, then extraction of DNA from either strain and exposure of the other strain to the extracted DNA should result in the transfer of the DNA molecules. By selecting one of the mutations as a selective marker and measuring cotransformation frequencies between the selective marker and the other genes individually, the frequency of the transfer will hint toward the mechanism of transfer.

Finally, if the transfer is by transduction, then by exposing the one cell type to extracted DNA from the other cell type, transfer of the genes would not be expected. Potential cotransduction frequencies could be measured similarly to the cotransformation frequency. Also, the presence of plaques might be evident.

Chapter Nine: Chromosome Variation

COMPREHENSION QUESTIONS

Section 9.1

*1. List the different types of chromosome mutations and define each one.
Chromosome rearrangements:

> *Deletion: loss of a portion of a chromosome.*
> *Duplication: addition of an extra copy of a portion of a chromosome.*
> *Inversion: a portion of the chromosome is reversed in orientation.*
> *Translocation: a portion of one chromosome becomes incorporated into a different (nonhomologous) chromosome.*

Aneupoloidy: loss or gain of one or more chromosomes so that the chromosome number deviates from 2n *or the normal euploid complement.*
Polyploidy: Gain of entire sets of chromosomes so the chromosome number changes from 2n *to* 3n *(triploid),* 4n *(tetraploid), and so on.*

Section 9.2

*2. Why do extra copies of genes sometimes cause drastic phenotypic effects?
The expression of some genes is balanced with the expression of other genes; the ratios of their gene products, usually proteins, must be maintained within a narrow range for proper cell function. Extra copies of one of these genes cause that gene to be expressed at proportionately higher levels, thereby upsetting the balance of gene products.

3. Draw a pair of chromosomes as they would appear during synapsis in prophase I of meiosis in an individual heterozygous for a chromosome duplication.
In the figure below, adapted from Figure 9.6b, the vertical dashed lines denote the locations of the genes labeled A, B, C,... G. The lower chromosome has duplicated a segment containing genes C, D, and E.

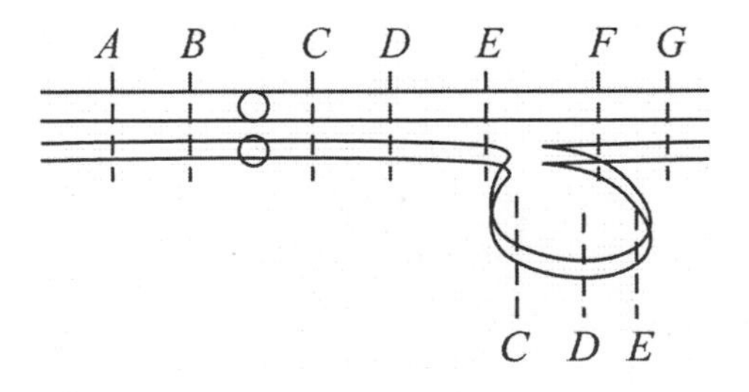

4. What is haploinsufficiency?
Haploinsufficiency is the term for a condition where having only one copy of a wild-type gene does not produce a wild-type phenotype in an otherwise diploid

organism. For haploinsufficient genes, the relative amounts of gene product is important.

*5. What is the difference between a paracentric and a pericentric inversion?
A paracentric inversion does not include the centromere; a pericentric inversion includes the centromere.

6. How do inversions cause phenotypic effects?
Although inversions do not result in loss or duplication of chromosomal material, inversions can have phenotypic consequences if the inversion disrupts a gene at one of its breakpoints or if a gene near a breakpoint is altered in its expression because of a change in its chromosomal environment, such as relocation to a heterochromatic region. Such effects on gene expression are called position effects.

7. Explain, with the aid of a drawing, how a dicentric bridge is produced when crossing over takes place in an individual heterozygous for a paracentric inversion.
The diagram below illustrates synapsis of chromosomes in an individual heterozygous for an inversion involving loci E, F, and G.

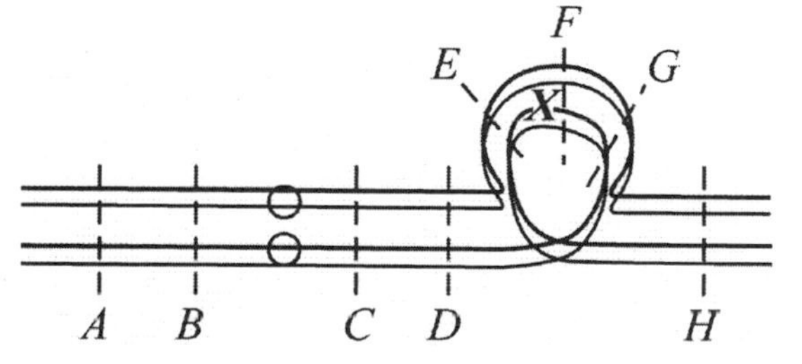

Suppose a single crossover event occurs with in the inversion loop, between E and F, involving the two chromatids depicted with thicker lines. The normal chromosome will then have one normal chromatid and one chromatid that crossed over with an inversion chromatid. Likewise, the inversion chromosome will have one inversion chromatid and one crossover chromatid.

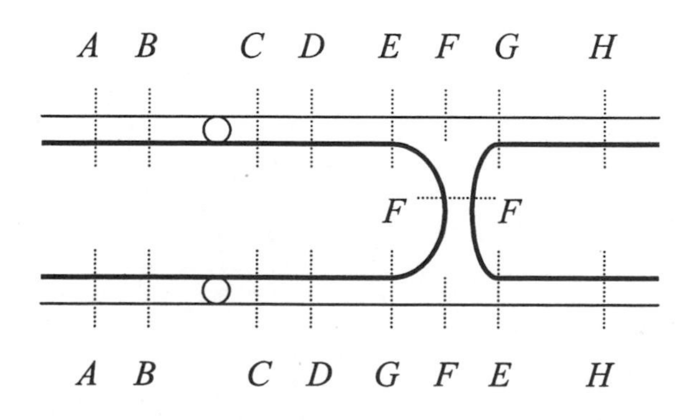

The chromatids involved in the crossover form a dicentric chromatid and a fragment without a centromere (acentric).

8. Explain why recombination is suppressed in individuals heterozygous for paracentric and pericentric inversions.
A crossover within a paracentric inversion produces a dicentric and an acentric recombinant chromatid. The acentric fragment is lost, and the dicentric fragment breaks, resulting in chromatids with large deletions that lead to nonviable gametes or embryonic lethality. A crossover within a pericentric inversion produces recombinant chromatids that have duplications or deletions. Again, gametes with these recombinant chromatids do not lead to viable progeny.

*9. How do translocations produce phenotypic effects?
Like inversions, translocations can produce phenotypic effects if the translocation breakpoint disrupts a gene or if a gene near the breakpoint is altered in its expression because of relocation to a different chromosomal environment (a position effect).

10. Sketch the chromosome pairing and the different segregation patterns that can arise in an individual heterozygous for a reciprocal translocation.

Chromosome pairing

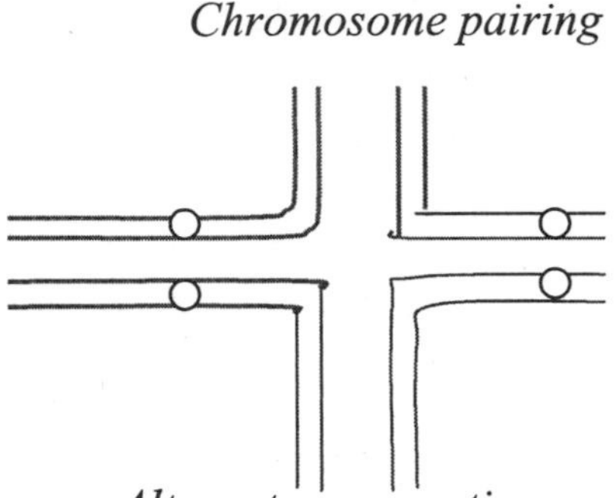

Alternate segregation

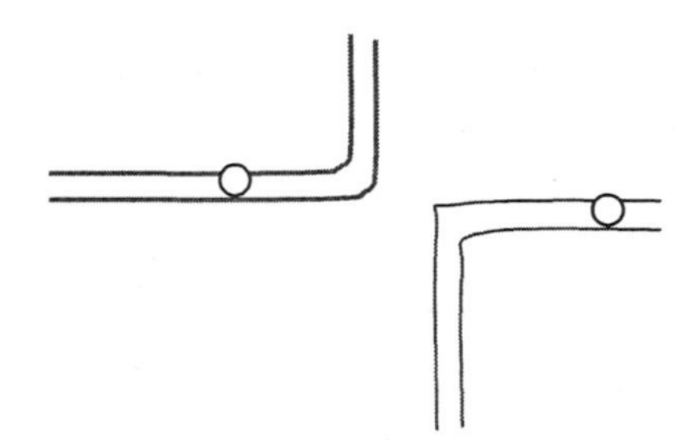

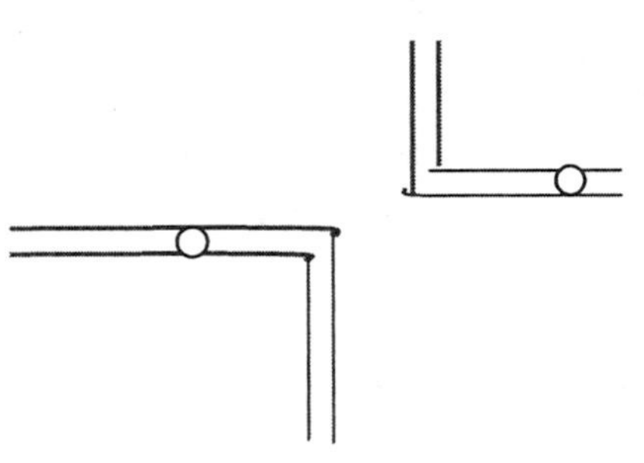

Adjacent-1 segregation

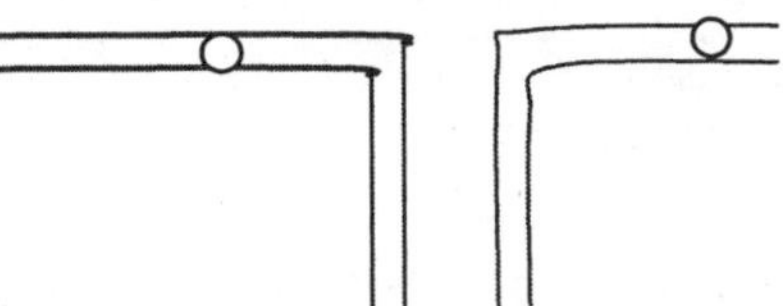

Adjacent-2 segregation

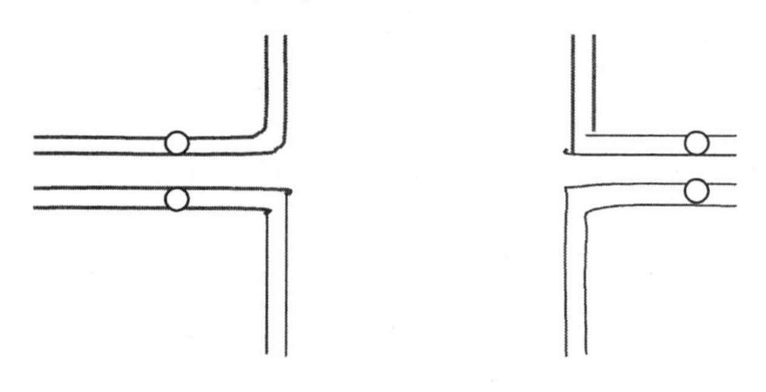

11. What is a Robertsonian translocation?
    The long arms of two acrocentric chromosomes are joined to a common centromere through translocation, resulting in a large metacentric chromosome and a very small chromosome with two very short arms. The very small chromosome may be lost.

Section 9.3

12. List four major types of aneuploidy.
    Nullisomy: having no copies of a chromosome.
    Monosomy: having only one copy of a chromosome.
    Trisomy: having three copies of a chromosome.
    Tetrasomy: having four copies of a chromosome.

*13. What is the difference between primary Down syndrome and familial Down syndrome? How does each arise?
    Primary Down syndrome is caused by spontaneous, random nondisjunction of chromosome 21, leading to trisomy 21. Familial Down syndrome most frequently arises as a result of a Robertsonian translocation of chromosome 21 with another chromosome, usually chromosome 14. Translocation carriers do not have Down syndrome, but their children have an increased incidence of Down syndrome. If the translocated chromosome segregates with the normal chromosome 21, the gamete will have two copies of chromosome 21 and result in a child with familial Down syndrome.

*14. What is uniparental disomy and how does it arise?
    Uniparental disomy refers to the inheritance of both copies of a chromosome from the same parent. This may arise originally from a trisomy condition in which the early embryo loses one of the three chromosomes, and the two remaining copies are from the same parent.

15. What is mosaicism and how does it arise?
    Mosaicism is a condition in which an individual has patches of cells that are genetically different from other cells. Mosaicism may arise from mitotic

nondisjunction during early embryonic divisions, X-inactivation in a heterozygous female, fusion of two zygotes into a single embryo, and other mechanisms.

Section 9.4

*16. What is the difference between autopolyploidy and allopolyploidy? How does each arise?
In autopolyploidy, all sets of chromosomes are from the same species. Autopolyploids typically arise from mitotic nondisjunction of all the chromosomes in an early 2n *embryo, resulting in an autotetraploid, or from meiotic nondisjunction that results in a* 2n *gamete fusing with a* 1n *gamete to form an autotriploid. In allopolyploidy, the chromosomes of two different species are contained in one individual through the hybridization of two related species followed by mitotic nondisjunction. Fusion of gametes from two different (but usually related) species results in a hybrid with a haploid set of chromosomes from each parent. If an early embryonic cell then undergoes mitotic nondisjunction and doubles each chromosome, then a fertile* 4n *allotetraploid individual having two copies of each chromosome from each species may result.*

17. Explain why autopolyploids are usually sterile, while allopolyploids are often fertile.
Autopolyploids arise from duplication of their own chromosomes. During meiosis, the presence of more than two homologous chromosomes results in faulty alignment of homologues in prophase I, and subsequent faulty segregation of the homologues in anaphase I. The resulting gametes have an uneven distribution of chromosomes and are genetically unbalanced. These gametes usually produce lethal chromosome imbalances in the zygote. Allopolyploids, however, have chromosomes from different species. As long as they have a diploid set of chromosomes from each species, as in an allotetraploid or even an allohexaploid, the homologous chromosome pairs from each species can align and segregate properly during meiosis. Their gametes will be balanced and will produce viable zygotes when fused with other gametes from the same type of allopolyploid individual.

APPLICATION QUESTIONS AND PROBLEMS

Section 9.1

*18. Which types of chromosome mutations:
- **a.** increase the amount of genetic material on a particular chromosome?
  Duplications
- **b.** increase the amount of genetic material for all chromosomes?
  Polyploidy
- **c.** decrease the amount of genetic material on a particular chromosome?
  Deletions

d. change the position of DNA sequences on a single chromosome without changing the amount of genetic material?
Inversions

e. move DNA from one chromosome to a nonhomologous chromosome?
Translocations

Section 9.2

*19. A chromosome has the following segments, where • represents the centromere.
AB • CDEFG

What types of chromosome mutations are required to change this chromosome into each of the following chromosomes? (In some cases, more than one chromosome mutation may be required.)

a. A B A B • C D E F G: *Tandem duplication of AB*
b. A B • C D E A B F G: *Displaced duplication of AB*
c. A B • C F E D G: *Paracentric inversion of DEF*
d. A • C D E F G: *Deletion of B*
e. A B • C D E: *Deletion of FG*
f. A B • E D C F G: *Paracentric inversion of CDE*
g. C • B A D E F G: *Pericentric inversion of ABC*
h. A B • C F E D F E D G: *Duplication and inversion of DEF*
i. A B • C D E F C D F E G: *Duplication of CDEF, inversion of EF*

20. A chromosome initially has the following segments:
AB • CDEFG

Draw and label the chromosome that would result from each of the following mutations:

a. Tandem duplication of DEF: *A B • C D E F D E F G*
b. Displaced duplication of DEF: *A B • C D E F G D E F*
c. Deletion of FG: *A B • C D E*
d. Deletion of CD: *A B • E F G*
e. Paracentric inversion that includes DEFG: *A B • C G F E D*
f. Pericentric inversion of BCDE: *A E D C • B F G*

21. The following diagrams represent two nonhomologous chromosomes:
AB • CDEFG
RS • TUVWX

What type of chromosome mutation would produce the following chromosomes?

a. AB • CD
RS • TUVWXEFG
Nonreciprocal translocation of E F G

b. AUVB • CDEFG
RS • TWX
Nonreciprocal translocation of U V

c. A B • T U V F G
R S • C D E W X
Reciprocal translocation of C D E and T U V

d. A B • C W G
R S • T U V D E F X
Reciprocal translocation of D E F and W

*22. The *Notch* mutation is a deletion on the X chromosome of *Drosophila melanogaster*. Female flies heterozygous for *Notch* have an indentation on the margins of their wings; *Notch* is lethal in the homozygous and hemizygous conditions. The *Notch* deletion covers the region of the X chromosome that contains the locus for white eyes, an X-linked recessive trait. Give the phenotypes and proportions of progeny produced in the following crosses:

a. A red-eyed, *Notch* female is mated with a white-eyed male.
Let X^N represent the Notch allele, X^w the white eye allele and X^+ the wild-type allele.

$X^NX^+ \times X^wY \rightarrow$
¼ X^NX^w Notch *female with white eyes*
¼ X^NY *lethal*
¼ X^+X^w *female with red eyes (wild-type)*
¼ X^+Y *wild-type male*

Overall, ⅓ of live progeny will be Notch *females with white eyes, ⅓ will be wild-type females, and ⅓ will be wild-type males.*

b. A white-eyed, *Notch* female is mated with a red-eyed male.

$X^NX^w \times X^+Y \rightarrow$
¼ X^NX^+ *Notch females*
¼ X^NY *lethal*
¼ X^+X^w *wild-type females*
¼ X^wY *white-eyed males*

The surviving progeny will therefore be ⅓ Notch *red-eyed females, ⅓ wild-type females, and ⅓ white-eyed males.*

c. A white-eyed, *Notch* female is mated with a white-eyed male.

$X^NX^w \times X^wY \rightarrow$
¼ X^NX^w *Notch, white-eyed females*
¼ X^NY *lethal*
¼ X^wX^w *white-eyed females*
¼ X^wY *white-eyed males*

Viable progeny are ⅓ Notch *white-eyed females, ⅓ white-eyed females, and ⅓ white-eyed males.*

23. The green-nose fly normally has six chromosomes, two metacentric and four acrocentric. A geneticist examines the chromosomes of an odd-looking green-nose fly and discovers that it has only five chromosomes; three of them are metacentric and two are acrocentric. Explain how this change in chromosome number might have occurred.
A Robertsonian translocation between two of the acrocentric chromosomes would result in a new metacentric chromosome and a very small chromosome that may have been lost.

*24. A wild-type chromosome has the following segments:
ABC • DEFGHI.

An individual is heterozygous for the following chromosome mutations. For each mutation, sketch how the wild-type and mutated chromosomes would pair in prophase I of meiosis, showing all chromosome strands.

a. ABC • DEFDEFGHI

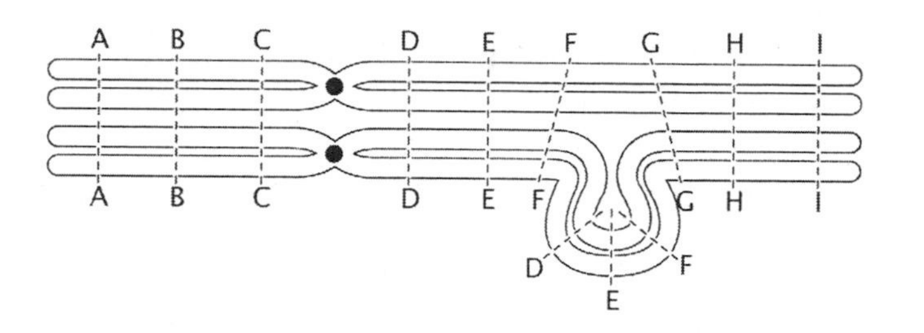

b. ABC • DHI

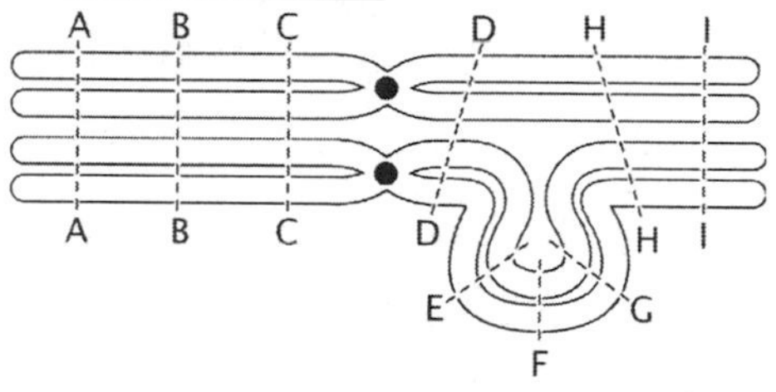

c. ABC • DGFEHI

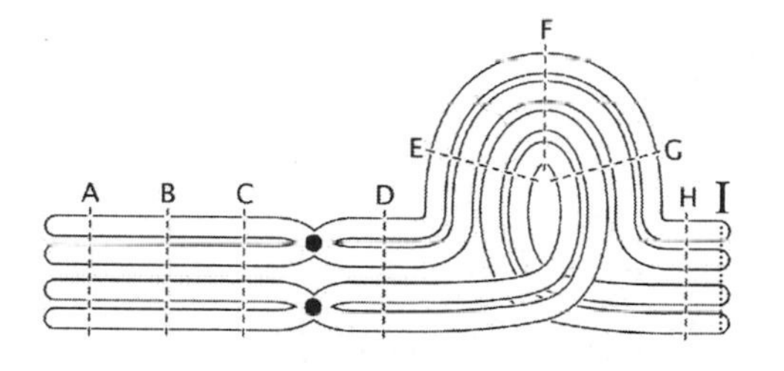

d. ABED • CFGHI

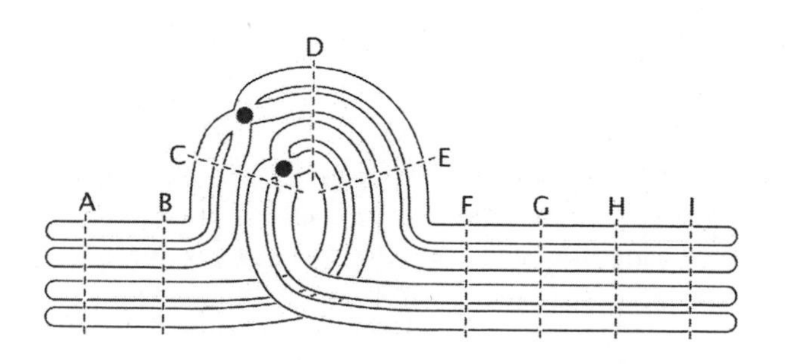

25. Draw the chromatids that would result from a two-strand double crossover between E and F in problem 24c.

    A B C•D E F G H I
    A B C•D E F G H I
    A B C•D G F E H I
    A B C•D G F E H I
    Here, all four chromatids will be functional.

26. As discussed in this chapter, crossing over within a pericentric inversion produces chromosomes that have extra copies of some genes and no copies of other genes. The fertilization of gametes containing such duplication/deficient chromosomes often results in children with syndromes characterized by developmental delay, mental retardation, abnormal development of organ systems, and early death. Using a special two-color FISH analysis that revealed the presence of crossing over within pericentric inversions, Maarit Jaarola and colleagues examined individual sperm cells of a male who was heterozygous for a pericentric inversion on chromosome 8 and determined that crossing over took place within the pericentric inversion in 26% of the meiotic divisions (M. Jaarola, R. H. Martin, and T. Ashley. 1998. *American Journal of Human Genetics* 63:218–224).

    Assume that you are a genetic counselor and that a couple seeks genetic counseling from you. Both the man and the woman are phenotypically normal, but the woman is heterozygous for a pericentric inversion on chromosome 8. The man is karyotypically normal. What is the probability that this couple will produce a child with a debilitating syndrome as the result of crossing over within the pericentric inversion?
    Each crossover event results in two recombinant and two nonrecombinant gametes. If one crossover occurs in 100% of meioses, the result would be 50% recombinant gametes. If crossing over occurs within the pericentric inversion at a rate of 26% of meioses, then 13% of the woman's oocytes will have duplication/deficient chromosome 8. If all of these oocytes form viable eggs, and if they do not result in early miscarriage after fertilization, then the probability of the couple having a child with a syndrome caused by the crossing over is 13%.

*27. An individual heterozygous for a reciprocal translocation possesses the following chromosomes.
    A B • C D E F G
    A B • C D V W X
    R S • T U E F G
    R S • T U V W X

a. Draw the pairing arrangement of these chromosomes in prophase I of meiosis.

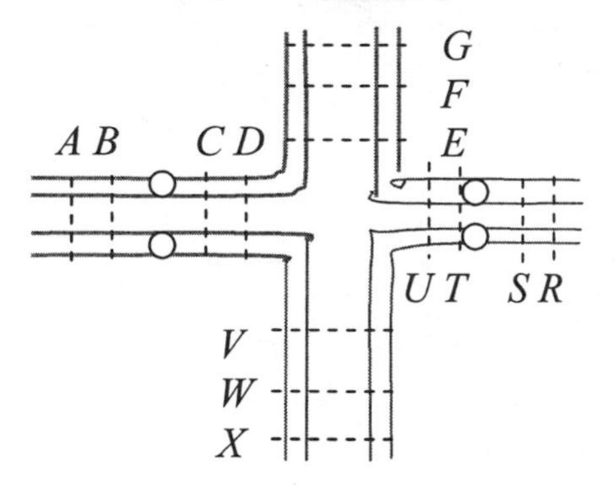

b. Diagram the alternate, adjacent-1, and adjacent-2 segregation patterns in anaphase I of meiosis.

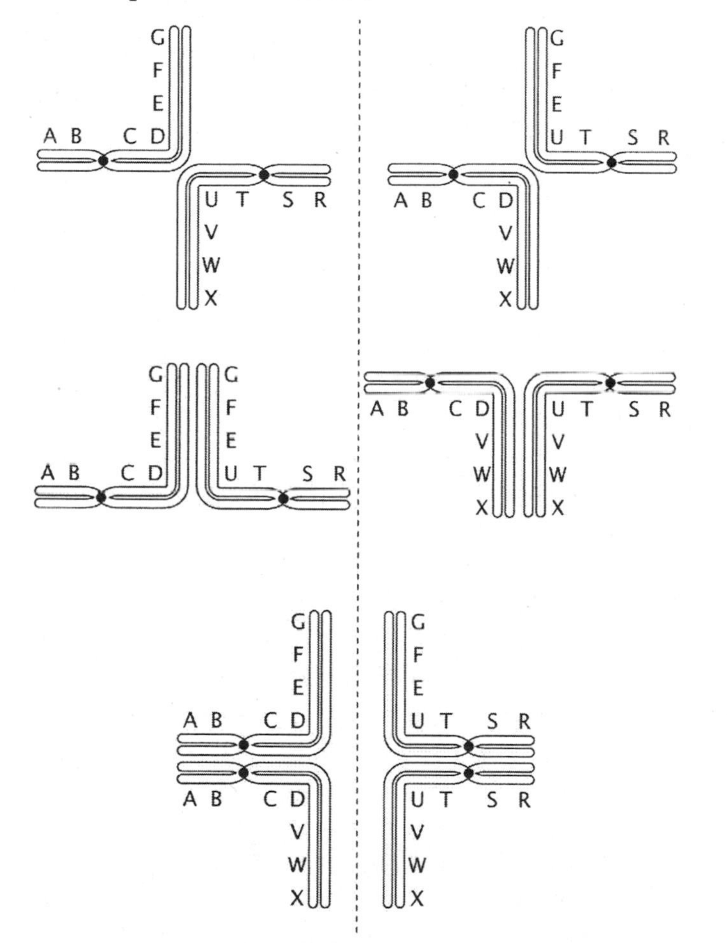

c. Give the products that result from alternate, adjacent-1, and adjacent-2 segregation.
Alternate: Gametes contain either both normal or both translocation chromosomes and are all viable.
AB•CDEFG + RS•TUVWX and
AB•CDVWX + RS•TUEFG

Adjacent-1: Gametes contain one normal and one translocation chromosome, resulting in duplication of some genes and deficiency for others.
A B • C D E F G + *R S • T U E F G* *and*
A B • C D V W X + *R S • T U V W X*

Adjacent-2 (rare): Gametes also contain one normal and one translocation chromosome, with duplication of some genes and deficiency for others.
A B • C D E F G + *A B • C D V W X* *and*
R S • T U V W X + *R S • T U E F G*

Section 9.3

28. Red-green color blindness is a human X-linked recessive disorder. A young man with a 47,XXY karyotype (Klinefelter syndrome) is color blind. His 46,XY brother also is color blind. Both parents have normal color vision. Where did the nondisjunction occur that gave rise to the young man with Klinefelter syndrome?
Because the father has normal color vision, the mother must be the carrier for color blindness. The color-blind young man with Klinefelter syndrome must have inherited two copies of the color-blind X chromosome from his mother. If no crossover occurred between the color blindness allele and the centromere, the nondisjunction event took place in meiosis II of the egg.

29. Junctional epidermolysis bullosa (JEB) is a severe skin disorder that results in blisters over the entire body. The disorder is caused by autosomal recessive mutations at any one of three loci that help to encode laminin 5, a major component in the dermal–epidermal basement membrane. Leena Pulkkinen and colleagues described a male newborn who was born with JEB and died at 2 months of age (L. Pulkkinen et al. 1997. *American Journal of Human Genetics* 61:611–619); the child had healthy unrelated parents. Chromosome analysis revealed that the infant had 46 normal-appearing chromosomes. Analysis of DNA showed that his mother was heterozygous for a JEB-causing allele at the LAMB3 locus, which is on chromosome 1. The father had two normal alleles at this locus. DNA fingerprinting demonstrated that the male assumed to be the father had, in fact, conceived the child.
   a. Assuming that no new mutations occurred in this family, explain the presence of an autosomal recessive disease in the child when the mother is heterozygous and the father is homozygous normal.
   One possibility is that the infant inherited both copies of chromosome 1 from the mother. A nondisjunction in meiosis II could result in a gamete with two copies of maternal chromosome 1 bearing the recessive JEB allele. Fertilization by a normal male sperm would normally result in trisomy 1, with early embryonic lethality. In this case, the paternal chromosome 1 may have been lost in the first mitotic division, resulting in an embryo with a normal karyotype, but carrying two maternal copies of chromosome 1 (disomy) and homozygosity for the JEB allele.

b. How might you go about proving your explanation? Assume that a number of genetic markers are available for each chromosome.
We could test a number of markers on chromosome 1, to find markers that distinguish maternal and paternal chromosomes. The absence of paternal chromosome 1 markers (and the presence of paternal markers for other chromosomes), would prove that both copies of chromosome 1 were maternally inherited. Markers close to the LABM3 locus should be homozygous and correspond to the maternal chromosome carrying the JEB allele.

30. Some people with Turner syndrome are 45,X/46,XY mosaics. Explain how this mosaicism could arise.
Such mosaicism could arise from mitotic nondisjunction early in embryogenesis, in which the Y chromosome fails to segregate and is lost. All the mitotic descendents of the resulting 45,X embryonic cell will also be 45,X. The fetus will then consist of a mosaic of patches of 45,X cells and patches of normal 46,XY cells. If the gonads are derived from 45,X cells, the individual will develop as a female with Turner Syndrome.

*31. Bill and Betty have had two children with Down syndrome. Bill's brother has Down syndrome and his sister has two children with Down syndrome. On the basis of these observations, which of the following statements is most likely correct? Explain your reasoning.

a. Bill has 47 chromosomes.
b. Betty has 47 chromosomes.
c. Bill and Betty's children each have 47 chromosomes.
d. Bill's sister has 45 chromosomes.
e. Bill has 46 chromosomes.
f. Betty has 45 chromosomes.
g. Bill's brother has 45 chromosomes.

The high incidence of Down syndrome in Bill's family and among Bill's relatives is consistent with familial Down syndrome, caused by a Robertsonian translocation involving chromosome 21. Bill and his sister, who are unaffected, are phenotypically normal carriers of the translocation and have 45 chromosomes. Their children and Bill's brother, who have Down syndrome, have 46 chromosomes, but one of these chromosomes is the translocation that has an extra copy of the long arm of chromosome 21. From the information given, there is no reason to suspect that Bill's wife Betty has any chromosomal abnormalities. Therefore, statement (d) is most likely correct.

*32. In mammals, sex-chromosome aneuploids are more common than autosomal aneuploids, but in fishes, sex-chromosome aneuploids and autosomal aneuploids are found with equal frequency. Offer an explanation for these differences in mammals and fishes.
In mammals, the higher frequency of sex chromosome aneuploids compared to autosomal aneuploids is due to X-chromosome inactivation and the lack of essential genes on the Y chromosome. If fishes do not have chromosome inactivation and

both their sex chromosomes have numerous essential genes, then the frequency of aneuploids should be similar for both sex chromosomes and autosomes.

*33. A young couple is planning to have children. Knowing that there have been a substantial number of stillbirths, miscarriages, and fertility problems on the husband's side of the family, they see a genetic counselor. A chromosome analysis reveals that, whereas the woman has a normal karyotype, the man possesses only 45 chromosomes and is a carrier for a Robertsonian translocation between chromosomes 22 and 13.

a. List all the different types of gametes that might be produced by the man.
If the translocation segregates away from the normal chromosomes 22 and 13, then the resulting gametes will have either (1) normal chromosome 13 and normal chromosome 22 or (2) translocated chromosome 13 + 22.

If the translocation and chromosome 22 segregate away from chromosome 13, then the resulting gametes will have either (3) translocated chromosome 13 + 22 and normal chromosome 22 or (4) normal chromosome 13.

If the translocation and chromosome 13 segregate away from chromosome 22 (probably rare), then the gametes will have either (5) normal chromosome 13 and translocated chromosome 13 + 22 or (6) normal chromosome 22.

b. What types of zygotes will develop when each of the gametes produced by the man fuses with a normal gamete produced by the woman?
(1) 13, 13, 22, 22; normal
(2) 13, 13 + 22, 22; translocation carrier
(3) 13, 13 + 22, 22, 22; trisomy 22
(4) 13, 13, 22; monosomy 22
(5) 13, 13, 13 + 22, 22; trisomy 13
(6) 13, 22, 22; monosomy 13

c. If trisomies and monosomies involving chromosome 13 and 22 are lethal, what proportion of the surviving offspring will be carriers of the translocation?
*Half, or 50%, from the answer in part **b**.*

Section 9.4

*34. A species has $2n = 16$ chromosomes. How many chromosomes will be found per cell in each of the following mutants in this species?

a. Monosomic: *15*
b. Autotriploid: *24*
c. Autotetraploid: *32*
d. Trisomic: *17*
e. Double monosomic: *14*
f. Nullisomic: *14*
g. Autopentaploid: *40*
h. Tetrasomic: *18*

35. Species I is diploid ($2n = 8$) with chromosomes AABBCCDD; related species II is diploid ($2n = 8$) with chromosomes MMNNOOPP. Individuals with the following sets of chromosomes represent what types of chromosome mutations?
   a. AAABBCCDD: *Trisomy A*
   b. MMNNOOOOPP: *Tetrasomy O*
   c. AABBCDD: *Monosomy C*
   d. AAABBBCCCDDD: *Triploidy*
   e. AAABBCCDDD: *Ditrisomy A and D*
   f. AABBDD: *Nullisomy C*
   g. AABBCCDDMMNNOOPP: *Allotetraploidy*
   h. AABBCCDDMNOP: *Allotriploidy*

36. Species I has $2n = 8$ chromosomes and species II has $2n = 14$ chromosomes. What would be the expected chromosome numbers in individual organisms with the following chromosome mutations? Give all possible answers.
   a. Allotriploidy including species I and II
   *Such allotriploids could have 1*n *from species I and 2*n *from species II for 3*n *= 18; alternatively they could have 2*n *from species I and 1*n *from species II for 3*n *= 15.*
   b. Autotetraploidy in species II
   *4*n *= 28*
   c. Trisomy in species I
   *2*n *+ 1 = 9*
   d. Monosomy in species II
   *2*n *– 1 = 13*
   e. Tetrasomy of species I
   *2*n *+ 2 = 10*
   f. Allotetraploid of species I and II
   *2*n *+ 2*n *= 22; 1*n *+ 3*n *= 25; 3*n *+ 1*n *= 19*

37. Consider a diploid cell that has $2n = 4$ chromosomes—one pair of metacentric chromosomes and one pair of acrocentric chromosomes. Suppose this cell undergoes nondisjunction, giving rise to an autotriploid cell ($3n$). The triploid cell then undergoes meiosis. Draw the different types of gametes that may result from meiosis in the triploid cell, showing the chromosomes present in each type. To distinguish between the different metacentric and acrocentric chromosomes, use a different color to draw each metacentric chromosome; similarly, use a different color to draw each acrocentric chromosome. (Hint: See Figure 9.27).

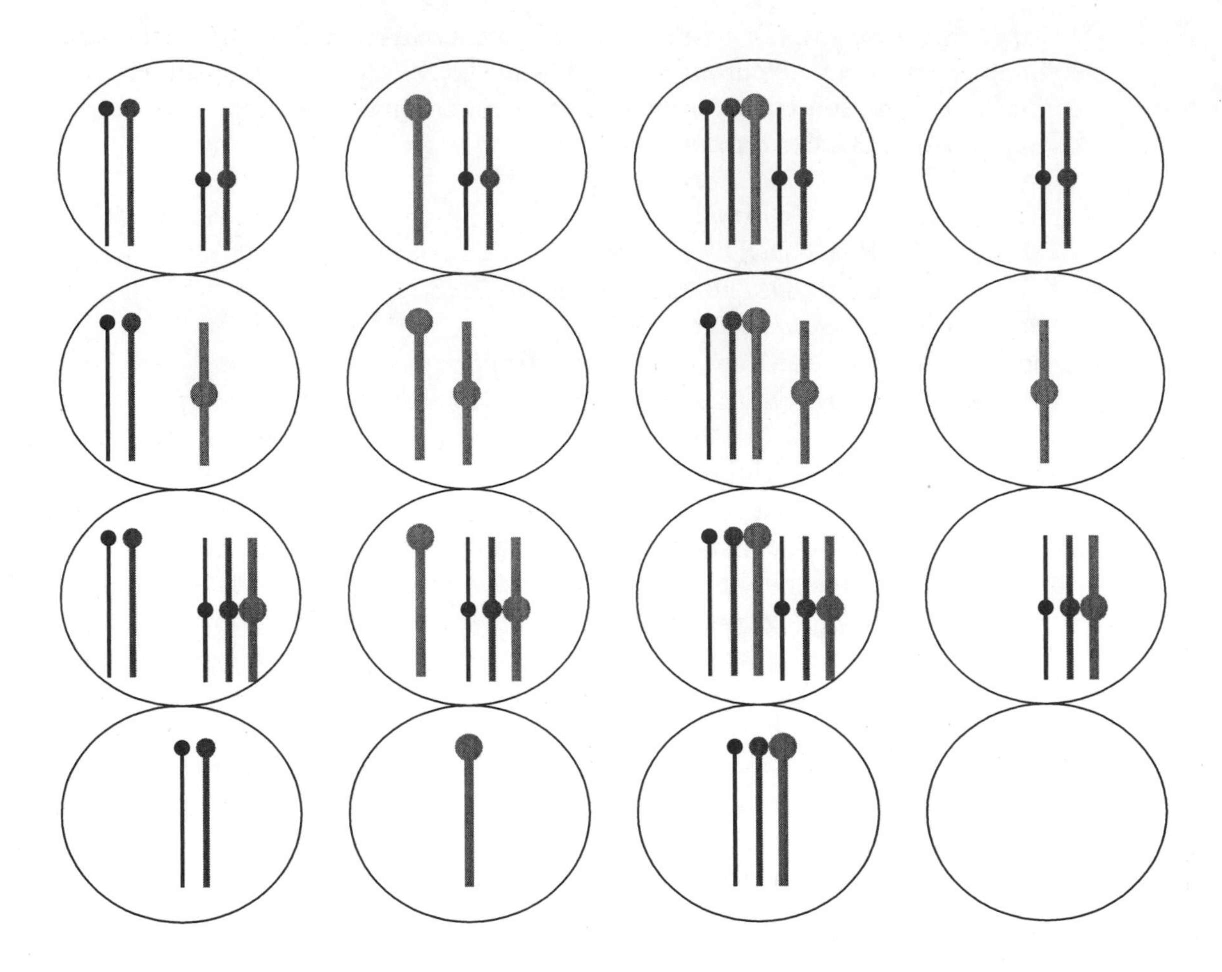

For each chromosome, the types of gametes that can result will have 1, 2, 3, or 0. Given two chromosomes, there are 16 basic combinations of the two chromosomes in the gametes. If the homologues are genetically distinct (as represented by different colors in the figure, there are many more possible genetic combinations of each chromosome (different colors combinations). Note that these combinations will not arise with equal frequencies; most trivalents will segregate 2:1, so combinations that result from 3:0 segregation will be less frequent.

38. *Nicotiana glutinosa* ($2n = 24$) and *N. tabacum* ($2n = 48$) are two closely related plant that can be intercrossed, but the F_1 hybrid plants that result are usually sterile. In 1925, Roy Clausen and Thomas Goodspeed crossed *N. glutinosa* and *N. tabacum*, and obtained one fertile F_1 plant (R. E. Clausen and T. H. Goodspeed. 1925 Genetics 10:278–284). They were able to self-pollinate the flowers of this plant to produce an F_2 generation. Surprisingly, the F_2 plants were fully fertile and produced viable seed. When Clausen and Goodspeed examined the chromosomes of the F_2 plants, they observed 36 pairs of chromosomes in metaphase I and 36 individual chromosomes in metaphase II. Explain the origin of the F_2 plants obtained by Clausen and Goodspeed and the numbers of chromosomes observed.
The rare fertile F_1 plant is most likely an allotetraploid. The initial fusion of haploid gametes forms a hybrid containing 12 glutinosa *chromosomes and 24*

tabacum *chromosomes. A mitotic nondisjunction results in an allotetraploid containing 12 pairs of* glutinosa *chromosomes and 24 pairs of* tabacum *chromosomes. In the allotetraploid, meiosis proceeds normally, since each chromosome can pair with a homologue.*

39. What would be the chromosome number of progeny resulting from the following crosses in wheat (see Figure 9.30)? What type of polyploidy (allotriploid, allotetraploid, etc.) would result from each cross?
   a. Einkorn wheat and Emmer wheat.
   Einkorn is T. monococcum *(*$2n = 14$*); Emmer is* T. turgidum *(*$4n = 28$*). Gametes from Einkorn have* $n = 7$ *and gametes from Emmer have* $2n = 14$ *chromosomes. The progeny from this cross would be* $3n = 21$ *and allotriploid.*
   b. Bread wheat and Emmer wheat.
   *Bread wheat (*T. aestivum*) is allohexaploid (*$6n = 42$*) and produce gametes with* $3n = 21$ *chromosomes. Progeny from this cross would have (*$3n = 21$*) + (*$2n = 14$*), or* $5n = 35$ *chromosomes, and would be allopentaploid.*
   c. Einkorn wheat and Bread wheat.
   ($n = 7$*) + (*$3n = 21$*) produces progeny with* $4n = 28$ *chromosomes, and are allotetraploid.*

40. Karl and Hally Sax crossed *Aegilops cylindrical* ($2n = 28$), a wild grass found in the Mediterranean region, with *Triticum vulgare* ($2n = 42$), a type of wheat (K. Sax and H. J. Sax. 1924. *Genetics* 9:454–464). The resulting F_1 plants from this cross had 35 chromosomes. Examination of metaphase I in the F_1 plants revealed the presence of 7 pairs of chromosomes (bivalents) and 21 unpaired chromosomes (univalents).
   a. If the unpaired chromosomes segregate randomly, what possible chromosome numbers will appear in the gametes of the F_1 plants?
   Anywhere from 7 to 28. Each gamete will get 7 chromosomes from segregation of the bivalents, plus 0 to 21 univalents.
   b. What does the appearance of the bivalents in the F_1 hybrids suggest about the origin of *Triticum vulgare* wheat?
   The 7 bivalents suggest that Triticum vulgare *has a common ancestor with* Aegilops *and that* Triticum *obtained 14 additional chromosomes from other ancestors not shared with* Aegilops.

CHALLENGE QUESTIONS

Section 9.3

41. Red-green color blindness is a human X-linked recessive disorder. Jill has normal color vision, but her father is color blind. Jill marries Tom, who also has normal color vision. Jill and Tom have a daughter who has Turner syndrome and is color blind.
   a. How did the daughter inherit color blindness?
   The daughter, with Turner syndrome, is 45,XO. A normal egg cell with a color-blind X chromosome was fertilized by a sperm carrying no sex chromosome.

Such a sperm could have been produced by a nondisjunction event during spermatogenesis.

b. Did the daughter inherit her X chromosome from Jill or from Tom?
The color-blind daughter must have inherited her X chromosome from Jill. Tom is not color blind and therefore could not have a color-blind allele on his single X chromosome. Jill's father was color blind, so Jill must have inherited a color-blind X chromosome from him and passed it on to her daughter.

42. Progeny of triploid tomato plants often contain parts of an extra chromosome, in addition to the normal complement of 24 chromosomes (J. W. Lesley and M. M. Lesley. 1929. *Genetics* 14:321–336). Mutants with a part of an extra chromosome are referred to as secondaries. James and Margaret Lesley observed that secondaries arise from triploid ($3n$), trisomic ($3n + 1$), and double trisomic ($3n + 1 + 1$) parents, but never from diploids ($2n$). Suggest one or more possible reasons that secondaries arise from parents that have unpaired chromosomes but not from parents that are normal diploids.
In triploids and other karyotypes with unpaired chromosomes, synapsis during meiotic prophase often involves three homologous chromosomes. Crossing over can then occur in such a way that one chromosome may cross over at one region with one homologue, and at another region with the other homologue. In metaphase, these trivalents are associated through chiasma. Segregation of such trivalents may cause a situation where one homologue is pulled to one pole, a second homologue is pulled to the other, and the third homologue that has formed chiasma with both homologues form a chromosomal bridge in anaphase, ultimately leading to chromosomal breakage.

43. Mules result from a cross between a horse ($2n = 64$) and a donkey ($2n = 62$), have 63 chromosomes and are almost always sterile. However, in the summer of 1985, a female mule named Krause who was pastured with a male donkey gave birth to a newborn foal (O. A. Ryder et al. 1985. *Journal of Heredity* 76:379–381). Blood tests established that the male foal, appropriately named Blue Moon, was the offspring of Krause and that Krause was indeed a mule. Both Blue Moon and Krause were fathered by the same donkey (see the illustration). The foal, like his mother, had 63 chromosomes—half of them horse chromosomes and the other half donkey chromosomes. Analyses of genetic markers showed that, remarkably, Blue Moon seemed to have inherited a complete set of horse chromosomes from his mother, instead of a random mixture of horse and donkey chromosomes that would be expected with normal meiosis. Thus, Blue Moon and Krause were not only mother and son, but also brother and sister.

a. With the use of a diagram, show how, if Blue Moon inherited only horse chromosomes from his mother, Blue Moon and Krause are both mother and son as well as brother and sister.

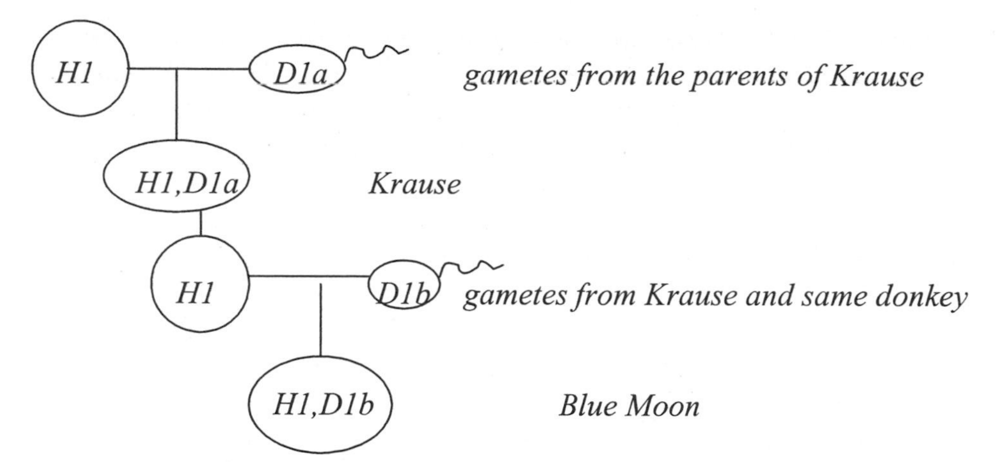

H1 represents the set of maternal horse chromosomes from Krause's mother; D1a represents the set of paternal donkey chromosomes from Krause's father. D1b represents a different set of paternal donkey chromosomes from Krause's (and Blue Moon's) father.

Krause and Blue Moon have the same father, and the same genetic mother, so they are siblings, although Krause is the mother of Blue Moon. In fact, they are closer than full siblings because Krause and Blue Moon have the exact same maternal horse chromosomes—they are half-clones, differing only in which set of chromosomes they received from the common father.

b. Although rare, additional cases of fertile mules giving births to offspring have been reported. In these cases, when a female mule mates with a male horse, the offspring is horselike in appearance but, when a female mule mates with a male donkey, the offspring is mulelike in appearance. Is this observation consistent with the idea that the offspring of fertile female mules inherit only a set of horse chromosomes from their mule mothers? Explain your reasoning.
Yes. Mules result when a donkey sperm fertilizes a horse egg. The reverse, a horse sperm fertilizing a donkey egg, produces a hinny, which is similar in appearance to a mule. The production of horse-like progeny when a female mule is mated with a male horse suggests that the mule's eggs largely lack donkey chromosomes. The observation that mating of the fertile female mule with a male donkey results in mule-like foals again suggests that the maternal chromosomes must be largely of horse origin.

c. Can you suggest a possible mechanism for how the offspring of fertile female mules might pass on a complete set of horse chromosomes to their offspring?
During oogenesis, meiotic divisions are unequal, resulting in a large egg cell and three small polar bodies. The key event is the first meiotic division: somehow a complete haploid set of horse chromosomes all segregate to the oocyte and not into the polar body. One possible mechanism is that horse chromosomes undergo an extra round of pre-meiotic replication, so the pre-meiotic cell has 2n horse chromosomes and 1n donkey chromosomes. Then the horse chromosomes pair and segregate, resulting in a complete set of horse chromosomes in the oocyte. Another possibility is that horse and donkey chromosomes pair, and the meiotic spindle apparatus interacts differentially with horse and donkey chromosomes so that horse chromosomes preferentially

segregate to the oocyte. A third possibility is that horse and donkey chromosomes do not synapse with each other, and horse chromosomes end up in the oocyte because the oocyte volume is so much larger. Donkey chromosomes are lost in the polar body because they have a more peripheral localization than horse chromosomes.

Section 9.5

44. Humans and many complex organisms are diploid, possessing two sets of genes, one inherited from the mother and one from the father. However, there are a number of eukaryotic organisms that spend the majority of their life cycle in a haploid state. Many of these, such as *Neurospora* (a fungus) and yeast, still undergo meiosis and sexual reproduction, but most of the cells that make up the organism are haploid.

    Considering that haploid organisms are fully capable of sexual reproduction and generating genetic variation, why are most complex organisms diploid? In other words, what might be the evolutionary advantage of existing in a diploid state instead of a haploid state? And why might a few organisms, such as *Neurospora* and yeast, exist as haploids?
    The most obvious advantage to being a diploid is genetic redundancy. Most mutations reduce or eliminate gene function, and are recessive. Having two copies means that the organism, and its component cells, are able to survive the vast majority of mutations that would be lethal or deleterious in a haploid state. This redundancy is especially important for organisms that have large, complex genomes, complex development, and relatively lengthy life cycles. Another advantage to diploidy is that gene expression levels can be higher than in haploid cells, often leading to larger cell sizes and larger, more robust organisms. Indeed, many cultivated polyploid crop plants have higher growth rates and yields than their diploid relatives. A third advantage is that the ability to carry recessive mutations in a masked state allows diploid populations to accumulate and harbor much more genetic diversity. Variant forms of genes that may be harmful to the organism and selected against in the current environment may prove advantageous when the environment changes.

    These advantages are less important for organisms that have small genomes and simpler, shorter life cycles. Yeast and Neurospora genomes are highly adapted to their ecological niches, and haploid growth exposes and weeds out less favorable genetic variants. The abililty to replicate their haploid genomes more quickly may give them a selective advantage over diploids. Indeed, these species become diploid and undergo meiosis only when conditions become less favorable for growth.

Chapter Ten: DNA: The Chemical Nature of the Gene

COMPREHENSION QUESTIONS

Section 10.1

*1. What three general characteristics must the genetic material possess?
(1) The genetic material must contain complex information that encodes the phenotype.
(2) The genetic material must replicate or be replicated faithfully.
(3) The genetic material must be able to mutate to generate diversity.

Section 10.2

2. Briefly outline the history of our knowledge of the structure of DNA until the time of Watson and Crick. Which do you think were the principal contributions and developments?
1869: Johann Friedrich Miescher isolates nuclei from white blood cells and extracts a substance that was slightly acidic and rich in phosphorous. He calls it nuclein.
Late 1800s: Albrecht Kossel determines that DNA contains the four nitrogenous bases: adenine, guanine, cytosine, and thymine.
1920s: Phoebus Aaron Levine discovers that DNA consists of repeating units, each consisting of a sugar, a phosphate, and a nitrogenous base.
1950: Erwin Chargaff formulates Chargaff's rules (A = T and G = C).
1947: William Ashbury begins studying DNA structure using X-ray diffraction.
1951–1953: Rosalind Franklin, working in Maurice Wilkins' lab, obtains higher resolution pictures of DNA structure using X-ray diffraction techniques.
1953: Watson and Crick propose the model of DNA structure.

All of these scientists contributed information that helped Watson and Crick determine the structure of the DNA double helix. Erwin Chargaff and Rosalind Franklin made two important contributions that directly led to the discoveries by Watson and Crick. By combining Chargaff's rules with Rosalind Franklin's X-ray diffraction data, Watson and Crick were able to predict accurately the structure of the DNA double helix.

*3. What experiments demonstrated that DNA is the genetic material?
Experiments by Hershey and Chase in the 1950s using the bacteriophage T2 and E. coli *cells demonstrated that DNA is the genetic material of the bacteriophage. Also, the experiments by Avery, Macleod, and McCarty demonstrated that the transforming material initially identified by Griffiths was DNA.*

4. What is transformation? How did Avery and his colleagues demonstrate that the transforming principle is DNA?
Transformation occurs when a transforming material (or DNA) genetically alters the bacterium that absorbs the transforming material. Avery and his colleagues

demonstrated that DNA is the transforming material by using enzymes that destroyed the different classes of biological molecules. Enzymes that destroyed proteins or nucleic acids had no effect on the activity of the transforming material. However, enzymes that destroyed DNA eliminated the biological activity of the transforming material. Avery and his colleagues were also able to isolate the transforming material and demonstrate that it had chemical properties similar to DNA.

*5. How did Hershey and Chase show that DNA is passed to new phages in phage reproduction?
Hershey and Chase used the radioactive isotope ^{32}P to demonstrate that DNA is passed to new phage particles during phage reproduction. The progeny phage released from bacteria infected with ^{32}P-labeled phages emitted radioactivity from ^{32}P. The presence of the ^{32}P in the progeny phage indicated that the infecting phage had passed DNA on to the progeny phage.

6. Why was Watson and Crick's discovery so important?
By deciphering the structure of the DNA molecule, Watson and Crick provided the foundation for molecular studies of the genetic material or DNA, allowing scientists to discern how genes function to produce phenotypes. Their model also suggested a possible mechanism for the replication of DNA that would ensure the fidelity of the replicated copies.

Section 10.3

*7. Draw and label the three parts of a DNA nucleotide.
The three parts of a DNA nucleotide are phosphate, deoxyribose sugar, and a nitrogenous base.

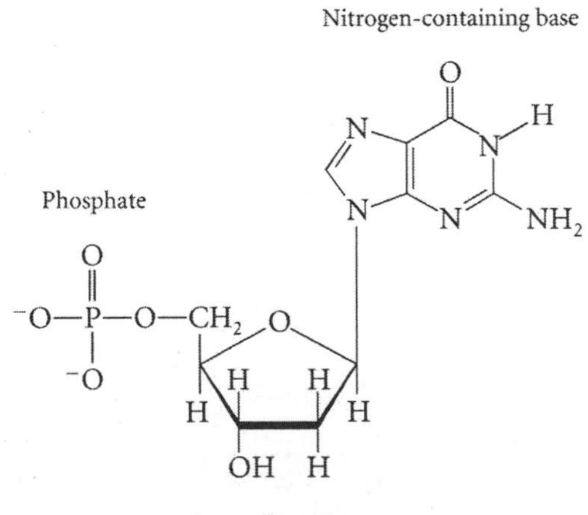

8. How does an RNA nucleotide differ from a DNA nucleotide?
DNA nucleotides, or deoxyribonucleotides, have a deoxyribose sugar that lacks an oxygen molecule at the 2′ carbon of the sugar molecule. Ribonucleotides, or RNA

nucleotides, have a ribose sugar with an oxygen linked to the 2′ carbon of the sugar molecule. Ribonucleotides may contain the nitrogenous base uracil, but not thymine. DNA nucleotides contain thymine, but not uracil.

9. How does a purine differ from a pyrimidine? What purines and pyrimidines are found in DNA and RNA?
A purine consists of a six-sided ring attached to a five-sided ring. A pyrimidine consists of only a six-sided ring. In both DNA and RNA, the purines found are adenine and guanine. DNA and RNA differ in their pyrimidine content. The pyrimidine cytosine is found in both RNA and DNA. However, DNA contains the pyrimidine thymine, whereas RNA contains the pyrimidine uracil but not thymine.

*10. Draw a short segment of a single polynucleotide strand, including at least three nucleotides. Indicate the polarity of the strand by labeling the 5′ end and the 3′ end.

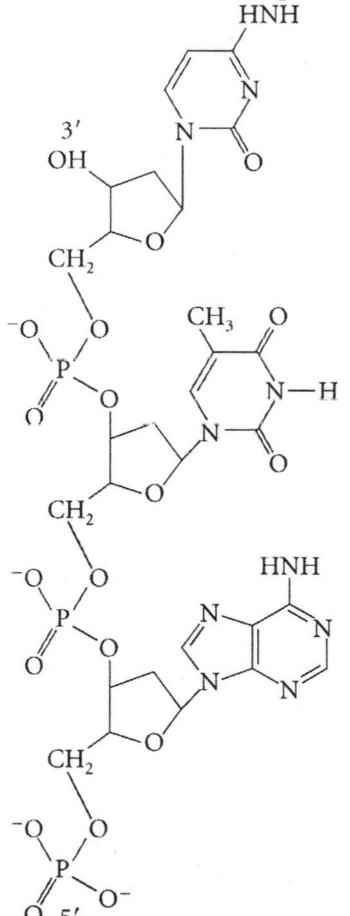

11. Which bases are capable of forming hydrogen bonds with each another?
Adenine is capable of forming two hydrogen bonds with thymine. Guanine is capable of forming three hydrogen bonds with cytosine.

12. What different types of chemical bonds are found in DNA and where do they occur?
The deoxyribonucleotides in a single chain or strand of DNA are held by covalent bonds called phosphodiester linkages between the 3′ end of the deoxyribose sugar of a nucleotide and the 5′ end of the deoxyribose sugar of the next nucleotide in the chain. Two chains of deoxyribonucleotides are held together by hydrogen bonds between the complementary nitrogenous bases of the nucleotides in each chain.

13. What are some of the important genetic implications of the DNA structure?
Referring back to Question 1, the structure of DNA gives insight into the three fundamental genetic processes. The Watson and Crick model suggests that the genetic information or instructions are encoded in the nucleotide sequences. The complementary polynucleotide strands indicate how faithful replication of the genetic material is possible. Finally, the arrangement of the nucleotides is such that they specify the primary structure or amino acid sequence of protein molecules.

*14. What are the major transfers of genetic information?
The major transfers of genetic information are replication, transcription, and translation. These are the components of the central dogma of molecular biology.

Section 10.4

15. What are hairpins and how do they form?
Hairpins are a type of secondary structure found in single strands of nucleotides. The formation of hairpins occurs when sequences of nucleotides on the single strand are inverted complementary repeats of one another.

16. What is DNA methylation?
DNA methylation is the addition of methyl groups ($-CH_3$) to certain positions on the nitrogenous bases on the nucleotide.

APPLICATION QUESTIONS AND PROBLEMS

Section 10.2

17. A student mixes some heat-killed-type IIS *Streptococcus pneumonia* bacteria with live-type IIR bacteria and injects the mixture into a mouse. The mouse develops pneumonia and dies. The student recovers some type IIS bacteria from the dead mouse. It is the only experiment conducted by the student. Has the student demonstrated that transformation has taken place? What other explanations might explain the presence of the type IIS bacteria in the dead mouse?
No, the student has not demonstrated that transformation has taken place. Unlike Griffiths, who used strains IIR and IIIS to demonstrate transformation, the student is using strains IIR and IIS. A mutation in the IIR strain injected into the mouse could be sufficient to convert the IIR strain into the virulent IIS strain. By not conducting

the appropriate control of injecting IIR bacteria only, the student cannot determine whether the conversion from IIR to IIS is due to transformation or to a mutation.

Although heat may have killed all the IIS bacteria, the student has not demonstrated that the heat was sufficient to kill all the IIS bacteria. A second useful control experiment would have been to inject the heat-killed IIS into mice and see if any of the IIS bacteria survived the heat treatment.

*18. Explain how heat-killed-type IIIS bacteria in Griffith's experiment genetically altered the live-type IIR bacteria? (Hint: See the discussion of transformation in Chapter 8.)
The IIR strain of Streptococcus *pneumonia must have been naturally competent or in other words, was capable of taking up DNA from the environment. The heat-killed strains of IIIS bacteria lysed releasing their DNA into the environment allowing for IIIS chromosomal DNA fragments to come in contact with IIR cells. The IIIS DNA responsible for the virulence of the IIIS strain was taken up by a IIR cell and integrated into the IIR cell's chromosome, thus "transforming" the IIR cell into a virulent IIIS cell.*

19. What results would you expect if the Hershey and Chase experiment were conducted on tobacco mosaic virus?
Infection by TMV results in the protein coat and RNA genome entering the host cell. Inside the plant cell, the TMV protein coat unwinds, releasing the viral genome, which initiates infection. If Hershey and Chase had used ^{32}P *and* ^{35}S *to label TMV particles, the RNA molecules would have been labeled with the* ^{32}P *and the viral proteins would have been labeled with* ^{35}S*. However, the protein coat and the RNA genome would have entered the cell, so "radioactive ghost proteins" would not have been located outside the cell. Newly synthesized viral RNAs would have contained measurable levels of* ^{32}P*.*

Section 10.3

20. DNA molecules of different size are often separated using a technique called electrophoresis (see Chapter 18). With this technique, DNA molecules are placed in a gel, an electrical current is applied to the gel, and the DNA molecules migrate toward the positive (+) pole of the current. What aspect of its structure causes DNA molecules to migrate toward to the positive pole?
The phosphate backbone of DNA molecules typically carries a negative charge, thus making the DNA molecules attractive to the positive pole of the current.

*21. Each nucleotide pair of a DNA double helix weighs about 1×10^{-21} g. The human body contains approximately 0.5 g of DNA. How many nucleotide pairs of DNA are in the human body? If you assume that all the DNA in human cells is in the B-DNA form, how far would the DNA reach if stretched end to end?
If each nucleotide pair of a DNA double helix weighs approximately 1×10^{-21} *g and the human body contains 0.5 grams of DNA, then the number of nucleotide pairs can*

be estimated as: (0.5 g DNA / human)/(1 × 10^{-21} g / nucleotide) = 5 × 10^{20} nucleotides pairs/human.

DNA that is in B form has an average distance of .34 nm between each nucleotide pair. If a human possesses 5 × 10^{20} nucleotide pairs, then that DNA stretched end to end would reach: (5 × 10^{20} nucleotides / human) × (0.34 nm/nucleotide pair) = 1.7 × 10^{20} nm, or 1.7 × 10^{8} km.

22. What aspects of its structure contribute to the stability of the DNA molecule? Why is RNA less stable than DNA?
Several aspects contribute to the stability of the DNA molecule. The relatively strong phosphodiester linkages connect the nucleotides of a given strand of DNA. The helical nature of the double-stranded DNA molecule results in the negatively charged phosphates of each strand being arranged to the outside and away from each other. The complementary nature of the nitrogenous bases of the nucleotides helps hold the two strands of polynucleotides together. The stacking interactions of the bases, which allow for any base to follow another in a given strand, also play a major role in holding the two strands together. Finally, the ability of DNA to have local variations in its secondary structure contributes to its stability.

RNA nucleotides or ribonucleotides contain an extra oxygen at the 2′ carbon. This extra oxygen at each nucleotide makes RNA a less stable molecule.

*23. Edwin Chargaff collected data on the proportions of nucleotide bases from the DNA of a variety of different organisms and tissues (E. Chargaff, in *The Nucleic Acids: Chemistry and Biology*, vol. 1, E. Chargaff and J. N. Davidson, Eds. New York: Academic Press, 1955). Data from the DNA of several organisms analyzed by Chargaff are presented here.

	Percent			
Organism and tissue	**A**	**G**	**C**	**T**
Sheep thymus	29.3	21.4	21.0	28.3
Pig liver	29.4	20.5	20.5	29.7
Human thymus	30.9	19.9	19.8	29.4
Rat bone marrow	28.6	21.4	20.4	28.4
Hen erythrocytes	28.8	20.5	21.5	29.2
Yeast	31.7	18.3	17.4	32.6
E. coli	26.0	24.9	25.2	23.9
Human sperm	30.9	19.1	18.4	31.6
Salmon sperm	29.7	20.8	20.4	29.1
Herring sperm	27.8	22.1	20.7	27.5

a. For each organism, compute the ratio of (A + G)/(T + C) and the ratio of (A + T)/(C + G).

Organism	*(A + G)/(C + T)*	*(A + T)/(C + G)*
Sheep Thymus	1.03	1.36
Pig liver	0.99	1.44
Human thymus	1.03	1.52
Rat bone marrow	1.02	1.36
Hen erythrocytes	0.97	1.38
Yeast	1.00	1.80
E. coli	1.04	1.00
Human sperm	1.00	1.67
Salmon sperm	1.02	1.43
Herring sperm	1.04	1.29

b. Are these ratios constant or do they vary among the organisms? Explain why.
The ratios for the (A + G)/(T + C) are constant at approximately 1.0 for the different organisms. Each of these organisms contains a double-stranded genome. The percentages of guanine and cytosine are almost equal to each other and the percentages of adenine and thymine are almost equal to each other as well. In other words, the percentage of purines should be equal to the percentage of pyrimidines for double-stranded DNA. This means that (A + G) = (C + T). The (A + T)/(C + G) ratios are not constant. The relative numbers of AT base pairs and GC base pairs are unique to each organism and can vary between the different species.

c. Is the (A + G)/(T + C) ratio different for the sperm samples? Would you expect it to be? Why or why not?
The ratios for the two sperm samples are essentially the same. The equal ratio should be expected. As stated in the answer to part ***b.*** *of this question, the percentage of purines should equal the percentage of pyrimindines.*

24. Boris Magasanik collected data on the amounts of the bases of RNA isolated from a number of sources (shown here), expressed relative to a value of 10 for adenine (B. Magasanik, in *The Nucleic Acids: Chemistry and Biology*, vol. 1, E Chargaff and J. N. Davidson, Eds. New York: Academic Press, 1955).

		Percent		
Organism and tissue	**A**	**G**	**C**	**U**
Rat liver nuclei	10	14.8	14.3	12.9
Rabbit liver nuclei	10	13.6	13.1	14.0
Cat brain	10	14.7	12.0	9.5
Carp muscle	10	21.0	19.0	11.0
Yeast	10	12.0	8.0	9.8

a. For each organism, compute the ratio of (A + G)/(U + C).

Organism and tissue	*(A + G)/ (U + C)*
Rat liver nuclei	0.91
Rabbit liver nuclei	0.87
Cat brain	1.15
Carp muscle	1.03
Yeast	1.24

b. How do these ratios compare with the (A + G)/(T + C) ratio found in DNA (see Problem 23)? Explain.
The ratios are not as similar to each other or as close to the value of 1.0 as found for the (A + G)/(T + C) ratio in DNA. Many RNA molecules are single-stranded and do not have large regions of complementary sequences as we would expect to find in DNA.

*25. Which of the following relations will be found in the percentages of bases of a double-stranded DNA molecule?
A double-stranded DNA molecule will contain equal percentages of A and T nucleotides and equal percentages of G and C nucleotides. The combined percentage of A and T bases added to the combined percentage of the G and C bases should equal 100.

a. $A + T = G + C$ *No*
b. $A + G = T + C$ *Yes*
c. $A + C = G + T$ *Yes*

d. $\frac{A + T}{C + G} = 1.0$ *No*

e. $\frac{A + G}{C + T} = 1.0$ *Yes*

f. $\frac{A}{C} = \frac{G}{T}$ *No*

g. $\frac{A}{G} = \frac{T}{C}$ *Yes*

h. $\frac{A}{T} = \frac{G}{C}$ *Yes*

*26. If a double-stranded DNA molecule has 15% thymine, what are the percentages of all the other bases?
The percentage of thymine (15%) should be approximately equal to the percentage of adenine (15%). The remaining percentage of DNA bases will consist of cytosine

and guanine bases (100% – 15% – 15% = 70%); these should be in equal amounts (70%/2 = 35%). Therefore, the percentages of each of the other bases if the thymine content is 15% are adenine = 15%; guanine = 35%; and cytosine = 35%.

27. Suppose that each of the bases in DNA were capable of pairing with any other base. What effect would this have on DNA's capacity to serve as the source of genetic information?
DNA's ability to be replicated faithfully and to encode phenotypes would be destroyed. If each base could pair with any other base, the result during replication would be changes in the DNA sequences of the newly replicated strands. The two new molecules of DNA would not be identical to the original molecule or to each other because different bases would be inserted in each newly synthesized strand. If the DNA base sequence was constantly changing due to the random pairing of bases, then no consistent "code" could be maintained. This lack of a code would inhibit the ability of a DNA molecule to faithfully code for any particular protein.

28. Heinz Shuster collected the following data on the base composition of ribgrass virus (H. Shuster, in *The Nucleic Acids: Chemistry and Biology,* vol. 3, E. Chargaff and J. N. Davidson, Eds. New York: Academic Press, 1955). On the basis of this information, is the hereditary information of the ribgrass virus RNA or DNA? Is it likely to be single stranded or double stranded?

	Percent				
	A	**G**	**C**	**T**	**U**
Ribgrass virus	29.3	25.8	18.0	0.0	27.0

Most likely, the ribgrass viral genome is a single-stranded RNA. The presence of uracil indicates that the viral genome is RNA. For the molecule to be double-stranded RNA, we would predict equal percentages of adenine and uracil bases and equal percentages of guanine and cytosine bases. Neither the percentages of adenine and uracil bases nor the percentages of guanine and cytosine bases are equal, indicating that the viral genome is likely single-stranded.

*29. The relative amounts of each nucleotide base are shown here for four different viruses. For each virus, indicate whether its genetic material is DNA or RNA and whether it is single-stranded or double-stranded. Explain your reasoning.

Virus	**T**	**C**	**U**	**G**	**A**
I	0	12	9	12	9
II	23	16	0	16	23
III	34	42	0	18	39
IV	0	24	35	27	17

Virus I is a double-stranded RNA virus. Uracil is present indicating an RNA genome and we see equal percentages of adenine and uracil and equal percentages of guanine and cytosine, which we would expect if it is a double-stranded genome.

Virus II is a doubled-stranded DNA virus. The presence of thymine indicates that the viral genome is DNA. As expected for a double-stranded DNA molecule, we see equal percentages of adenine and thymine bases and equal percentages of guanine and cytosine bases.

Virus III is a single-stranded DNA virus. Thymine is present suggesting a DNA genome. However, we see unequal percentages of thymine and adenine and unequal percentages of guanine and cytosine, which suggest a single-stranded DNA molecule.

Virus IV is a single-stranded RNA virus. Uracil is present indicating an RNA genome. However, the percentage of adenine does not equal the percentage of uracil, and the percentage of guanine does not equal the percentage of cytosine. These unequal amounts suggest a single-stranded genome.

*30. A B-DNA molecule has 1 million nucleotide pairs.

a. How many complete turns are there in this molecule?
B-form DNA contains approximately 10 nucleotides per turn of the helix. A B-DNA molecule of 1 million nucleotide pairs will have about the following number of complete turns: (1,000,000 nucleotides) / 10 nucleotides/turn) = 100,000 complete turns.

b. If this same molecule were in the Z-DNA configuration, how many complete turns would it have?
If the same DNA molecule assumes a Z-DNA configuration, then each turn would consist of about 12 nucleotides. The determination of the number of complete turns in the 1 million nucleotide molecule is:
(1,000,000 nucleotides) / (12 nucleotides / turn) = 83333.3, or 83333 complete turns.

31. For entertainment on a Friday night, a genetics professor proposed that his children diagram a polynucleotide strand of DNA. Having learned about DNA in preschool, his 5-year-old daughter was able to draw a polynucleotide strand, but she made several mistakes. The daughter's diagram (represented here) contained at least 10 mistakes.

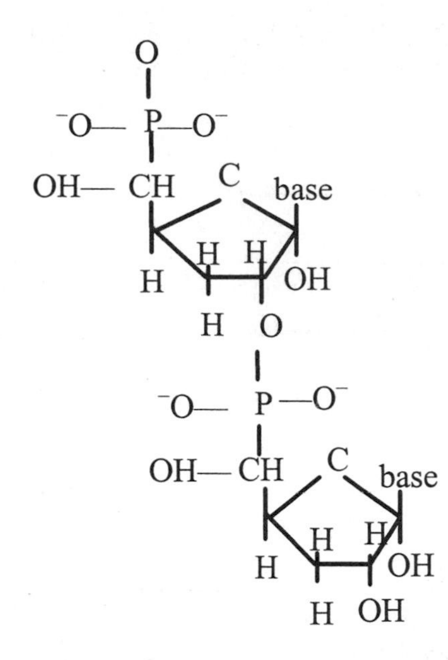

a. Make a list of all the mistakes in the structure of this DNA polynucleotide strand.

(1) Neither 5′ carbon of the two sugars is directly linked to phosphorous.

(2) Neither 5′ carbon of the two sugars has an OH group attached.

(3) Neither sugar molecule has oxygen in its ring structure between the 1′ and 4′ carbons.

(4) In both sugars, the 2′ carbon has an –OH group attached, which does not occur in deoxyribonucleotides.

(5) At the 3′ position in both sugars, only hydrogen is attached, as opposed to an –OH group.

(6) The 1′ carbon of both sugars has an –OH group, as opposed to just a hydrogen attached.

b. Draw the correct structure for the polynucleotide strand.

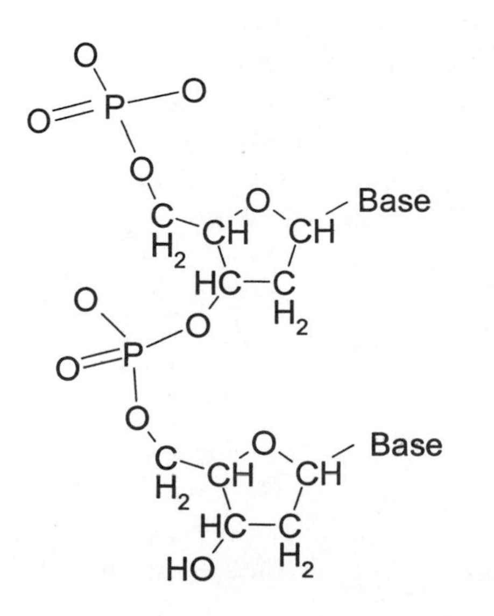

*32. Chapter 1 considered the theory of the inheritance of acquired characteristics and noted that this theory is no longer accepted. Is the central dogma consistent with the theory of the inheritance of acquired characteristics? Why or why not?
The central dogma of molecular biology is not consistent with the theory of inheritance of acquired characteristics. The flow of information predicted by the central dogma is:

DNA ⟶ *RNA* ⟶ *Protein*

One exception to the central dogma is reverse transcription, whereby RNA codes for DNA. However, biologists currently do not know of a process that will allow for the flow of information from proteins back to DNA. The theory of inheritance of acquired characteristics necessitates such a flow of information from proteins back to the DNA.

Section 10.4

33. Write a sequence of bases in an RNA molecule that would produce a hairpin structure.
For a hairpin structure to form in a RNA molecule, an inverted complementary RNA sequence separated by a region of noncomplementary sequence is necessary. The inverted complements form the stem structure, and the loop of the hairpin is formed by the noncomplementary sequences.

5′—UGCAU—3′ ...unpaired nucleotides...5′—AUGCA—3′

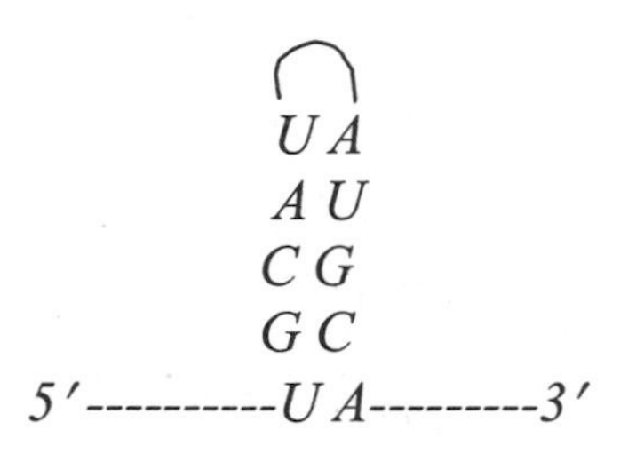

34. Write out a sequence of nucleotides on a strand of DNA that will form a hairpin structure.
For a hairpin structure to form in a strand of DNA, the DNA strand must contain inverted complementary DNA sequences separated by a region of noncomplementary sequence. The inverted complements form the stem structure, and the loop of the hairpin is formed by the noncomplementary sequences:

5′—TGCATTACTCAATGCA—3′

or

C T
A C
T A
G C
C G
A T
T A
5′ T A 3′

CHALLENGE QUESTIONS

Section 10.1

*35. Suppose that an automated, unmanned probe is sent into deep space to search for extraterrestrial life. After wandering for many light-years among the far reaches of the universe, this probe arrives on a distance planet and detects life. The chemical composition of life on this planet is completely different from that of life on Earth, and its genetic material is not composed of nucleic acids. What predictions can you make about the chemical properties of the genetic material on this planet?

Although the chemical composition of the genetic material may be different DNA, it more than likely will have similar properties to DNA. As discussed earlier in the chapter, the genetic material must possess three general characteristics:

(1) It must contain complex information.

(2) It must replicate or be replicated faithfully.

(3) It must encode the phenotype.

Even if the material is not DNA, it must meet these criteria. For instance, if the material could not be replicated or duplicated faithfully, then life on that planet could not continue because ultimately no offspring could be produced. A lack of fidelity would result in the loss of information. Genetic material from any lifeform has to store the information necessary for the survival of that organism. Also, the genetic material will need to be stable. Unstable molecules will not allow for long-term storage of information, resulting in the loss of information and change in phenotype. In addition, this material must be able to mutate, and the mutant form must be as stable as the original form, or else species will not be able to evolve.

Section 10.2

36. How might ^{32}P and ^{35}S be used to demonstrate that the transforming principle is DNA? Briefly outline an experiment that would show that DNA and not protein is the transforming principle.

The first step would be to label the DNA and proteins of the donor bacteria cells with ^{35}S and ^{32}P. The DNA could be labeled by growing a culture of bacteria in the presence of ^{32}P. The cells as they replicate ultimately will incorporate radioactive phosphorous into their DNA. A second culture of bacteria should be grown in the presence of ^{35}S, which ultimately will be incorporated into proteins.

Material from each culture should be used to transform bacteria cells that previously had not been exposed to the radioactive isotopes. Transformed cells (or colonies) that would be identified by the acquisition of a new phenotype should contain low levels of the radioactive material due to the uptake of the labeled molecules. If the transforming material were protein, then cells transformed by the material from the ^{35}S exposed bacterial cultures would also contain ^{35}S. If the transforming material were DNA, then the cells transformed by the material from the ^{32}P exposed bacterial cultures would also contain ^{32}P.

Section 10.3

37. Researchers have proposed that early life on Earth used RNA as its source of genetic information and that DNA eventually replaced RNA as the source of genetic information. What aspects of DNA structure might make it better suited than RNA to be the genetic material?
Due to the lack of an attached oxygen molecule at the 2' carbon position of the sugar molecule in deoxynucleotides, DNA molecules are more stable and less reactive than RNA molecules. The double-helical nature of the DNA molecule provides a greater opportunity for DNA repair and fidelity during replication. If mistakes occur in one strand, the complementary strand can serve as a template for corrections.

38. Scientists have reportedly isolated short fragments of DNA from fossilized dinosaur bones hundreds of millions of years old. The technique used to isolate this DNA is the polymerase chain reaction (PCR), which is capable of amplifying very small amounts of DNA a million fold (see Chapter 16). Critics have claimed that the DNA isolated from dinosaur bones is not of ancient origin but instead represents contamination of the samples with DNA from present-day organisms, such as bacteria, mold, or humans. What precautions, analyses, and control experiments could be carried out to ensure that DNA recovered from fossils is truly of ancient origin?
An initial precaution would be to handle all the material in the most sterile manner possible. People handling the samples should wear gloves and masks to help keep the area as devoid of extraneous DNA as possible. Instruments used in the sampling should be sterilized to eliminate any contamination by bacteria, viruses, fungi, and so on. They should also be treated to remove trace DNAs. In addition, the source material surrounding the bones should be treated to remove contaminating DNAs.

Controls also need to be conducted. The DNA from people involved in the procedure should be tested to see if amplification occurs. Material at various locations around the site and isolated bugs and microorganisms from the area should be sampled to see if similar amplification patterns emerge. The design of the primers used for amplification should be considered carefully and should be executed considering the sequences of potential dinosaur descendants, such as birds or reptiles, in an attempt to limit random amplifications. Furthermore, every experiment should be reproducible.

Chapter Eleven: Chromosome Structure and Transposable Elements

COMPREHENSION QUESTIONS

Section 11.1

*1. How does supercoiling arise? What is the difference between positive and negative supercoiling?
Supercoiling arises from overwinding (positive supercoiling) or underwinding (negative supercoiling) the DNA double helix. Supercoiling may occur:
(1) when the DNA molecule does not have free ends, as in circular DNA molecules, or
(2) when the ends of the DNA molecule are bound to proteins that prevent them from rotating about each other.

2. What functions does supercoiling serve for the cell?
Supercoiling compacts the DNA. Negative supercoiling helps to unwind the DNA duplex for replication and transcription.

Section 11.3

*3. Describe the composition and structure of the nucleosome. How do core particles differ from chromatosomes?
The nucleosome core particle contains two molecules each of histones H2A, H2B, H3, and H4, which form a protein core with 145–147 bp of DNA wound around the core. Chromatosomes contain the nucleosome core with a molecule of histone H1.

4. Describe in steps how the double helix of DNA, which is 2 nm in width, gives rise to a chromosome that is 700 nm in width.
DNA is first packaged into nucleosomes; the nucleosomes are packed to form a 30-nm fiber. The 30-nm fiber forms a series of loops that pack to form a 250 nm fiber, which in turn coils to form a 700-nm chromatid.

5. What are polytene chromosomes and chromosomal puffs?
Polytene chromosomes are giant chromosomes formed by repeated rounds of DNA replication without nuclear division, found in the larval salivary glands of Drosophila *and a few other species of flies. Certain regions of polytene chromosomes can become less condensed, resulting in localized swelling, or chromosomal puffs, because of intense transcriptional activity at the site.*

*6. Describe the function and molecular structure of the centromere.
Centromeres are the points of attachment for mitotic and meiotic spindle fibers and are required for the movement of chromatids to the poles in anaphase. Centromeres have distinct centromeric DNA sequences where the kinetochore proteins bind. For some species, such as yeast, the centromere is compact, consisting of only 125 bp. For other species, including Drosophila *and mammals, the centromere is larger,*

ranging from several thousands to hundreds of thousands of basepairs of DNA sequence.

*7. Describe the function and molecular structure of a telomere.
Telomeres are the ends of the linear chromosomes in eukaryotes. They cap and stabilize the ends of the chromosomes to prevent degradation by exonucleases or joining of the ends. Telomeres also enable replication of the ends of the chromosome. Telomeric DNA sequences consist of repeats of a simple sequence, usually in the form of 5′$C_n(A/T)_m$.

8. What is the difference between euchromatin and heterochromatin?
Euchromatin undergoes regular cycles of condensation during mitosis and decondensation during interphase, whereas heterochromatin remains highly condensed throughout the cell cycle, except transiently during replication. Nearly all transcription takes place in euchromatic regions, with little or no transcription of heterochromatin.

Section 11.4

9. What is the C value of an organism?
The C value is the amount of DNA per cell of an organism.

*10. Describe the different types of DNA sequences that exist in eukaryotes.
Unique-sequence DNA, present in only one or a few copies per haploid genome, represents most of the protein coding sequences, plus a great deal of sequences with unknown function.

Moderately repetitive sequences, a few hundred to a few thousand base pairs long, are present in up to several thousand copies per haploid genome. Some moderately repetitive DNA consists of functional genes that code for rRNAs and tRNAs, but most is made up of transposable elements and remnants of transposable elements.

Highly repetitive DNA, or satellite DNA, consists of clusters of tandem repeats of short (often less than 10 base pairs) sequences present in hundreds of thousands to millions of copies per haploid genome.

Section 11.5

*11. What general characteristics are found in many transposable elements? Describe the differences between replicative and nonreplicative transposition.
Most transposable elements have terminal inverted repeats and are flanked by short direct repeats. Replicative transposons use a copy-and-paste mechanism in which the transposon is replicated and inserted in a new location, leaving the original transposon in place. Nonreplicative transposons use a cut-and-paste mechanism in which the original transposon is excised and moved to a new location.

*12. What is a retrotransposon and how does it move?

A retrotransposon is a transposable element that relocates through an RNA intermediate. First, it is transcribed into RNA. A reverse transcriptase encoded by the retrotransposon then reverse transcribes the RNA template into a DNA copy of the transposon, which then integrates into a new location in the host genome.

*13. Describe the process of replicative transposition through DNA intermediates. What enzymes are involved?

First, a transposase makes single-stranded nicks on either side of the transposon and on either side of the target sequence. Second, the free ends of the transposon are joined by a DNA ligase to the free ends of the DNA at the target site. Third, the free 3′ ends of DNA on either side of the transposon are used to replicate the transposon sequence, forming the cointegrate. The enzymes normally required for DNA replication are required for this step, including DNA polymerase. The cointegrate has two copies of the transposon and the target site sequence on one side of each copy. Fourth, the cointegrate undergoes resolution, which involves a crossing over within the transposon, by resolvase enzymes such as those used in homologous recombination.

Section 11.6

*14. Draw the structure of a typical insertion sequence and identify its parts.

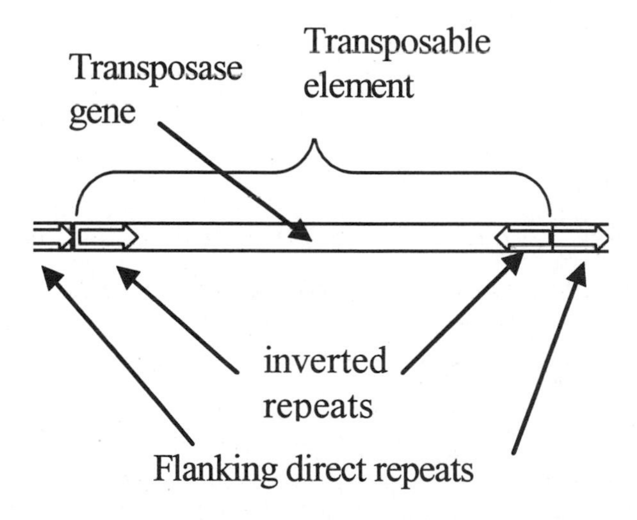

15. Draw the structure of a typical composite transposon in bacteria and identify its parts.

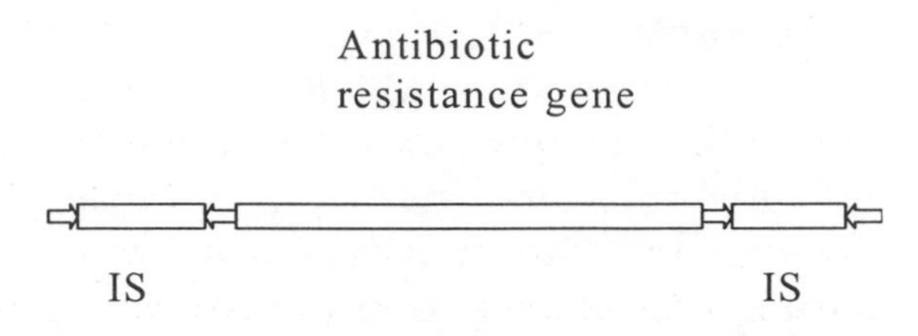

16. How are composite transposons and retrotransposons alike and how are they different?
Composite transposons and retrotransposons are similar in their complexity and some aspects of structure. Both have long, flanking repeat sequences: IS elements for composite transposons, long terminal repeat sequences for retrotransposons. Both generate direct duplications of their target site sequences.

However, the IS elements in composite transposons may be either direct repeats or inverted, whereas retrotransposons have direct long terminal repeats. Furthermore, retrotransposons transpose through an RNA intermediate and depend on reverse transcriptase for transposition, whereas composite transposons transpose through DNA intermediates by the action of transposases encoded by one of their insertion sequences.

17. Explain how *Ac* and *Ds* elements produce variegated corn kernels.
A mutation caused by insertion of either an Ac *or* Ds *element in the pigment-producing gene causes the corn kernel to be colorless. During development of the kernel, the* Ac *or* Ds *element may transpose out of the pigment gene and pigment production in the cell and its descendents will be restored. If the excision of the transposable element occurs early in kernel development, the kernel will have a relatively large sector of purple pigment. If the excision occurs later, the kernel will have relatively smaller sectors or specks of purple pigment.*

18. Briefly explain hybrid dysgenesis and how *P* elements lead to hybrid dysgenesis.
When a sperm containing P *elements from a* P^+ *male fly fertilizes an egg from a female that does not contain* P *elements, the resulting zygote undergoes hybrid dysgenesis: transposition of* P *elements causes chromosomal abnormalities and disrupts gamete formation. The hybrid progeny fly is therefore sterile. Hybrid dysgenesis does not occur if the female fly is* P^+ *because the* P *elements also encode a repressor that prevents* P *element transposition. The egg cells from a* P^+ *female fly contain enough of this repressor protein in the cytoplasm to repress* P *element transposition in the progeny.*

19. What are some differences between LINEs and SINEs?
Although both LINEs (long interspersed elements) and SINEs (short interspersed elements) transpose through RNA intermediates, they have different origins. LINEs resemble retroviruses (encode reverse transcriptase), whereas SINEs resemble RNA Pol*III transcripts (*Alu *sequences resemble the 7S RNA gene). SINEs, with a typical*

length of 200–400 bp, are also significantly smaller than LINEs, which can range 6000 bp or longer for full length elements.

Section 11.7

*20. Briefly summarize three hypotheses for the widespread occurrence of transposable elements.

The cellular function hypothesis proposes that transposable elements have a function in the cell or organism, such as regulation of gene expression.

The genetic variation hypothesis suggests that transposable elements serve to generate genomic variation. A larger pool of genomic variants would accelerate evolution by natural selection.

The selfish DNA hypothesis suggests that transposable elements are simply parasites, serving only to replicate and spread themselves.

APPLICATION QUESTIONS AND PROBLEMS

Section 11.3

*21. Compare and contrast prokaryotic and eukaryotic chromosomes. How are they alike and how do they differ?

Prokaryotic chromosomes are usually circular, whereas eukaryotic chromosomes are linear. Prokaryotic chromosomes generally contain the entire genome, whereas each eukaryotic chromosome has only a portion of the genome: the eukaryotic genome is divided into multiple chromosomes. Prokaryotic chromosomes are generally smaller and have only a single origin of DNA replication. Eukaryotic chromosomes are often many times larger than prokaryotic chromosomes and contain multiple origins of DNA replication. Prokaryotic chromosomes are typically condensed into nucleoids, which have loops of DNA compacted into a dense body. Eukaryotic chromosomes contain DNA packaged into nucleosomes, which are further coiled and packaged into successively higher-order structures. The condensation state of eukaryotic chromosomes varies with the cell cycle.

22. **a.** In a typical eukaryotic cell, would you expect to find more molecules of the H1 histone or more molecules of the H2A histone? Explain your reasoning.

Because each nucleosome contains two molecules of histone H2A and only one molecule of histone H1, eukaryotic cells will have more H2A than H1.

b. Would you expect to find more molecules of H2A or more molecules of H3? Explain your reasoning.

Because each nucleosome contains two molecules of H2A and two molecules of H3, eukaryotic cells should have equal amounts of these two histones.

23. Suppose you examined polytene chromosomes from the salivary glands of fruit fly larvae and counted the number of chromosomal puffs observed in different regions of DNA.

a. Would you expect to observe more puffs from euchromatin or from heterochromatin? Explain your answer.
Euchromatin is less condensed and capable of being transcribed, whereas heterochromatin is highly condensed and rarely transcribed. Because chromosomal puffs are sites of active transcription, they should occur primarily in euchromatin.

b. Would you expect to observe more puffs in unique-sequence DNA, moderately repetitive DNA, or repetitive DNA? Why?
Highly repetitive DNA consists of simple tandem repeats usually found in heterochromatic regions and are rarely transcribed. Moderately repetitive DNA comprises transposons and remnants of transposons. Again, with the exception of the rDNA cluster, these sequences are rarely transcribed or transcribed at low levels. The most actively transcribed genes occur as single-copy sequences, or as small gene families. Therefore, more chromosomal puffs would be observed in unique-sequence DNA than in moderately or highly repetitive DNA.

*24. A diploid human cell contains approximately 6.4 billion base pairs of DNA.

a. How many nucleosomes are present in such a cell? (Assume that the linker DNA encompasses 40 bp.)
Given that each nucleosome contains about 140 bp of DNA tightly associated with the core histone octamer, another 20 bp associated with histone H1, and 40 bp in the linker region, then one nucleosome occurs for every 200 bp of DNA.

6.4×10^9 *bp divided by* 2×10^2 *bp/nucleosome* = 3.2×10^7 *nucleosomes (32 million).*

b. How many histone proteins are complexed to this DNA?
Each nucleosome contains two of each of the following histones: H2A, H2B, H3, and H4. A nucleosome plus one molecule of histone H1 constitute the chromatosome. Therefore, nine histone protein molecules occur for every nucleosome.

3.2×10^7 *nucleosomes* $\times$ *9 histones* = 2.9×10^8 *molecules of histones are complexed to 6.4 billion bp of DNA.*

*25. Would you expect to see more or less acetylation in regions of DNA that are sensitive to digestion by DNase I? Why?
More acetylation. Regions of DNase I sensitivity are less condensed than DNA that is not sensitive to DNase I, the sensitive DNA is less tightly associated with nucleosomes, and it is in a more open state. Such a state is associated with acetylation of lysine residues in the N-terminal histone tails. Acetylation eliminates the positive charge of the lysine residue and reduces the affinity of the histone for the negatively charged phosphates of the DNA backbone.

26. Gunter Korge examined several proteins that are secreted from the salivary glands of *Drosophila melanogaster* during larval development (G. Korge. 1975. *Proceedings of the National Academy of Sciences of the United States of America* 72:4550–4554). One protein, called protein fraction 4, was encoded by a gene found

by deletion mapping to be located on the X chromosome at position 3C. Korge observed that, about 5 hours before the first synthesis of protein fraction 4, an expanded and puffed-out region formed on the X chromosome at position 3C. This chromosome puff disappeared before the end of the third larval instar stage, when the synthesis of protein fraction 4 ceased. He observed that there was no puff at position 3C in a special strain of flies that lacked secretion of protein fraction 4. Explain these results. What is the chromosome puff at region 3, and why does its appearance and disappearance roughly coincide with the secretion of protein fraction 4?
Chromosomal puffs correspond to relaxation of chromatin structure and transcriptional activity at the locus. The puff at region 3C indicates active transcription of that region, including the gene for protein fraction 4.

27. Suppose a chemist develops a new drug that neutralizes the positive charges on the tails of histone proteins. What would be the most likely effect of this new drug on chromatin structure? Would you predict that this drug would have any effect on gene expression? Explain your answers.
Such a drug would disrupt the ionic interactions between the histone tails and the phosphate backbone of DNA and thereby cause a loosening of the DNA from the nucleosome. The drug may mimic the effects of histone acetylation, which neutralizes the positively charged lysine residues. Changes in chromatin structure would result from the altered nucleosome-DNA packing and possible changes in interaction with other chromatin modifying enzymes and proteins. Changes in transcription would result because DNA may be more accessible to transcription factors.

Section 11.4

*28. Which of the following two molecules of DNA has the lower melting temperature? Why?

AGTTACTAAAGCAATACATC	AGGCGGGTAGGCACCCTTA
TCAATGATTTCGTTATGTAG	TCCGCCCATCCGTGGGAAT

The molecule on the left, with a higher percentage of A–T base pairs, will have a lower melting temperature than the molecule on the right, which has mostly G–C base pairs. A–T base pairs have two hydrogen bonds, and thus less stability, than G–C base pairs, which have three hydrogen bonds.

29. In a DNA hybridization study, DNA was isolated from a particular species, labeled with ^{32}P, and sheared into small fragments (S. K. Dutta et al. 1967. *Genetics* 57:719–727). Hybridizations between these labeled fragments and denatured DNA from different species were then compared. The following table gives the percentages of labeled wheat DNA that hybridized to DNA molecules of wheat, corn, radish, and cabbage.

Species	Percentage of bound wheat DNA hybridized relative to wheat
Wheat	100
Cabbage	23
Corn	63
Radish	30

What do these results indicate about the evolutionary differences among these organisms?

The extent of DNA hybridization correlates with DNA sequence similarity. Corn DNA is more similar to wheat than radish DNA. Cabbage DNA is least similar. Therefore, corn is most closely related to wheat, followed by radish; and cabbage is the most evolutionarily distant from wheat.

Section 11.5

*30. A particular transposable element generates flanking direct repeats that are 4 bp long. Give the sequence that will be found on both sides of the transposable element if this transposable element inserts at the position indicated on each of the following sequences:

a. 5′—ATTCGAAC**TGAC**(transposable element)**TGAC**CGATCA—3′

b. 5′—ATT**CGAA**(transposable element)**CGAA**CTGACCGATCA—3′

For (a) and (b) the target site duplication is indicated in bold.

*31. White eyes in *Drosophila melanogaster* result from an X-linked recessive mutation. Occasionally, white-eye mutants give rise to offspring that possess white eyes with small red spots. The number, distribution, and size of the red spots are variable. Explain how a transposable element could be responsible for this spotting phenomenon.

Such a fly may carry an allele of the white-eye locus that contains a transposon insertion. The eye cells in these flies cannot make red pigment. During eye development, the transposon may spontaneously transpose out of the white-eye locus, restoring function to this gene so the cell and its mitotic progeny can make red pigment. Depending on how early during eye development the transposition occurs, the number and size of red spots in the eyes will be variable.

*32. What factor do you think determines the length of the flanking direct repeats that are produced in transposition?

The length of the flanking direct repeats that are generated depends on the number of base pairs between the staggered single-stranded nicks made at the target site by the transposase.

Section 11.6

33. Which of the following pairs of sequences might be found at the ends of an insertion sequence?
    a. 5′—GGGCCAATT—3′ and 5′—CCCGGTTAA—3′.
    b. 5′—AAACCCTTT—3′ and 5′—AAAGGGTTT—3′.
    c. 5′—TTTCGAC—3′ and 5′—CAGCTTT—3′.
    d. 5′—ACGTACG—3′and 5′—CGTACGT—3′.
    e. 5′—GCCCCAT—3′ and 5′—GCCCAT—3′.
    The pairs of sequences in (b) and (d) are inverted repeats because they are both reversed and complementary and might be found at the ends of insertion sequences. Sequences in (a), (c), and (e) would not be expected at the ends of an insertion seqence. The sequences in (a) are complementary, but not inverted. The sequences in (c) are reversed, but not complementary. The sequences in (e) are imperfect direct repeats.

34. Two different strains of *Drosophila melanogaster* are mated in reciprocal crosses. When strain A males are crossed with strain B females, the progeny are normal. However, when strain A females are crossed with strain B males, many mutations and chromosome rearrangements occur in the gametes of the F_1 progeny and they are effectively sterile. Explain these results.
    These results could be explained by hybrid dysgenesis, with strain B harboring P elements and strain A having no P elements. When sperm from strain A males fertilize eggs with P elements from strain B females, the progeny are normal because the strain B egg cytoplasm contains a repressor of P element transposition. However, when P^+ sperm cells from strain B fertilize P^- eggs from strain A, the P elements undergo a burst of transposition in the embryo because the P^- egg cytoplasm lacks the repressor.

*35. An insertion sequence contains a large deletion in its transposase gene. Under what circumstances would this insertion sequence be able to transpose?
    Without a functional transposase gene of its own, the transposon would be able to transpose only if another transposon of the same type were in the cell and able to express a functional transposase enzyme. This transposase enzyme will recognize the inverted repeats and transpose its own element as well as other nonautonomous copies of the transposon with the same inverted repeats.

36. A transposable element is found to encode a transposase enzyme. On the basis of this information, what conclusions can you make about the likely structure and method of transposition of this element?
    This element probably has short inverted terminal repeats, and transposes through a DNA intermediate, using either a cut-and-paste nonreplicative mechanism or a copy-and-paste replicative mechanism. Because it does not encode a reverse transcriptase, it is not likely to be a retrotransposon.

37. Zidovudine (AZT) is a drug used to treat patients with AIDS. AZT works by blocking the reverse transcriptase enzyme used by human immunodeficiency virus (HIV), the causative agent of AIDS. Do you expect that AZT would have any effect on transposable elements? If so, what type of transposable elements would be affected and what would be the most likely effect?
AZT should affect retrotransposons because they transpose through an RNA intermediate that is reverse transcribed to DNA by reverse transcriptase. If endogenous reverse transcriptases in human cells have similar sensitivity to AZT as HIV reverse transcriptase, then AZT should inhibit retrotransposons.

38. A transposable element is found to encode a reverse transcriptase enzyme. On the basis of this information, what conclusions can you make about the likely structure and method of transposition of this element?
Like other retrotransposons, this element probably has long terminal direct repeats and transposes through an RNA intermediate that is reverse transcribed to DNA.

39. A geneticist examines an ear of corn in which most kernels are yellow, but she finds a few kernels with purple spots, as shown here. Give a possible explanation for the appearance of the purple spots in these otherwise yellow kernels, accounting for their different sizes. (Hint: See section on *Ac* and *Ds* elements in maize on pp. 304-305).

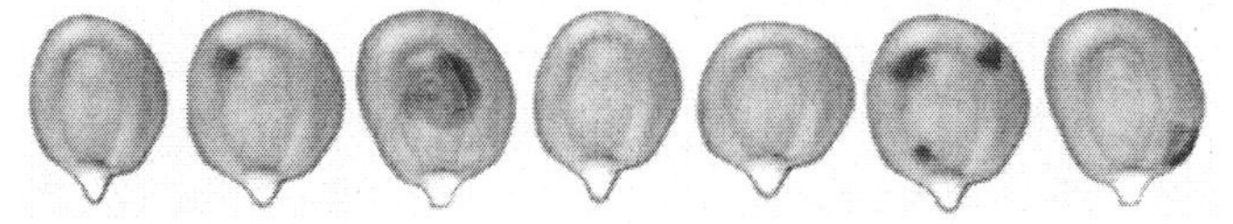

The appearance of purple spots of varying sizes in these few yellow corn kernels could be explained by transposition. The yellow kernels may be due to inactivation of a pigment gene by insertion of a Ds *element in the plant bearing this ear. Because the* Ds *element cannot transpose on its own, the mutant allele is stable in the absence of* Ac *and the plant produces yellow kernels when fertilized by pollen from the same strain (lacking* Ac*). However, a few kernels may have been fertilized by pollen from a different strain with an active* Ac *element. The* Ac *element can then mobilize transposition of the* Ds *element out of the pigment gene, restoring pigment gene function. Excision of the* Ds *element earlier in kernel development will produce larger clones of cells producing purple pigment. Excision later in kernel development will produce smaller clones of purple cells.*

40. A geneticist studying the DNA of the Japanese bottle fly finds many copies of a particular sequence that appears similar to the *copia* transposable element in *Drosophila* (see Table 11.6). Using recombinant DNA techniques, the geneticist places an intron into a copy of this DNA sequence and inserts it into the genome of a Japanese bottle fly. If the sequence is a transposable element similar to *copia*, what prediction would you make concerning the fate of the introduced sequence in the genomes of offspring of the fly receiving it?
Because copia *is a retrotransposon, the* copia*-like element probably transposes through an RNA intermediate. The recombinant transposon containing an intron*

will be transcribed into an RNA molecule containing the intron. However, the intron will be removed by splicing. Reverse transcription of the spliced RNA will generate copies of the transposon that lack the intron. Therefore, the daughter transposons will lack the intron sequence.

CHALLENGE QUESTIONS

Section 11.3

41. An explorer discovers a strange new species of plant and sends some of the plant tissue to a geneticist to study. The geneticist isolates chromatin from the plant and examines it with the electron microscope. She observes what appear to be beads on a string. She then adds a small amount of nuclease, which cleaves the string into individual beads that each contain 280 bp of DNA. After digestion with more nuclease, she finds that a 120 bp fragment of DNA remains attached to a core of histone proteins. Analysis of the histonc core reveals histones in the following proportions:

H1	12.5%
H2A	25%
H2B	25%
H3	0%
H4	25%
H7 (a new histone)	12.5%

On the basis of these observations, what conclusions could the geneticist makc about the probable structure of the nucleosome in the chromatin of this plant?
The 120 bp of DNA associated with the histone core is smaller than the 140 bp associated with typical nucleosomes. The new plant also is lacking histone H3. The new histone H7 apparently does not replace histone H3 in the nucleosome core because it is present in the same ratio as histone H1, or half of the ratios of nucleosomal core histones H2A, H2B, and H4. Finally, the 280 bp fragments with limited DNase digestion are larger than the 200 bp fragments seen with typical eukaryotic chromatin.

These observations suggest a model in which the nucleosome core consists of just six histones, two each of H2A, H2B, and H4, explaining the lack of H3 and the smaller amount of DNA. The longer DNA per nucleosome can be explained in part by a molecule of H7 either in the spacer between nucleosomes, or perhaps helping to cap nucleosomes in conjunction with H1.

Section 11.4

42. Although highly repetitive DNA is common in eukaryotic chromosomes, it does not code for proteins; in fact, it is probably never transcribed into RNA. If highly repetitive DNA does not code for RNA or proteins, why is it present in eukaryotic genomes? Suggest some possible reasons for the widespread presence of highly repetitive DNA.

Highly repetitive DNA may have important structural roles for eukaryotic chromosomes. Highly repetitive DNA is present in regions of the chromosome that are heterochromatic, near the centromere and near the telomeres. Highly repetitive DNA clusters may play important roles in facilitating chromatin condensation in mitotic or meiotic prophase. They may serve to insulate critical regions from chromatin decondensation or from transcription. Near the telomeres, they may serve as buffers between the ends of the chromosome and essential protein coding genes, to minimize deleterious effects from loss of telomeric DNA. Heterochromatic sequences may also regulate the expression of genes located in the heterochromatic region of the chromosome or near the heterochromatin/euchromatin border.

43. In DNA hybridization experiments on six species of plants in the genus *Vicia*, DNA was isolated from each of the six species, denatured by heating, and sheared into small fragments (W. Y. Chooi. 1971. *Genetics* 68:213–230). In one experiment, DNA from each species and *E. coli* was allowed to renature. The adjoining graph shows the results of this renaturation experiment.

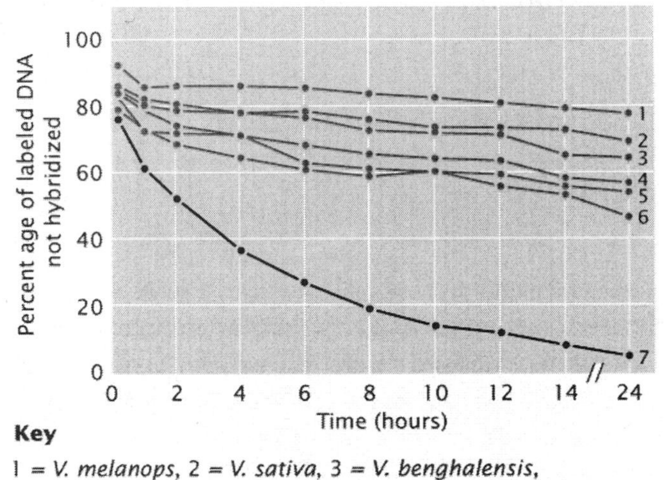

Key
1 = *V. melanops*, 2 = *V. sativa*, 3 = *V. benghalensis*,
4 = *V. atropurpurea*, 5 = *V. faba*, 6 = *V. narbonensis*, 7 = *E. coli*

a. Can you explain why the *E. coli* DNA renatures at a much faster rate than DNA from all of the *Vicia* species?
The E. coli *genome is far smaller than the genomes of the plant species, and therefore has lower complexity. Given the same concentration of DNA, the* E. coli *DNA sequences are at higher copy number than the plant DNA sequences, and therefore renature faster.*

b. Notice that, for the *Vicia* species, the rate of renaturation is much faster in the first hour and then slows down. What might cause this initial rapid renaturation and the subsequent slowdown?
The most repetitive sequences in the Vicia *genome will renature fastest. The single-copy sequences renature most slowly. The rapidly renaturing sequences are the highly repetitive sequences.*

Section 11.6

44. Marilyn Houck and Margaret Kidwell proposed that *P* elements were carried from *Drosophila willistoni* to *D. melanogaster* by mites that fed on fruit flies (M. A. Houck et al. 1991. *Science* 253:1125-1129). What evidence do you think would be required to demonstrate that *D. melanogaster* acquired *P* elements in this way? Propose a series of experiments to provide such evidence.
    This hypothesis requires not only that mites pick up P *elements when they feed on* P^+ *fruit flies, but also that they can transmit the* P *elements to a host that does not have them.*

    Mites that have infected a laboratory colony of D. willistoni *should be isolated and tested for the presence of* P *elements. If* P *elements are present in these mites, these mites should then be allowed to infect a colony of* D. melanogaster *that is free of* P *elements. After several generations the colony of* D. melanogaster *should be tested for the presence of* P *elements after they have been disinfected of the mites. One way to test for the presence of* P *elements would be to mate females from this test colony with males that are* P^+. *Fertile progeny, testifying to a lack of hybrid dysgenesis, would indicate that these females are* P^+. *These experiments would show whether mites are capable of transmitting* P *elements from one species to another.*

Chapter Twelve: DNA Replication and Recombination

COMPREHENSION QUESTIONS

Section 12.2

1. What is semiconservative replication?
In semiconservative replication, the original two strands of the double helix serve as templates for new strands of DNA. When replication is complete, two double-stranded DNA molecules will be present. Each will consist of one original template strand and one newly synthesized strand that is complementary to the template.

*2. How did Meselson and Stahl demonstrate that replication in *E. coli* takes place in a semiconservative manner?
Meselson and Stahl grew E. coli *cells in a medium containing the heavy isotope of nitrogen (^{15}N) for several generations. The ^{15}N was incorporated in the DNA of the* E. coli *cells. The* E. coli *cells were then switched to a medium containing the common form of nitrogen (^{14}N) and allowed to proceed through a few cycles of cellular generations. Samples of the bacteria were removed at each cellular generation. Using equilibrium density gradient centrifugation, Meselson and Stahl were able to distinguish DNAs that contained only ^{15}N from DNAs that contained only ^{14}N or a mixture of ^{15}N and ^{14}N because DNAs containing the ^{15}N isotope are "heavier." The more ^{15}N a DNA molecule contains, the further it will sediment during equilibrium density gradient centrifugation. DNA from cells grown in the ^{15}N medium produced only a single band at the expected position during centrifugation. After one round of replication in the ^{14}N medium, one band was present following centrifugation, but the band was located at a position intermediate to that of a DNA band containing only ^{15}N and a DNA band containing only ^{14}N. After two rounds of replication, two bands of DNA were present. One band was located at a position intermediate to that of a DNA band containing only ^{15}N and a DNA band containing only ^{14}N, while the other band was at a position expected for DNA containing only ^{14}N. These results were consistent with the predictions of semiconservative replication and incompatible with the predictions of conservative and dispersive replication.*

*3. Draw a molecule of DNA undergoing theta replication. On your drawing, identify (1) origin, (2) polarity (5′ and 3′ ends) of all template strands and newly synthesized strands, (3) leading and lagging strands, (4) Okazaki fragments, and (5) location of primers.

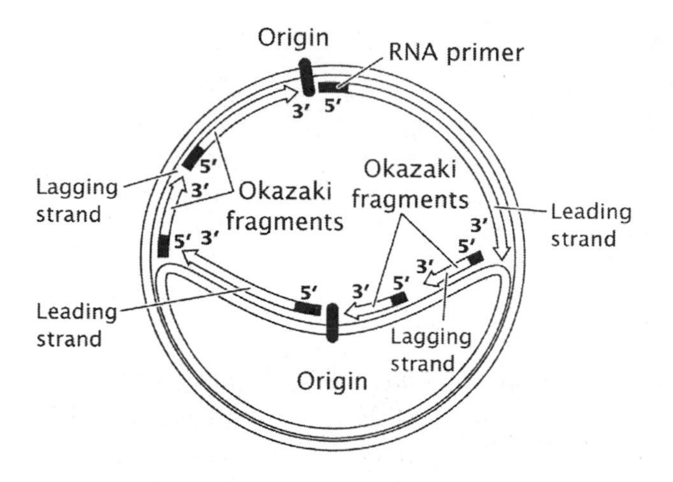

4. Draw a molecule of DNA undergoing rolling-circle replication. On your drawing, identify (1) origin, (2) polarity (5′ and 3′ ends) of all template strands and newly synthesized strands, (3) leading and lagging strands, (4) Okazaki fragments, and (5) location of primers.

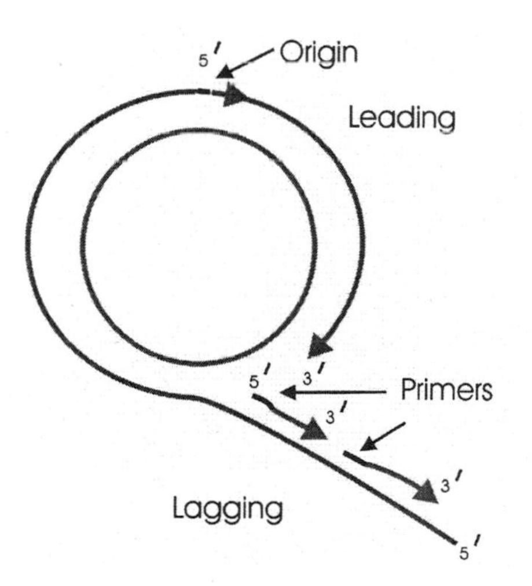

5. Draw a molecule of DNA undergoing eukaryotic linear replication. On your drawing, identify (1) origin, (2) polarity (5′ and 3′ ends) of all template strands and newly synthesized strands, (3) leading and lagging strands, (4) Okazaki fragments, and (5) location of primers.

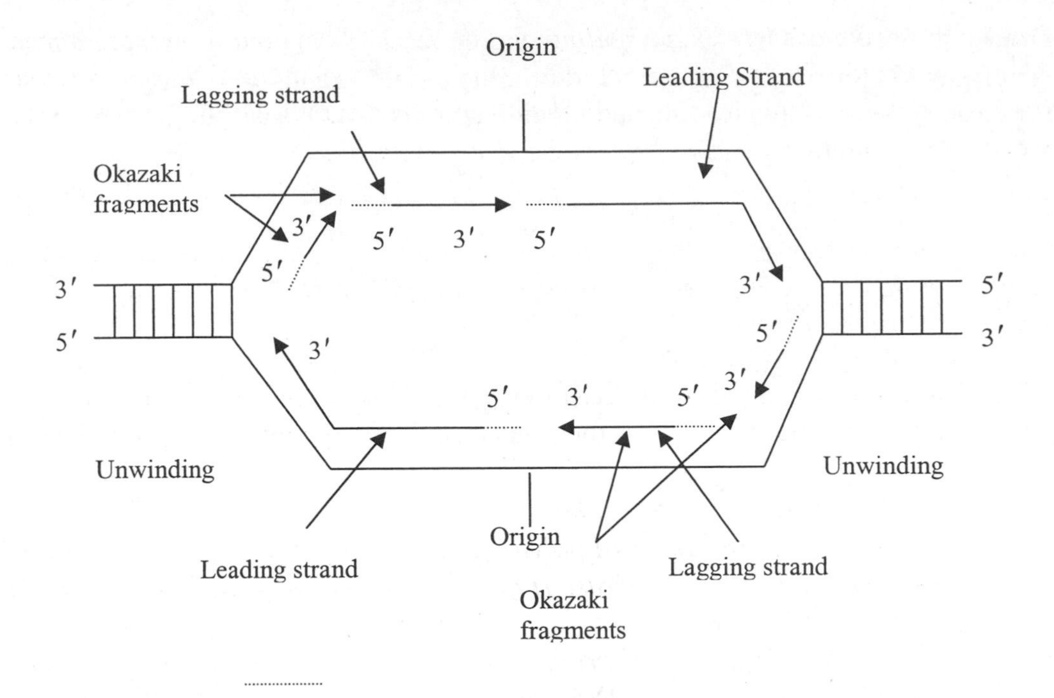

6. What are three major requirements of replication?
   (1) A single-stranded DNA template
   (2) Nucleotide substrates for synthesis of the new polynucleotide strand
   (3) Enzymes and other proteins associated with replication to assemble the nucleotide substrates into a new DNA molecule

*7. What substrates are used in the DNA synthesis reaction?
   The substrates for DNA synthesis are the four types of deoxyribonucleoside triphosphates: deoxyadenosine triphosphate, deoxyguanosine triphosphate, deoxycytosine triphosphate, and deoxythymidine triphosphate.

Section 12.3

8. List the different proteins and enzymes taking part in bacterial replication. Give the function of each in the replication process.
   DNA polymerase III is the primary replication polymerase. It elongates a new nucleotide strand from the 3′–OH of the primer.

   DNA polymerase I removes the RNA nucleotides of the primers and replaces them with DNA nucleotides.

   DNA ligase connects Okazaki fragments by sealing nicks in the sugar phosphate backbone.

   DNA primase synthesizes the RNA primers that provide the 3′–OH group needed for DNA polymerase III to initiate DNA synthesis.

   DNA helicase unwinds the double helix by breaking the hydrogen bonding between the two strands at the replication fork.

DNA gyrase reduces DNA supercoiling and torsional strain that is created ahead of the replication fork by making double-stranded breaks in the DNA and passing another segment of the helix through the break before resealing it. Gyrase is also called topoisomerase II.

Initiator proteins bind to the replication origin and unwind short regions of DNA.

Single-stranded binding protein (SSB protein) stabilizes single-stranded DNA prior to replication by binding to it, thus preventing the DNA from pairing with complementary sequences.

9. What similarities and differences exist in the enzymatic activities of DNA polymerases I, II, and III? What is the function of each type of DNA polymerase in bacterial cells?

   Each of the three DNA polymerases has a 5′ to 3′ polymerase activity. They differ in their exonuclease activities. DNA polymerase I has a 3′ to 5′ as well as a 5′ to 3′ exonuclease activity. DNA polymerase II and DNA polymerase III have only a 3′ to 5′ exonuclease activity.
   (1) DNA polymerase I carries out proofreading. It also removes and replaces the RNA primers used to initiate DNA synthesis.
   (2) DNA polymerase II functions as a DNA repair polymerase. It restarts replication after DNA damage has halted replication. It has proofreading activity.
   (3) DNA polymerase III is the primary replication enzyme and also has a proofreading function in replication.

*10. Why is primase required for replication?

   Primase is a DNA-dependent RNA polymerase. Primase synthesizes the short RNA molecules, or primers, that have a free 3′–OH to which DNA polymerase can attach deoxyribonucleotides in replication initiation. The DNA polymerases require a free 3′–OH to which they add nucleotides, and therefore they cannot initiate replication. Primase does not have this requirement.

11. What three mechanisms ensure the accuracy of replication in bacteria?
   (1) Highly accurate nucleotide selection by the DNA polymerases when pairing bases.
   (2) The proofreading function of DNA polymerase, which removes incorrectly inserted bases.
   (3) A mismatch repair apparatus that repairs mistakes after replication is complete.

12. How does replication licensing ensure that DNA is replicated only once at each origin per cell cycle?

   Only replication origins to which replication licensing factor (RPF) has bound can undergo initiation. Shortly after the completion of mitosis, RPF binds the origin during G_1 and is removed by the replication machinery during S phase.

*13. In what ways is eukaryotic replication similar to bacterial replication, and in what ways is it different?

Eukaryotic and bacterial replication of DNA replication share some basic principles:

(1) Semiconservative replication.
(2) Replication origins serve as starting points for replication.
(3) Short segments of RNA called primers provide a 3'–OH for DNA polymerases to begin synthesis of the new strands.
(4) Synthesis occurs in a 5' to 3' direction.
(5) The template strand is read in a 3' to 5' direction.
(6) Deoxyribonucleoside triphosphates are the substrates.
(7) Replication is continuous on the leading strand and discontinuous on the lagging strand.

Eukaryotic DNA replication differs from bacterial replication in that:
(1) It has multiple origins of replications per chromosome.
(2) It has several different DNA polymerases with different functions.
(3) Immediately following DNA replication, assembly of nucleosomes takes place.

14. What is the end-of-chromosome problem for replication? Why, in the absence of telomerase, do the ends of chromosomes get progressively shorter each time the DNA is replicated?

For DNA polymerases to work, they need the presence of a 3' OH group to which to add a nucleotide. At the ends of the chromosomes when the RNA primer is removed, there is no adjacent 3' OH group to which to add a nucleotide, thus no nucleotides are added leaving a gap at the end of the chromosome. Telomerase can extend the single stranded protruding end by pairing with the overhanging 3' end of the DNA and adding a repeated sequence of nucleotides. In the absence of telomerase, DNA polymerase will be unable to add nucleotides to the end of the strand. After multiple rounds of replication without a functional telomerase the chromosome will become progressively shorter.

15. Outline in words and pictures how telomeres at the end of eukaryotic chromosomes are replicated.

Telomeres are replicated by the enzyme telomerase. Telomerase, a ribonucleoprotein, consists of protein and an RNA molecule that is complementary to the 3' end of the DNA of a eukaryotic chromosome. The RNA molecule also serves as a template for the addition of nucleotides to the 3' end. After the 3' end has been extended, the 5' end of the DNA can be extended as well, possibly by lagging strand synthesis of a DNA polymerase using the extended 3' end as a template.

DNA replication of the linear eukaryotic chromosomes generates a 3' overhang. Part of the RNA sequence within telomerase is complementary to the overhang.

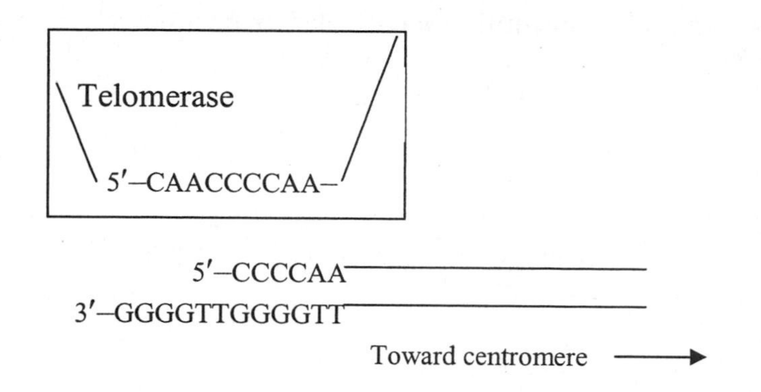

Telomerase RNA sequence pairs with the 3′ overhang and serves as a template for the addition of DNA nucleotides to the 3′ end of the DNA molecule, which serves to extend the 3′ end of the chromosome.

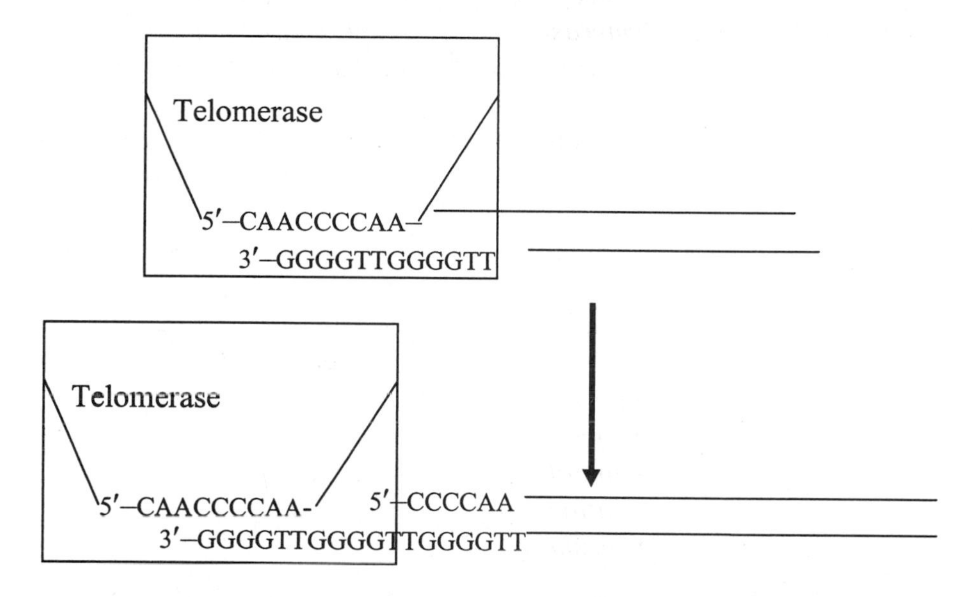

Additional nucleotides to the 5′ end are added by DNA synthesis using a DNA polymerase with priming by primase.

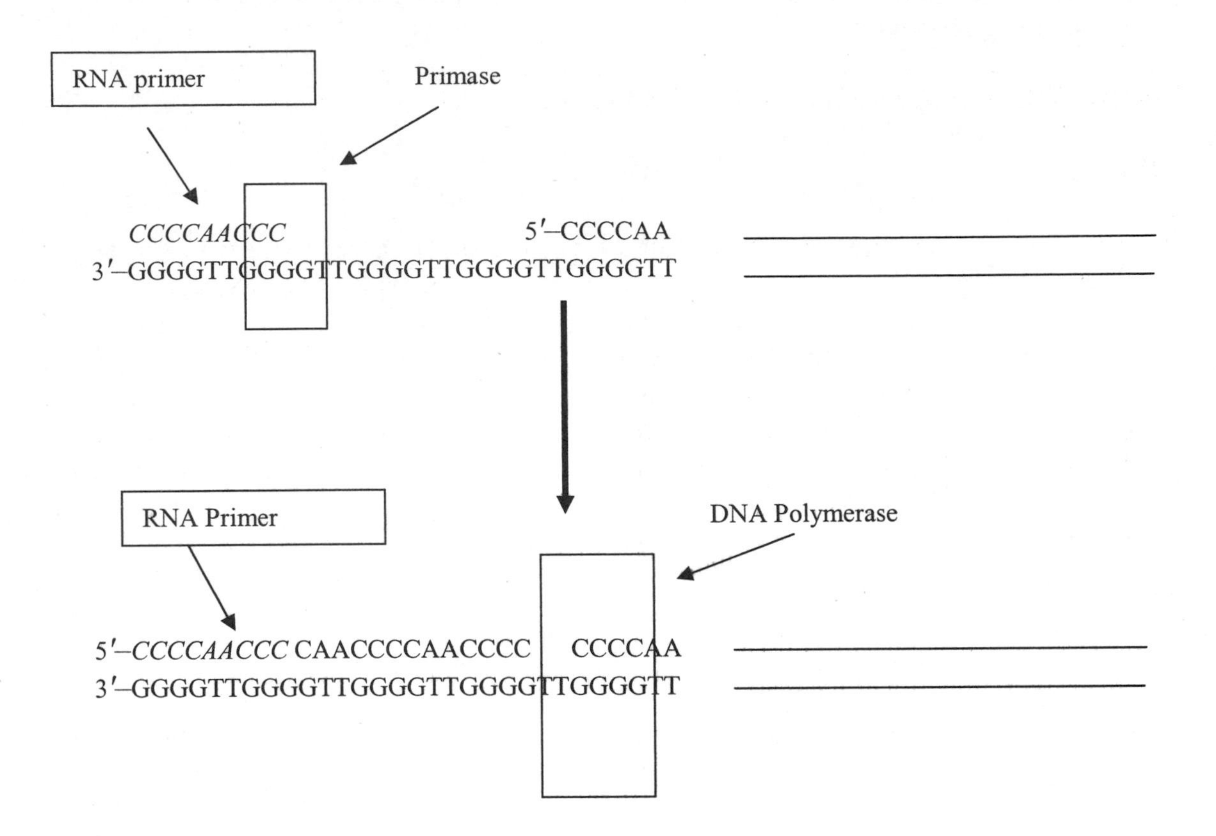

Section 12.4

*16. What are some of the enzymes taking part in recombination in *E. coli,* and what roles do they play?

(1) RecBCD protein unwinds double-stranded DNA and can cleave polynucleotide strands.
(2) RecA protein allows a single strand to invade a double-stranded DNA.
(3) RuvA and RuvB proteins promote branch migration during homologous recombination.
(4) RuvC protein is resolvase, a protein that resolves the Holliday structure by cleavage of the DNA.
(5) DNA ligase repairs nicks or cuts in the DNA generated during recombination.

17. What is gene conversion? How does it arise?

Gene conversion results from nonreciprocal genetic exchange between chromosomes and can result in abnormal ratios of gametes. Gene conversion occurs during the heteroduplex formation of recombination when a single-stranded DNA molecule of one chromosome pairs with a single-stranded molecule of another chromosome. If the two chromosomes contain different alleles at the duplex region, then a base-pairing mismatch will occur. Mismatch repair enzymes may repair the mismatch by excising nucleotides from one strand and using the complementary strand as a template essentially converting one allele into the other or in other words gene conversion.

APPLICATION QUESTIONS AND PROBLEMS

Section 12.2

*18. Suppose a future scientist explores a distant planet and discovers a novel form of double-stranded nucleic acid. When this nucleic acid is exposed to DNA polymerases from *E. coli,* replication takes place continuously on both strands. What conclusion can you make about the structure of this novel nucleic acid?
Each strand of the novel double-stranded nucleic acid must be oriented parallel to the other, as opposed to the antiparallel nature of earthly double-stranded DNA. Replication by E. coli *DNA polymerases can proceed continuously only in a 5′ to 3′ direction, which requires the template to be read in a 3′ to 5′ direction. If replication is continuous on both strands, the two strands must have the same direction and be parallel.*

*19. Phosphorous is required to synthesize the deoxyribonucleoside triphosphates used in DNA replication. A geneticist grows some *E. coli* in a medium containing nonradioactive phosphorous for many generations. A sample of the bacteria is then transferred to a medium that contains a radioactive isotope of phosphorus (^{32}P). Samples of the bacteria are removed immediately after the transfer and after one and two rounds of replication. What will be the distribution of radioactivity in the DNA of the bacteria in each sample? Will radioactivity be detected in neither, one, or both strands of the DNA?
In the initial sample removed immediately after transfer, no ^{32}P should be incorporated into the DNA because replication in the medium containing ^{32}P has not yet occurred. After one round of replication in the ^{32}P containing medium, one strand of each newly synthesized DNA molecule will contain ^{32}P, while the other strand will contain only nonradioactive phosphorous. After two rounds of replication in the ^{32}P containing medium, 50% of the DNA molecules will have ^{32}P in both strands, while the remaining 50% will contain ^{32}P in one strand and nonradioactive phosphorous in the other strand.

20. A line of mouse cells is grown for many generations in a medium with ^{15}N. Cells in G_1 are then switched to a new medium that contains ^{14}N. Draw a pair of homologous chromosomes from these cells at the following stages, showing the two strands of DNA molecules found in the chromosomes. Use different colors to represent strands with ^{14}N and ^{15}N.

a. Cells in G_1, before switching to medium with ^{14}N

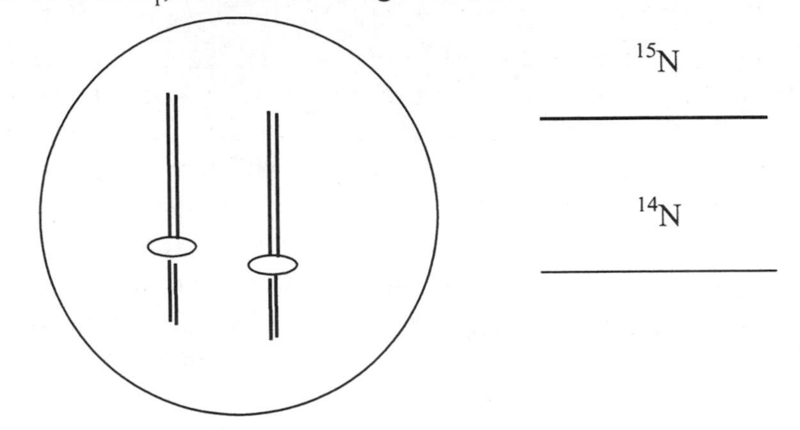

b. Cells in G_2, after switching to medium with ^{14}N

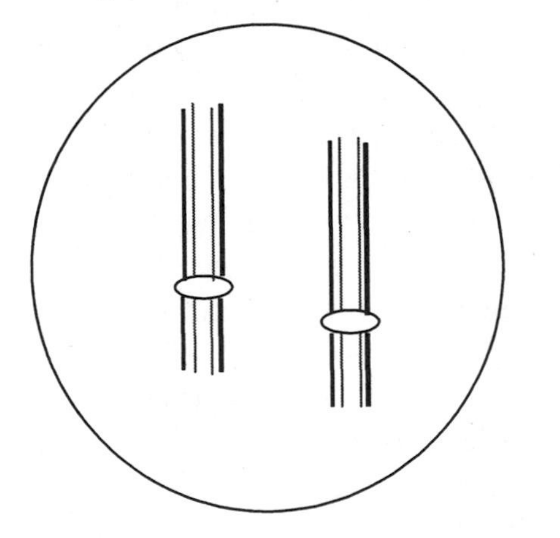

c. Cells in anaphase of mitosis, after switching to medium with ^{14}N

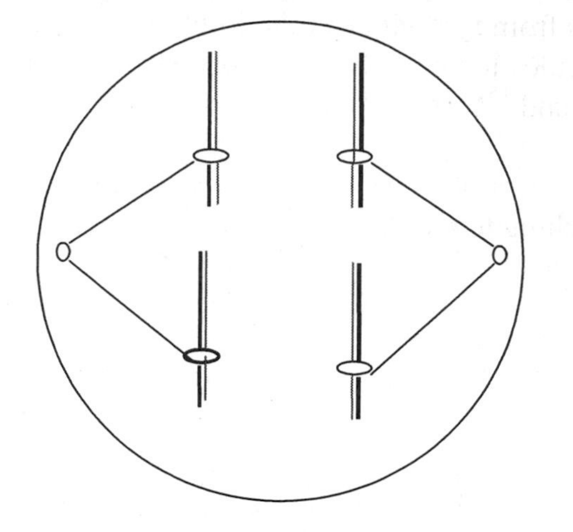

d. Cells in metaphase I of meiosis, after switching to medium with ^{14}N

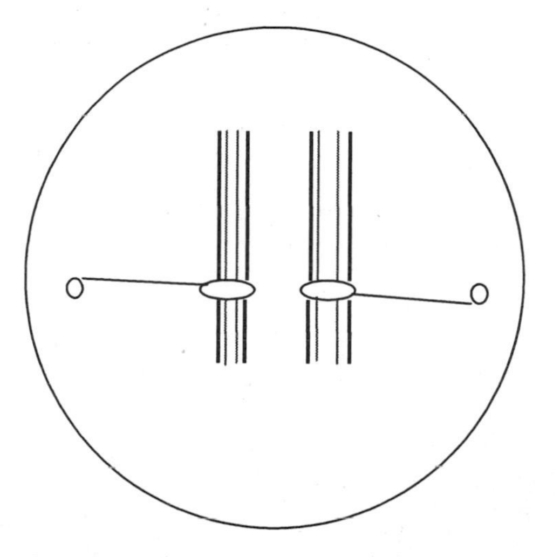

e. Cells in anaphase II of meiosis, after switching to medium with ^{14}N

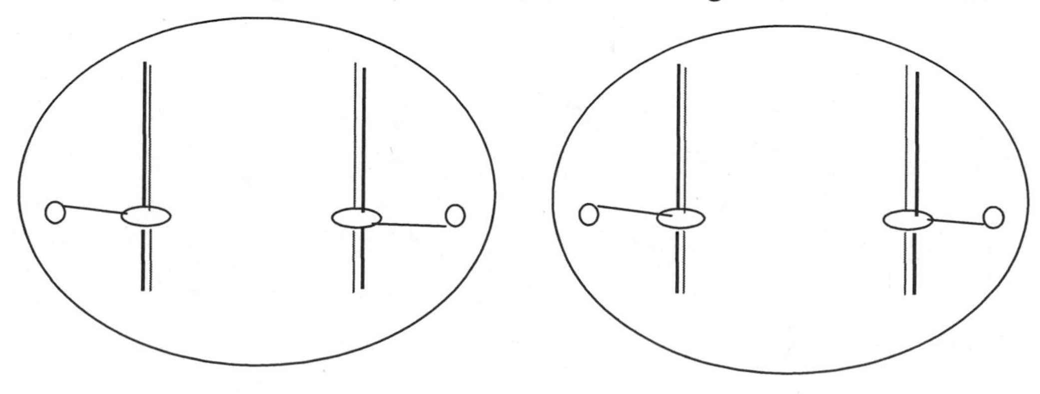

*21. A circular molecule of DNA contains 1 million base pairs. If DNA synthesis at a replication fork occurs at a rate of 100,000 nucleotides per minute, how long will theta replication require to completely replicate the molecule, assuming that theta replication is bidirectional? How long will replication of this circular chromosome take by rolling-circle replication? Ignore replication of the displaced strand in rolling-circle replication.
In bidirectional replication there are two replication forks, each proceeding at a rate of 100,000 nucleotides per minute. Therefore, it would require 5 minutes for the circular DNA molecule to be replicated by bidirectional replication because each fork could synthesize 500,000 nucleotides (5 minutes × 100,000 nucleotides per minute) within the time period. Because rolling-circle replication is unidirectional and thus has only one replication fork, 10 minutes will be required to replicate the entire circular molecule.

22. A bacterium synthesizes DNA at each replication fork at a rate of 1000 nucleotides per second. If this bacterium completely replicates its circular chromosome by theta replication in 30 minutes, how many base pairs of DNA will its chromosome contain?
Each replication complex is synthesizing DNA at each fork at a rate of 1000 nucleotides per second. So for each second, 2000 nucleotides are being synthesized by both forks (1000 nucleotides / second × 2 forks = 2000 nucleotides / second) or 120,000 nucleotides per minute. If the bacterium requires 30 minutes to replicate its chromosome, then the size of the chromosome is 3,600,00 nucleotides (120,000 nucleotides / minute × 30 minutes = 3,600,000).

Section 12.3

*23. The following diagram represents a DNA molecule that is undergoing replication. Draw in the strands of newly synthesized DNA and identify the following:
a. Polarity of newly synthesized strands
b. Leading and lagging strands
c. Okazaki fragments
d. RNA primers

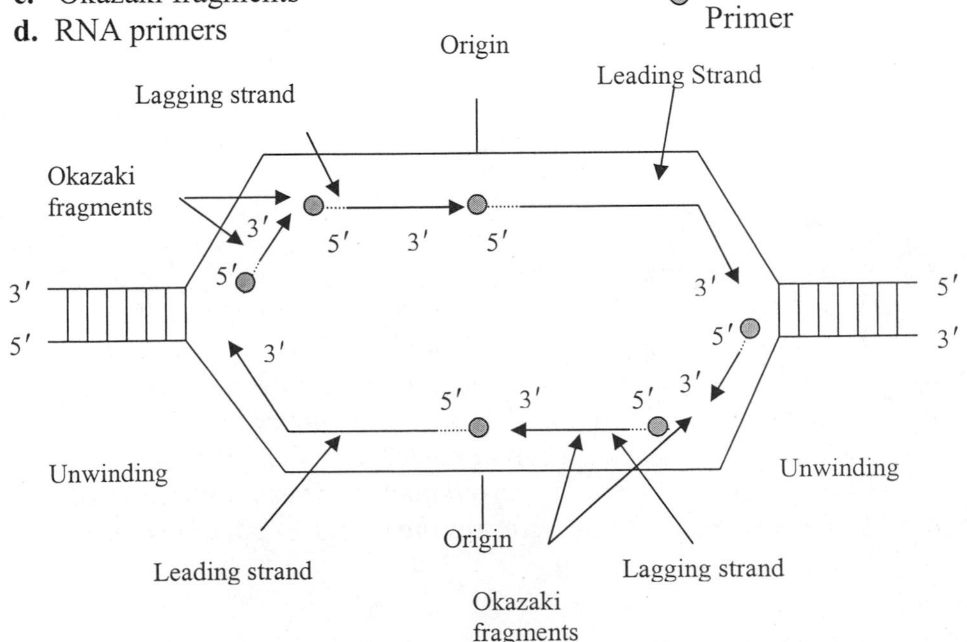

*24. What would be the effect on DNA replication of mutations that destroyed each of the following activities in DNA polymerase I?

a. 3′ → 5′ exonuclease activity
The 3′ → 5′ exonuclease activity is important for proofreading newly synthesized DNA. If the activity is nonfunctional, then the fidelity of replication by DNA polymerase I will decrease, resulting in more misincorporated bases in the DNA.

b. 5′ → 3′ exonuclease activity
Loss of the 5′ → 3′ exonuclease activity would result in the RNA primers used to initiate replication not being removed by DNA polymerase I.

c. 5′ → 3′ polymerase activity
DNA polymerase I would be unable to synthesize new DNA strands if the 5′ → 3′ polymerase activity was destroyed. RNA primers could be removed by DNA polymerase I using the 5′ → 3′ exonuclease activity, but new DNA sequences could not be added in their place by DNA polymerase I.

25. How would DNA replication be affected in a cell that is lacking topoisomerase?
Topoisomerase II or gyrase reduces the positive supercoiling or torsional strain that develops ahead of the replication fork due to the unwinding of the double helix. If the topoisomerase activity was lacking, then the torsional strain would continue to increase, making it more difficult to unwind the double helix. Ultimately, the increasing strain would lead to an inhibition of the replication fork movement.

26. If the gene for primase were mutated so that no functional primase was produced, what would be effect on theta replication? On rolling-circle replication?
Primase is required for replication initiation in theta form replication. If primase is nonfunctional then replication initiation would not take place resulting in no replication occurring. Rolling circle replication does not require primase. A single-stranded break within one strand provides a 3′ OH group to which nucleotides can be added so rolling circle replication could occur without a functional primase.

27. DNA polymerases are not able to prime replication, yet primase and other RNA polymerases can. Some geneticists have speculated that the inability of DNA polymerase to prime replication is due to its proofreading function. This hypothesis argues that proofreading is essential for faithful transmission of genetic information and that, because DNA polymerases have evolved the ability to proofread, they cannot prime DNA synthesis. Explain why proofreading and priming functions in the same enzyme might be incompatible?
If DNA polymerase inserts an incorrect nucleotide, the 3′ OH group of the mismatched nucleotide is improperly positioned within the active site for the addition of the next nucleotide, thus stalling the polymerase. The 3′ to 5′ exonuclease function of the DNA polymerase then removes the mismatch base and the polymerase can now insert the correct nucleotide. Primase does not add a nucleotide to the 3′ end of a previous paired nucleotide as it initiates replication. Proofreading by DNA polymerases requires the incorrect position of the 3′ OH

within the active site. Without the requirement for the correct positioning of the 3′ OH, then proofreading is likely not to occur.

*28. Marina Melixetian and her colleagues suppressed the expression of Geminin protein in human cells by treating the cells with small interfering RNAs (siRNAs) complementary to Geminin messenger RNA (M. Melixetian et al. 2004. *Journal of Cell Biology* 165:473–482). (Small interfering RNAs cleave mRNAs with which they are complementary, so there is no translation of the mRNA; see pp. 386–387 in Chapter 14) Forty-eight hours after treatment with siRNA, the Geminin-depleted cells were enlarged and contained a single giant nucleus. Analysis of DNA content showed that many of these Geminin-depleted cells were 4N or greater. Explain these results.
The Geminin protein is an inhibitor of replication licensing by the minichromosome maintenance complex (MCM). Licensing at the origins is necessary for the functioning of initiator proteins at the origins of replication. If MCM is allowed to bind DNA sequences at the origin, then replication can initiate. Typically, Geminin protein blocks MCM from binding the origin until after mitosis is completed when the Geminin protein is degraded. If expression of Geminin is prevented by addition of the siRNAs, then no Geminin protein will be produced, thus allowing MCM to bind at the origins without regulation. The continued and repeated licensing of the replication origin by MCM could lead to an increase in chromosome number within the nuclei. These polyploid nuclei were likely blocked from dividing through mitosis by cell cycle checkpoints generating a single nucleus with the multiple sets of chromosomes.

29. What results would be expected in the experiment outlined in Figure 12.17 if, during replication, all the original histone proteins remained on one strand of the DNA and new histones attached to the other strand?
Two distincts bands should be seen within the tube after centrifugation. The original histone proteins contained amino acids labeled with a heavy isotope while the newly synthesized histones contain amino acids labeled with a light isotope. If the original histones remained on one strand then we would expect after centrifugation to see that the original histones sedimented nearer the bottom of the tube in a distinct band. The newly synthesized histones would be lighter and should be in a distinct band higher in the tube.

30. A number of scientists who study ways to treat cancer have become interested in telomerase. Why would they be interested in telomerase? How might cancer drug therapies that target telomerase work?
Telomerase is an enzyme that functions in cells that undergo continuous cell division and may play a role in the lack of cellular aging in these cells. Cells that lack telomerase exhibit progressive shortening of the chromosomal ends or telomeres. This shortening leads to unstable chromosomes and ultimately to programmed cell death. Many tumor cells also express telomerase, which may assist these cells in becoming immortal. If the telomerase activity in these cancer cells could be inhibited, then cell division might be halted in the cancer cells thus controlling cancer cell growth. Chemically modified antisense RNAs or DNA oligonucleotides complementary to the telomerase RNA sequence might block the

telomerase activity by base pairing with the telomerase RNA, making it unavailable as a template. A second strategy would be target the DNA synthesis activity of the telomerase protein preventing the telomere DNA from being synthesized.

31. The enzyme telomerase is part protein and part RNA. What would be the most likely effect of a large deletion in the gene that encodes the RNA part of telomerase? How would the function of telomerase be affected?
The RNA portion of telomerase is needed to provide the template for synthesizing complementary DNA telomere sequences at the ends of the chromosomes. A large deletion would affect how the telomeres are synthesized at the ends of the chromosomes by telomerase and could potentially element telomere synthesis.

32. Dyskeratosis congenita (DKC) is a rare genetic disorder characterized by abnormal fingernails and skin pigmentation, the formation of white patches on the tongue and cheek, and progressive failure of the bone marrow. An autosomal dominant form of DKC results from mutations in the gene that encodes the RNA component of telomerase. Tom Vulliamy and his colleagues examined 15 families with autosomal dominant DKC (T. Vulliamy et al. 2004. *Nature Genetics* 36:447–449). They observed that the median age of onset of DKC in parents was 37 years, whereas the median age of onset in the children of affected parents was 14.5 years. Thus, DKC in these families arose at progressively younger ages in successive generations, a phenomenon known as anticipation (see p. 122 in Chapter 5). The researchers measured telomere length of members of these families; the measurements are given in the adjoining table. Telomere length normally shortens with age, and so telomere length was adjusted for age. Note that the age-adjusted telomere length of all members of these families is negative, indicating that their telomeres are shorter than normal. For age-adjusted telomere length, the more negative the number, the shorter the telomere.

Age-Adjusted Telomere Length in Children and Their Parents in Families with DKC

Parent telomere length	Child telomere length
−4.7	−6.1
	−6.6
	−6.0
−3.9	−0.6
−1.4	−2.2
−5.2	−5.4
−2.2	−3.6
−4.4	−2.0
−4.3	−6.8
−5.0	−3.8
−5.3	−6.4
−0.6	−2.5
−1.3	−5.1
	−3.9
−4.2	−5.9

a. How do the telomere lengths of parents compare with telomere length of their children?
In general, the telomere lengths of the parents are longer than the telomere lengths of the children (with few exceptions).

b. Explain why the telomeres of people with DKC are shorter than normal and why DKC arises at an earlier age in subsequent generations.
Individuals with DKC possess an allele that codes for a mutant form of the RNA component of telomerase. The RNA component serves as the template for telomere synthesis and is necessary for the synthesis of telomeres of appropriate lengths. The defective RNA component results in telomeres that are shortened. For individuals in subsequent generations who inherit the DKC, it is likely that these individuals also inherited shorter telomeres from the parent possessing the DKC allele as evidenced from the above data. Once telomeres become too short, cells normally undergo cell death (apoptosis), which likely leads to the symptoms of DKC. In subsequent generations, telomeres may reach that critical length earlier leading to an earlier onset of DKC.

CHALLENGE QUESTIONS

Section 12.3

33. Conditional mutations express their mutant phenotype only under certain conditions (the restrictive conditions) and express the normal phenotype under other conditions (the permissive conditions). One type of conditional mutation is a temperature-sensitive mutation, which expresses the mutant phenotype only at certain temperatures.

    Strains of *E. coli* have been isolated that contain temperature-sensitive mutations in the genes encoding different components of the replication machinery. In each of these strains, the protein produced by the mutated gene is nonfunctional under the restrictive conditions. These strains are grown under permissive conditions and then abruptly switched to the restrictive condition. After one round of replication under the restrictive condition, the DNA from each strain is isolated and analyzed. What characteristics would you expect to see in the DNA isolated from each strain with a temperature-sensitive mutation in its gene that encodes the following?

    Temperature-sensitive mutation in gene encoding:

    a. DNA ligase
    DNA ligase is required to seal the nicks left after DNA polymerase I removes the RNA primers used to begin DNA synthesis. If DNA ligase is not functioning, multiple nicks in the lagging strand will be expected and the Okazaki fragments will not be joined. However, DNA replication will take place.

    b. DNA polymerase I
    RNA primers are removed by DNA polymerase I. After one round of replication, the DNA molecule would still contain the RNA nucleotide primers since DNA polymerase I is not functioning. However, replication of the DNA will take place.

c. DNA polymerase III
No replication would be expected. DNA polymerase III is the primary replication enzyme. If it is not functioning, then neither lagging- nor leading-strand DNA synthesis will take place.

d. Primase
No replication would be expected. Primase synthesizes the short RNA molecules that act as primers for DNA synthesis. If the RNA primers are not synthesized, then no free 3′–OH will be available for DNA polymerase III to attach DNA nucleotides. Thus, DNA synthesis will not take place.

e. Initiator protein
Initiator proteins bind to the oriC *and unwind the DNA, allowing for the binding of DNA helicase and single-stranded binding proteins to the DNA. If these proteins are unable to bind, then DNA replication initiation will not occur, so DNA synthesis will not take place.*

34. DNA topoisomerases play important roles in DNA replication and supercoiling (see Chapter 11). These enzymes are also the targets for certain anticancer drugs. Eric Nelson and his colleagues studied m-AMSA, one of the anticancer compounds that acts on topisomerase enzymes (E. M. Nelson, K. M. Tewey, and L. F. Liu. 1984. *Proceedings of the National Academy of Sciences* 81:1361–1365). They found that m-AMSA stabilizes an intermediate produced in the course of the topoisomerase's action. The intermediate consisted of the topoisomerase bound to the broken ends of the DNA. Breaks in DNA that are produced by anticancer compounds such as m-AMSA inhibit the replication of the cellular DNA and thus stop cancer cells from proliferating. Propose a mechanism for how m-AMSA and other anticancer agents that target topoisomerase enzymes taking part in replication might lead to DNA breaks and chromosome rearrangments.
Compounds such as m-AMSA prevent topoisomerases from completing their functions. Type II topoisomerases (such as gyrase) function to reduce torsional strain generated ahead of the replication fork. The enzymes reduce the strain by generating a double-stranded break in a segment of the DNA and passing another segment of the DNA through the break, which is then followed by the resealing of the broken ends. The anticancer agent m-AMSA and related substances inhibit the function of topoisomerases by stabilizing the topoisomerase on the 5′ ends of the double-stranded break. This prevents the other segment from passing through and prevents resealing of the break. Essentially the compounds generate double-stranded breaks in the DNA due to the action of topoisomerase. Breaks in the DNA molecule may stimulate recombination and repair enzymes leading to chromosomal rearrangements.

*35. Regulation of replication is essential to genomic stability, and normally the DNA is replicated just once every cell cycle (during S phase). Normal cells produce protein A, which increases in concentration during S phase. In cells that have a mutated copy of the gene for protein A, however, replication occurs continuously throughout the cell cycle, with the result that cells may have 50 times the normal amount of DNA. Protein B is normally present in G_1 but disappears from the cell nucleus during S phase. In cells with a mutated copy of the gene for protein A, the levels of protein B fail to decrease during S phase and, instead, remain high throughout the cell cycle. When the gene for protein B is mutated, no replication occurs.

Propose a mechanism for how protein A and protein B might normally regulate replication so that each cell gets the proper amount of DNA. Explain how mutation of these genes produces the effects described above.
Protein B may be needed for replication to successfully initiate at replication origins. Protein B is present at the beginning of S phase but disappears later in the stage when initiation (but not replication) is complete. Protein A may be responsible for removing or inactivating protein B. As levels of protein A increase, the levels of protein B decrease preventing extra initiation events. When protein A is mutated, it can no longer inactivate protein B, thus successive rounds of replication can begin due to the high levels of protein B. When protein B is mutated, it cannot assist initiation and replication ceases.

1-8, 10-13, 15-21, 24, 26ac 27, 29, 31-33

Chapter Thirteen: Transcription

COMPREHENSION QUESTIONS

Section 13.1

*1. Draw an RNA nucleotide and a DNA nucleotide, highlighting the differences. How is the structure of RNA similar to that of DNA? How is it different?
RNA and DNA are polymers of nucleotides that are held together by phosphodiester bonds. An RNA nucleotide contains a ribose sugar, whereas a DNA nucleotide contains a deoxyribose sugar. Also, the pyrimidine base uracil is found in RNA but thymine is not. DNA, however, contains thymine but not uracil. Finally, an RNA polynucleotide is typically single-stranded even though RNA molecules can pair with other complementary sequences. DNA molecules are almost always double-stranded.

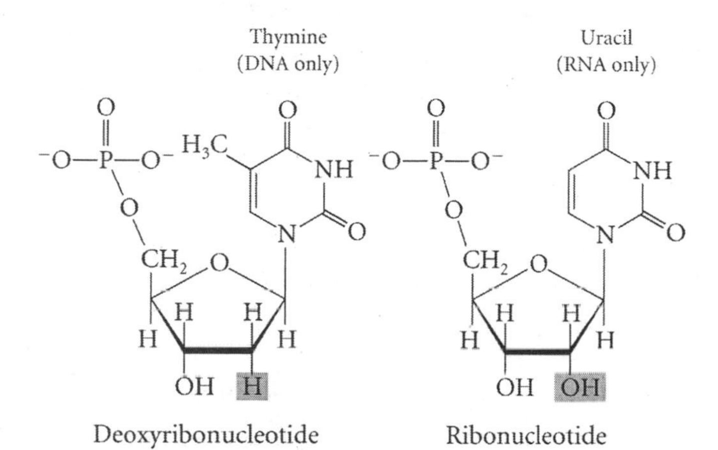

2. What are the major classes of cellular RNA? Where would you expect to find each class of RNA within eukaryotic cells?
Cellular RNA molecules are made up of six classes:
(1) Ribosomal RNA, or rRNA, is found in the cytoplasm.
(2) Transfer RNA, or tRNA, is found in the cytoplasm.
(3) Messenger RNA is found in the cytoplasm (however, pre-mRNA is found only in the nucleus).
(4) Small nuclear RNA, or snRNA, is found in the nucleus as part of riboproteins called snrps.
(5) Small nucleolar RNA, snoRNA, is found in the nucleus.
(6) Small cytoplasmic RNA, or scRNA, is found in the cytoplasm.

3. Why is DNA more stable than RNA?
The presence of the free 2′OH in the ribose sugar makes RNA more susceptible to degradation under alkaline conditions. DNA molecules contain the sugar deoxyribose and lack the 2′OH found in ribose sugars, so DNA is more stable.

Section 13.2

*4. What parts of DNA make up a transcription unit? Draw and label a typical transcription unit in a bacterial cell.

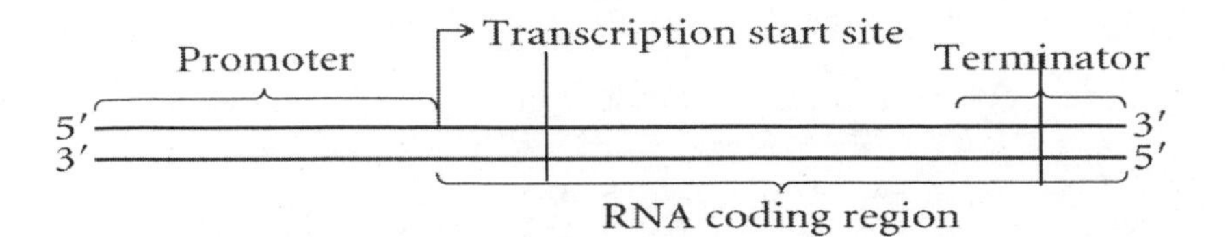

5. What is the substrate for RNA synthesis? How is this substrate modified and joined together to produce an RNA molecule?
Four ribonucleoside triphosphates serve as the substrate for RNA synthesis: adenosine triphosphate, guanosine triphosphate, cytosine triphosphate, and uridine monophosphate. The enzyme RNA polymerase uses a DNA polynucleotide strand as a template to synthesize a complementary RNA polynucleotide strand. The nucleotides are added to the RNA polynucleotide strand, one at time, at the 3′–OH of the RNA molecule. As each nucleoside triphosphate is added to the growing polynucleotide chain, two phosphates are removed from the 5′ end of the nucleotide. The remaining phosphate is linked to the 3′–OH of the RNA molecule to form the phosphodiester bond.

6. Describe the structure of bacterial RNA polymerase.
Bacterial RNA polymerase consists of several polypeptides. The RNA polymerase core enzyme is composed of four polypeptide subunits: two copies of the alpha subunit, the beta subunit, and the beta prime subunit. The addition of a sigma factor to the core enzyme forms the RNA polymerase holoenzyme.

*7. Give the names of the three RNA polymerases found in eukaryotic cells and the types of RNA they transcribe.
(1) RNA polymerase I transcribes rRNA.
(2) RNA polymerase II transcribes pre-mRNA and some snRNAs.
(3) RNA polymerase III transcribes small RNA molecules such as 5S rRNA, tRNAs, and some small nuclear RNAs.

Section 13.3

8. What are the three basic stages of transcription? Describe what happens at each stage.
(1) Initiation: Transcription proteins assemble at the promoter to form the basal transcription apparatus and begin synthesis of RNA.
(2) Elongation: RNA polymerase moves along the DNA template in a 3′ to 5′ direction unwinding the DNA and synthesizing RNA in a 5′ to 3′ direction.
(3) Termination: Synthesis of RNA is terminated, and the RNA molecule separates from the DNA template.

*9. Draw and label a typical bacterial promoter. Include any common consensus sequences.

```
                                              Transcription start site
                                                       ┌──►
5′—TTGACA---------------------TATAAT---┼---3′
3′—AACTGT---------------------ATATTA-------5′
     −35                        −10
    Region                     Region
```

The typical bacterial promoter consists of the −35 and −10 consensus sequences. An upstream element rich in AT sequences is found only in some bacterial promoters and is located upstream of the −35 consensus sequence

10. What are the two basic types of terminators found in bacterial cells? Describe the structure of each.
The two basic types of terminators in bacterial cells are rho-independent and rho-dependent terminators. Rho-independent terminators consist of inverted repeats that can form a hairpin structure. Immediately following the inverted repeats is a string of six adenine nucleotides. Rho-dependent terminators require the interaction of the protein rho with RNA polymerase. Two features are typical for rho-dependent termination: (1) variable DNA sequences that cause RNA polymerase to pause during transcription and (2) upstream from variable region lies DNA sequence that encodes RNA devoid of secondary structure, which also serves as the rho binding site.

Section 13.4

11. How is the process of transcription in eukaryotic cells different from that in bacterial cells?
Eukaryotic transcription requires the action of three RNA polymerases. Each type of polymerase recognizes and transcribes from different types of promoters. Binding to the promoter and initiation from the promoter requires the action of many protein factors; different promoters require different sets of protein factors. The RNA molecule produced by transcription in eukaryotic cells usually requires extensive processing, such as the addition of a 5′ cap, a 3′ poly(A) tail, and the removal of introns prior to becoming functional. Bacterial promoters tend to be more uniform in composition, and only one RNA polymerase does transcription. Bacterial RNAs are typically functional once transcription has taken place.

12. Compare the roles of general transcription factors and transcriptional activator proteins.
General transcription factors form the basal transcription apparatus together with RNA polymerase and are needed to initiate minimal levels of transcription. Transcriptional activator proteins bring about higher levels of transcription by stimulating the assembly of the basal transcriptional apparatus at the start site.

*13. Compare and contrast transcription and replication. How are these processes similar and how are they different?

Common characteristics of transcription and replication:

(1) Utilize a DNA template.

(2) Synthesize molecules in a 5' to 3' direction.

(3) Synthesize molecules that are antiparallel and complementary to the template.

(4) Use nucleotide triphosphates as substrates.

(5) Involve complexes of proteins and enzymes necessary for catalysis.

Unique characteristics of transcription:

(1) Unidirectional synthesis of only a single strand of nucleic acid.

(2) Initiation does not require a primer.

(3) Subject to numerous regulatory mechanisms.

(4) Each gene is transcribed separately.

Unique characteristics of replication:

(1) Bidirectional synthesis of two strands of nucleic acid.

(2) Initiates from replication origins.

14. How are the processes of transcription in archaeans and eukaryotes different? How are they similar?

A comparison of transcription between eukaryotes and archaea shows that transcription in eukaryotes shares more similarities with archaeal transcription than they do with transcription in bacteria. Organisms in domain archaea use a single RNA polymerase for transcription while eukaryotes have three RNA polymerases for transcribing nuclear genes. Archaea lack a nucleus so transcription occurs within the cytoplasm. The RNA polymerase from archaea is similar to RNA polymerases of eukaryotes. Archaea possess a TATA-binding protein, which is also found in eukaryotes and is critical for all three nuclear RNA polymerases. The TATA-binding protein in archaea binds the TATA box with the assistance of TFIIB, which is found in eukaryotes as well.

APPLICATION QUESTIONS AND PROBLEMS

Section 13.1

15. An RNA molecule has the following percentages of bases: A = 23%, U = 42%, C = 21%, G = 14%.

a. Is this RNA single-stranded or double-stranded? How can you tell?

The RNA molecule is likely to be single-stranded. If the molecule was double-stranded, we would expect nearly equal percentages of adenine and uracil, as well as equal percentages of guanine and cytosine. In this RNA molecule, the percentages of these potential base pairs are not equal, so the molecule is single-stranded.

b. What would be the percentages of bases in the template strand of the DNA that contains the gene for this RNA?

Because the DNA template strand is complementary to the RNA molecule, we would expect equal percentages for bases in the DNA complementary to the

RNA bases. Therefore, in the DNA we would expect A = 42%, T = 23%, C = 14%, and G = 21%.

Section 13.2

*16. The following diagram represents DNA that is part of the RNA-coding sequence of a transcription unit. The bottom strand is the template strand. Give the sequence found on the RNA molecule transcribed from this DNA and label the 5′ and 3′ ends of the RNA.

5′–A T A G G C G A T G C C A–3′
3′–T A T C C G C T A C G G T–5′ ← template strand

The RNA molecule would be complementary to the template strand, contain uracil, and be synthesized in an antiparallel fashion. The sequence would be:
5′–A U A G G C G A U G C C A–3′.

The RNA strand contains the same sequence as the nontemplate DNA strand except that the RNA strand contains uracil in place of thymine.

17. The following sequence of nucleotides is found in a single-stranded DNA template:
A T T G C C A G A T C A T C C C A A T A G A T
Assume that RNA polymerase proceeds along this template from left to right.

a. Which end of the DNA template is 5′ and which end is 3′?
RNA is synthesized in a 5′ to 3′ direction by RNA polymerase, which reads the DNA template in a 3′ to 5′ direction. So, if the polymerase is moving from left to right on the template then the 3′ end must be on the left and the 5′ end on the right.

3′–A T T G C C A G A T C A T C C C A A T A G A T–5′

b. Give the sequence and label the 5′ and 3′ ends of the RNA copied from this template.

5′–U A A C G G U C U A G U A G G G U U A U C U A–3′

18. RNA polymerases carry out transcription at a much slower rate than DNA polymerases carry out replication. Why is speed more important in replication than in transcription?
DNA polymerases are required to replicate much larger regions of DNA, such as entire chromosomes. Speed is essential to complete the replication process in a timely manner. RNA polymerases typically transcribe only small areas of the chromosomes. The speed required for replication by DNA polymerases is not needed by the RNA polymerases to transcribe these smaller regions.

Section 13.3

19. Write the consensus sequence for the following set of nucleotide sequences:

    A G G A G T T
    A G C T A T T
    T G C A A T A
    A C G A A A A
    T C C T A A T
    T G C A A T T

    The consensus sequence is identified by determining which nucleotide is used most frequently at each position. For the two nucleotides that occur at an equal frequency at the first position, both are listed at that position in the sequence and identified by a slash mark:
    T/A G C A A T T

*20. List at least five properties that DNA polymerases and RNA polymerases have in common. List at least three differences.
Similarities: (1) Both use DNA templates, (2) DNA templates are read in the 3′ to 5′ direction, (3) the complementary strand is synthesized in a 5′ to 3′ direction that is antiparallel to the template, (4) both use triphosphates as substrates, and (5) their actions are enhanced by accessory proteins.

Differences: (1) RNA polymerases use ribonucleoside triphosphates as substrates, whereas DNA polymerases use deoxyribonucleoside triphophates; (2) DNA polymerases require a primer that provides an available 3′–OH group where synthesis begins, whereas RNA polymerases do not require primers to begin synthesis; and (3) RNA polymerases synthesize a copy off only one of the DNA strands, whereas DNA polymerases can synthesize copies off both strands.

21. RNA molecules have *three* phosphates at their 5′ end, but DNA molecules never do. Explain this difference.
During initiation of DNA replication, DNA nucleotide triphosphates must be attached to a 3′–OH of a RNA molecule by DNA polymerase. This process removes the terminal two phosphates of the nucleotides. If the RNA molecule is subsequently removed, then a single phosphate would remain at the 5′ end of the DNA molecule. RNA polymerase does not require the 3′–OH to initiate synthesis of RNA molecules. Therefore, the 5′ end of a RNA molecule will retain all three of the phosphates from the original nucleotide triphosphate substrate.

22. Write out a hypothetical sequence of bases that might be found in the first 20 nucleotides of a promoter of a bacterial gene. Include both strands of DNA and label the 5′ and 3′ ends of both strands. Be sure to include the start site for transcription and any consensus sequences found in the promoter.

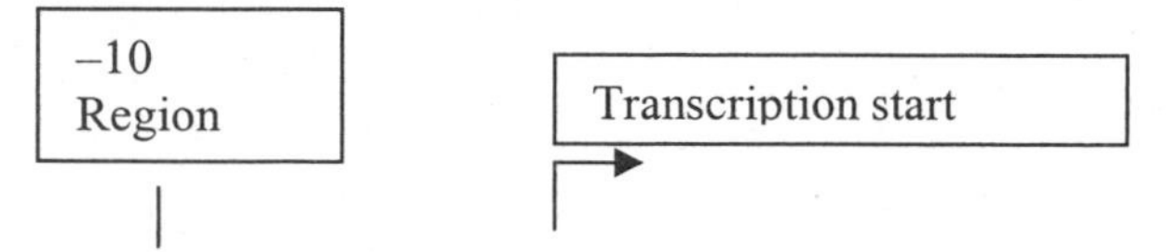

5′–GGACTATATGATGCGGCCCAT–3′

3′–CCTGATATACTACGCCGGGTA–5′

The –10 region, or Pribnow Box, has the consensus sequence of TATAAT. However, few bacterial promoters actually contain the exact consensus sequence. A common sequence at the transcription start site is 5′–CAT–3′ with transcription beginning at the "A."

*23. What would be the most likely effect of a mutation at the following locations in an *E. coli* gene?

a. –8
A mutation at the –8 position would probably affect the –10 consensus sequence (TATAAT), which is centered on position –10. This consensus sequence is necessary for binding of RNA polymerase. A mutation in there would most likely decrease transcription.

b. –35
A mutation in the –35 region could affect the binding of the sigma factor to the promoter. Deviations away from the consensus typically reduce transcription, so transcription is likely to be reduced or inhibited.

c. –20
The –20 region is located between the consensus sequences of an E. coli *promoter. Although the holoenzyme may cover the site, it is unlikely that a mutation will have any effect on transcription.*

d. Start site
A mutation in the start site would have little effect on transcription.

24. A strain of bacteria possesses a temperature-sensitive mutation in the gene that encodes the sigma factor. At elevated temperatures, the mutant bacteria produce a sigma factor that is unable to bind to RNA polymerase. What effect will this mutation have on the process of transcription when the bacteria are raised at elevated temperatures?
Binding of the sigma factor to the RNA polymerase core enzyme forms the RNA polymerase holoenzyme. Only the holoenzyme binds to the promoter. Without the sigma factor, RNA polymerase will be unable to bind the promoter and transcription initiation will not occur. Any RNA polymerase that has completed transcription inititation and has begun elongation will complete transcription

because the sigma factor is not needed for elongation. However, no further initiation will be possible at the elevated temperature.

*25. The following diagram represents a transcription unit on a DNA molecule.

a. Assume that this DNA molecule is from a bacterial cell. Draw in the approximate location of the promoter and terminator for this transcription unit.

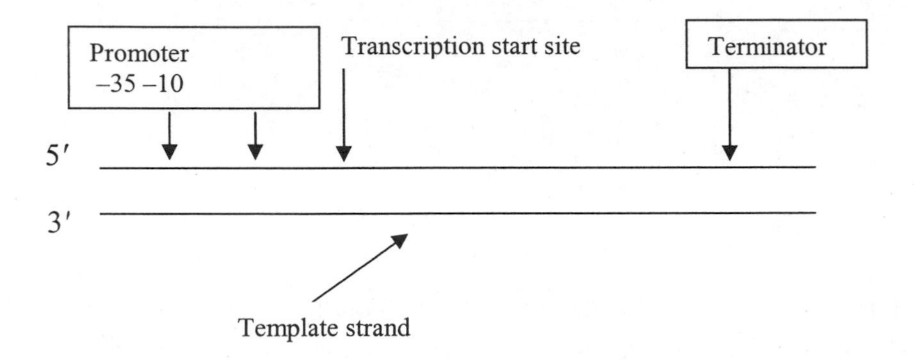

b. Assume that this DNA molecule is from a eukaryotic cell. Draw in the approximate location of an RNA polymerase II promoter.

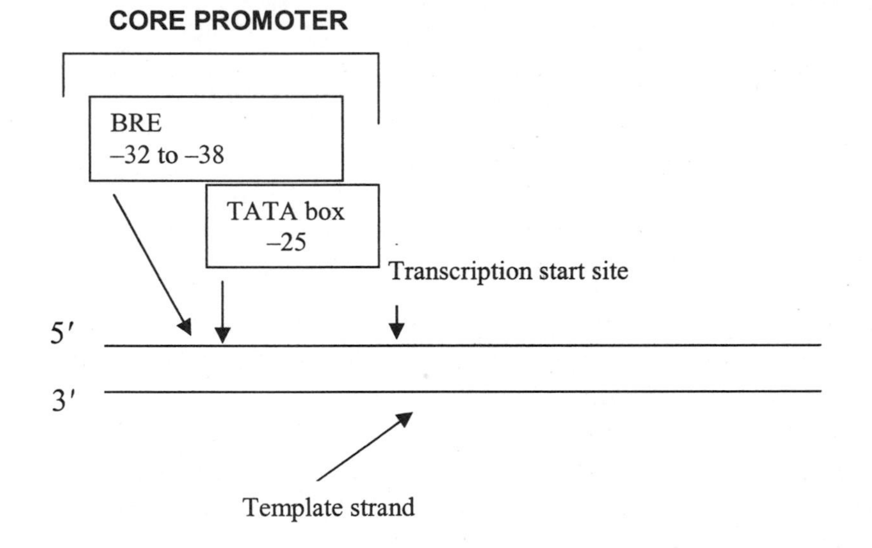

26. The following DNA nucleotides are found near the end of a bacterial transcription unit. Find the terminator in this sequence:
3′–AGCATACAGCAGACCGTTGGTCTGAAAAAAGCATACA–5′

a. Mark the point at which transcription will terminate.

Termination

3′–AGCATACAGCAGACCGTTGGTCTGAAAAAAGCATACA–5′

b. Is this terminator intrinsic or rho-dependent?
Based on the potential hairpin structure that can form and the run of U's that will be synthesized in the RNA, this terminator is an intrinsic terminator.

c. Draw a diagram of the RNA that will be transcribed from this DNA, including its nucleotide sequence and any secondary structures that form.

```
                 A
               C   A
                G C
                G C
                U A
                C G
                U A
5′–UCGUAUGUCG CUUUU–3′
```

*27. A strain of bacteria possesses a temperature-sensitive mutation in the gene that encodes the rho subunit of RNA polymerase. At high temperatures, rho is not functional. When these bacteria are raised at elevated temperatures, which of the following effects would you expect to see? Explain your reasoning for accepting or rejecting each of these five options.

a. Transcription does not take place.
No. Because the rho protein is involved in transcription termination, it should not affect transcription initiation or elongation. So you would expect to see transcription.

b. All RNA molecules are shorter than normal.
No. Without the rho protein, transcription would be expected to continue past the normal termination site of rho-dependent terminators producing some longer molecules than expected.

c. All RNA molecules are longer than normal.
No, only some will be. Only RNA molecules produced from genes using rho-dependent termination should be longer. Genes that are terminated through rho-independent termination should remain unaffected.

d. Some RNA molecules are longer than normal.
Yes. You would expect to see some RNA molecules that are longer than normal. Only genes that use rho-dependent termination would be expected to not terminate at the normal termination site, thus producing some RNA molecules that are longer than normal.

e. RNA is copied from both DNA strands.
No. RNA will be copied from only a single strand because rho protein does not affect transcription initiation or elongation.

28. The following diagram represents the Christmas-tree-like structure of active transcription observed by Miller, Hamkalo, and Thomas (see Figure 13.3). On the diagram, identify parts (a) through (i):

a. DNA molecule

b. 5′ and 3′ ends of the template strand of DNA

c. At least one RNA molecule
d. 5′ and 3′ ends of at least one RNA molecule
e. Direction of movement of the transcription apparatus on the DNA molecule
f. Approximate location of the promoter
g. Possible location of a terminator
h. Upstream and downstream directions
i. Molecules of RNA polymerase (use dots to represent these molecules)

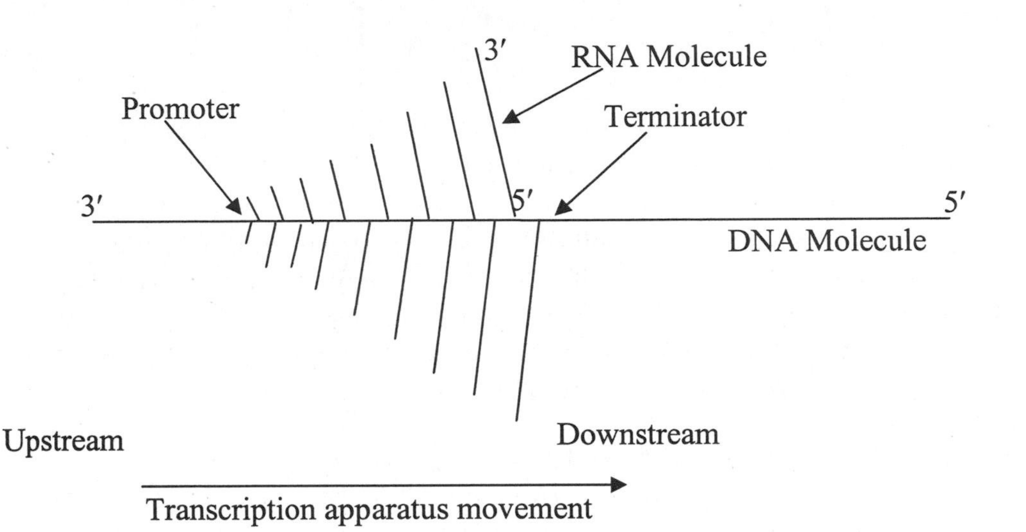

29. Suppose that the string of A's following the inverted repeat in a rho-independent terminator was deleted, but the inverted repeat was left intact. How would this deletion affect termination? What would happen when RNA polymerase reached this region?
Termination would not take place if the string of A's following the inverted repeat were deleted. Although RNA polymerase may stall briefly at the hairpin, the presence of the A–U base pairs is needed to destabilize the DNA-RNA interaction and to end transcription. If the string of A's is not present, then transcription by RNA polymerase would continue.

Section 13.4

30. The following diagram represents a transcription unit in a hypothetical DNA molecule:

5′...TTGACA...TATAAT...3′
3′...AACTGT...ATATTA...5′

a. On the basis of the information given, is this DNA from a bacterium or from a eukaryotic organism?
The DNA is from a bacterium as evidenced by the TATAAT sequence that corresponds to the –10 consensus sequence of a bacterial promoter and the

TTGACA sequence that is identical to the –35 consensus sequence of a bacterial promoter.

b. If this DNA molecule is transcribed, which strand will be the template strand and which will be the nontemplate strand?

5′...TTGACA...TATAAT...3′ Nontemplate
3′...AACTGT...ATATTA...5′ Template

c. Where, approximately, will the start site of transcription be?

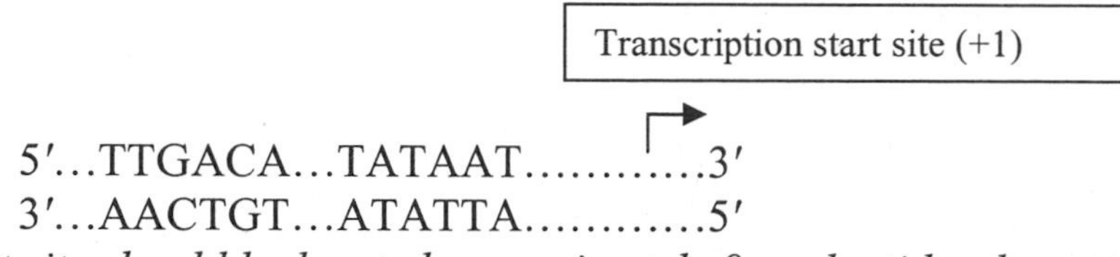

The start site should be located approximately 9 nucleotides downstream from the last nucleotide of the –10 consensus sequence (TATAAT) or the +1 nucleotide.

31. Computer programmers, working with molecular geneticists, have developed computer programs that can identify genes within long stretches of DNA sequences. Imagine that you are working with a computer programmer on such a project. Based on what you know about the process of transcription, what sequences might you suggest the computer program look for to identify the beginning and end of a gene?

The sequences recommended to the programmer will depend on the type of organism whose DNA is being studied. However, both bacteria and eukaryotic organisms have both promoter and termination consensus DNA sequences that could be recognized by the programmer's computer program as potential sites of interest for genes. The program should be able to recognize the consensus sequences as well as deviations from the consensus since the consensus sequence may only rarely be present. The identification by the program of both promoter sequences, terminator sequences and/or potential cleavage sites in a region of the DNA should provide evidence of a gene's presence in that region. For bacterial promoters, the –35 sequences and –10 sequences would be good starting points. At the end of bacterial genes are terminators. The program should be able to find inverted repeats and strings of A's, which are indicative of Rho-independent termination. For rho-dependent termination sequences may prove more difficult to identify. The presence of both the promoter sequences and a termination sequence in a region of DNA would be evidence of a bacterial gene. Similar promoter and termination sequences as well as 3′ cleavage site sequences *in eukaryotic cells could also be used. For RNA polymerase II genes, identifying sequences of the core promoter such as the TFIIB recognition element (BRE) sequences, the TATA box, the initiator element sequence, and the downstream core promoter element sequences might indicate the beginning of a gene. Although not every RNA polymerase II promoter contains each sequence, the core promoter will contain at least one of these sequences. Transcription terminators at the ends of RNA polymerase II transcribed genes could prove difficult to recognize. However, another potential sequence for identification would be the cleavage site consensus sequences at the 3′ regions of the gene.*

*32. Through genetic engineering, a geneticist mutates the gene that encodes TBP in cultured human cells. This mutation destroys the ability of TBP to bind to the TATA box. Predict the effect of this mutation on cells that possess it.
If TBP cannot bind the TATA box, then genes with these promoters will be transcribed at very low levels or not at all. Because the TATA box is the most common promoter element for RNA polymerase II transcription units and is found in some RNA polymerase III promoters, transcription will decline significantly. The lack of proteins encoded by these genes will most likely result in cell death.

33. Elaborate repair mechanisms are associated with replication to prevent permanent mutations in DNA, yet no similar repair is associated with transcription. Can you think of a reason for these differences in replication and transcription? (Hint: Think about the relative effects of a permanent mutation in a DNA molecule compared with one in an RNA molecule.)
RNA molecules are constantly being synthesized and subsequently degraded. Typically, RNA is produced regularly by new transcription. A defective RNA molecule would soon be replaced and would only briefly affect a cell. Furthermore, the defective RNA would not be passed to offspring of that cell. DNA mutations, however, are permanent. All RNAs transcribed from that gene would be affected by a change in the DNA sequence. The mutation in the DNA molecule would be passed to progeny cells and propagated. Finally, replication occurs only once during the cell cycle. Once the mutation occurs, there is no subsequent replication that could produce the correct sequence.

CHALLENGE QUESTIONS

Section 13.3

34. Many genes in bacteria and eukaryotes contain numerous sequences that potentially cause pauses or premature terminations of transcription. Nevertheless, transcription of these genes within a cell normally produces multiple RNA molecules thousands of nucleotides long without pausing or terminating prematurely. However, when a single round of transcription takes place on such templates in a test tube, RNA synthesis is frequently interrupted by pauses and premature terminations, which slow the rate at which transcription occurs and frequently reduces the length of mRNA molecules produced. The majority of these pauses and premature terminations occur when RNA polymerase temporarily backtracks (i.e., backs up) for one or two nucleotides along the DNA. Experimental studies have demonstrated that most transcriptional delays and premature terminations disappear if several RNA polymerases are simultaneously transcribing the DNA molecule. Propose an explanation for why transcription is more rapid and results in longer mRNA when the template DNA is being transcribed by multiple RNA polymerases.
When a single RNA polymerase is transcribing a template in a test tube, the backtracking often may lead to the RNA polymerase and the 3′ end of the RNA transcript losing contact with each other resulting in premature termination. If other RNA polymerases are present on the template, they may act cooperatively

with the leading RNA polymerase that has backtracked by pushing the leading RNA polymerase foward, thus continuing transcription before termination can occur.

Section 13.4

35. Enhancers are sequences that affect the initiation of the transcription of genes that are hundreds or thousands of nucleotides away. Enhancer-binding proteins usually interact directly with transcription factors at promoters by causing the intervening DNA to loop out. An enhancer of the bacteriophage T4 does not function by looping of the DNA (D. R. Herendeen, et al., 1992, *Science* 256:1298–1303). Propose some additional mechanisms (other than DNA looping) by which this enhancer might affect transcription at a gene thousands of nucleotides away.
For T4 late genes, the replication fork serves as a mobile enhancer. Three T4 replication proteins also serve as transcriptional activators from the enhancers. Two potential mechanisms by which these proteins can activate transcription are (1) DNA tracking by the enhancer proteins and (2) topological changes in the DNA induced by the binding of the T4 enhancer by the transcriptional activator proteins. In the DNA tracking model, the enhancer-binding proteins move along the DNA and activate the transcriptional apparatus at the promoter or, perhaps, the activator proteins deliver the transcriptional apparatus to the promoter. In the second model, binding of the enhancer by the activator proteins leads to topological changes in the DNA structure that make the promoter more accessible to the transcriptional apparatus.

***36.** The locations of the TATA box in the genes of two species of yeast, *Saccharomyces pombe* and *S. cerevisiae,* differ dramatically. The TATA box of *S. pombe* is about 30 nucleotides upstream of the start site, similar to the location for most other eukaryotic cells. However, the TATA box of *S. cerevisiae* is 40 to 120 nucleotides upstream of the start site.

To better understand what sets the start site in these organisms, researchers at Stanford University conducted a series of experiments to determine which components of the transcription apparatus of these two species could be interchanged (Y. Li et al. 1994. *Science* 263:805–807). In these experiments, different transcription factors and RNA polymerases were switched in *S. pombe* and *S. cerevisiae,* and the effects of the switch on the level of RNA synthesis and on the start point of transcription were observed in a series of tests. The results from one set of experiments are shown in the adjoining table. Components cTFIIB, cTFIIE, cTFIIF, and cTFIIH are transcription factors from *S. cerevisisae.* Components pTFIIB, pTFIIE, pTFIIF, and pTFIIH are transcription factors from *S. pombe.* Components cPol II and pPol II are RNA polymerase II from *S. cerevisiae* and *S. pombe,* respectively. The table indicates whether the component was present (+) or missing (–) in each of the tests (1–7). In the accompanying gel, the presence of a band indicates that RNA was produced and the position of the band indicates whether it was the length predicated when transcription begins 30 bp downstream from the TATA box or from 40 to 120 bp downstream from the TATA box.

Components	1	2	3	4	5	6	7
cTFIIE	+	–	+	+	+	+	–
cTFIIH + cTFIIF	+	–	+	+	+	–	+
cTFIIB + cPol II	+	–	–	–	–	–	–
pPol II	–	+	+	+	–	+	+
pTFIIB	–	+	+	–	+	+	+
pTFIIE + pTFIIH + pTFIIF	–	+	–	–	–	–	–

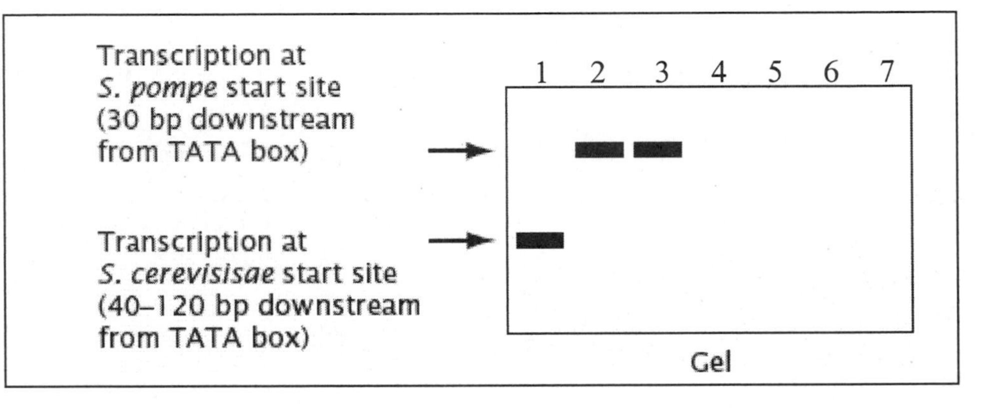

a. What conclusion can you draw from these data about what components determine the start site for transcription?
From reaction 3 and the corresponding lane in the gel, it appears that pTFIIB, along with the RNA polymerase from S. pombe *(pPolII), were sufficient to determine the start site of transcription. The* S. pombe *start site was used even though many of the other transcription factors were from* S. cervesiae.

b. What conclusions can you draw about the interactions of the different components of the transcription apparatus?
When TFIIE and TFIIH were individually exchanged, transcription was affected. However, the paired exchange of TFIIE and TFIIH allowed for transcription, suggesting that TFIIE and TFIIH have a species-specific interaction that is essential for transcription, the absence of this interaction will inhibit transcription. Similar results were observed in the paired exchange of TFIIB and RNA polymerase, which suggests species-specific interaction between TFIIB and RNA polymerase is needed for transcription.

c. Propose a mechanism for why the start site for transcription in *S. pombe* is about 30 bp is 30 nucleotides downstream from the TATA box, whereas the start site for transcription in *S. cerevisiae* is 40 to 120 bp downstream from the TATA box.
TFIIB likely interacts with both RNA polymerase and with other transcription factors necessary for initiation at the promoter. The data indicate that a requirement is for the TFIIB and the RNA polymerase II to be from the same species for transcription to occur. TFIIB potentially assembles downstream of

the TATA box but in association with other transcription factors located at or near the TATA box. Possibly, these downstream interactions between TFIIB and the other transcription factors are involved in stimulating conformation changes in RNA polymerase and the DNA sequence allowing for transcription to begin at the appropriate start site. In reaction 3, both the RNA polymerase and TFIIB were from S. pombe, *leading to the positioning of the RNA polymerase at the* S. pombe *transcription start site by pTFIIB.*

37. Glenn Croston and his colleagues studied the relation between chromatin structure and transcription activity. In one set of experiments, they measured the level of in vitro transcription of a *Drosophila* gene by RNA polymerase II with the use of DNA and various combinations of histone proteins (G. E. Croston et al. 1991. *Science* 251:643–649).

First, they measured the level of transcription for naked DNA, with no associated histone proteins. Then, they measured the level of transcription after nucleosome octamers (without H1) were added to the DNA. The addition of the octamers caused the level of transcription to drop by 50%. When both the nucleosome octamers and the H1 proteins were added to the DNA, transcription was greatly repressed, dropping to less than 1% of that obtained with naked DNA (see the following table).

Treatment	Relative amount of transcription
Naked DNA	100
DNA + octamers	50
DNA + octamers + H1	<1
DNA + GAL4-VP16	1000
DNA + octamers + GAL4-VP16	1000

GAL4-VP16 is a protein that binds to the DNA of certain eukaryotic genes. When GAL4-VP16 is added to DNA, the level of RNA polymerase II transcription is greatly elevated. Even in the presence of the H1 protein, GAL4-VP16 stimulates high levels of transcription.

Propose a mechanism for how the H1 protein represses transcription and how GAL4-VP16 overcomes this repression. Explain how your proposed mechanism would produce the results obtained in these experiments.

The addition of H1 in the presence of nucleosome octamers almost completely inhibits transcription, suggesting that the promoter is not available to the transcriptional apparatus. If GAL4-VP16 could remove and prevent H1 from blocking the promoter, then transcription initiation would occur. Also, the addition of GAL4-VP16 to the DNA increases transcription tenfold over naked DNA. So GAL4-VP16 is likely a transcriptional activator protein as well.

1-3 1,2,4,6a,c 8-13, 15, 16, 19-24, 26-30

Chapter Fourteen: RNA Molecules and RNA Processing

COMPREHENSION QUESTIONS

Section 14.1

*1. What is the concept of colinearity? In what way is this concept fulfilled in bacterial and eukaryotic cells?
Colinearity is the concept that the sequence of codons in the DNA of a gene is the same as the sequence of amino acids in the protein. If we examine the location of three codons for three amino acids in a protein, the order of the codons in the gene is always the same as the order of the amino acids in the protein. The presence of large regions of non coding DAN in introns means that DNA sequences that code for adjacent amino acids may be separated by many base pairs of DNA, but introns to not alter the order of codons in the gene.

2. What are some characteristics of introns?
Eukaryotic genes commonly contain introns. However, introns are rare in bacterial genes. The number of introns found in an organism's genome is typically related to complexity—more complex organisms possess more introns. Introns typically do not encode proteins and are usually larger in size than exons.

*3. What are the four basic types of introns? In which genes are they found?
(1) Group I introns are found in rRNA genes and some bacteriophage genes.
(2) Group II introns are found in protein-encoding genes of mitochondria, chloroplasts, and a few eubacteria.
(3) Nuclear pre-mRNA introns are found in the protein-encoding genes of the nucleus.
(4) Transfer RNA introns are found in tRNA genes.

Section 14.2

*4. What are the three principal elements in mRNA sequences in bacterial cells?
(1) The 5′ untranslated region, which contains the Shine-Dalgarno sequence
(2) The protein-encoding region
(3) The 3′ untranslated region

5. What is the function of the Shine-Dalgarno consensus sequence?
The Shine-Dalgarno consensus sequence functions as the ribosome-binding site on the mRNA molecule.

*6. **a.** What is the 5′ cap?
The 5′ end of eukaryotic mRNA is modified by the addition of the 5′ cap. The cap consists of an extra guanine nucleotide linked 5′ to 5′ to the mRNA molecule. This nucleotide is methylated at position 7 of the base. The ribose sugars of adjacent bases may be methylated at the 2′–OH.

b. How is the 5′ cap added to eukaryotic pre-mRNA?
Initially, the terminal phosphate of the three 5′ phosphates linked to the end of the mRNA molecule is removed. Subsequently, a guanine nucleotide is attached to the 5′ end of the mRNA using a 5′ to 5′ phosphate linkage. Next, a methyl group is attached to position 7 of the guanine base. Ribose sugars of adjacent nucleotides may also be methylated, but at the 2′–OH.

c. What is the function of the 5′ cap?
CAP binding proteins recognize the 5′ cap and stimulate binding of the ribosome to the 5′ cap and to the mRNA molecule. The 5′ cap may also increase mRNA stability in the cytoplasm. Finally, the 5′ cap is needed for efficient splicing of the intron that is nearest the 5′ end of the pre-mRNA molecule.

7. How is the poly(A) tail added to pre-mRNA? What is the purpose of the poly(A) tail?
Initially, a complex consisting of several proteins forms on the 3′ UTR of the pre-mRNA molecule. Cleavage and polyadenylation specificity factor (CPSF) bind to the AAUAAA consensus sequence, which is located upstream of the 3′ cleavage site. Another protein, cleavage stimulation factor (CsTF), binds downstream of the cleavage site. Two cleavage factors (CFI and CFII) and polyadenylate polymerase (PAP) also become part of the complex. Once the complex has formed, the pre-mRNA is cleaved. CsTF and the two cleavage factors leave the complex. PAP adds approximately 10 adenine nucleotides to the 3′ end of the pre-mRNA molecule. The addition of the short poly(A) tail allows for the binding of the poly(A) binding protein (PABII) to the tail. PABII increases the rate of polyadenylation, which subsequently allows for more PABII protein to bind the tail.

The presence of the poly(A) tail increases the stability of the mRNA molecule through the interaction of proteins at the poly(A) tail.

*8. What makes up the spliceosome? What is the function of the spliceosome?
The spliceosome consists of five small ribonucleoproteins (snRNPs). Each snRNP is composed of multiple proteins and a single small nuclear RNA molecule or snRNA. The snRNPs are identified by which snRNA (U1, U2, U3, U4, U5, or U6) each contains. Splicing of pre-mRNA nuclear introns takes place within the spliceosome.

9. Explain the process of pre-mRNA splicing in nuclear genes.
Removal of an intron from the pre-mRNA requires the assembly of the spliceosome complex on the pre-mRNA, cleavage at both the 5′ and 3′ splice sites of the intron, and two transesterification reactions ultimately leading to the joining of the two exons. Initially, snRNP U1 binds to the 5′ splice site through complementary base pairing of the U1 snRNA. Next, snRNP U2 binds to the branch point within the intron. The U5 and U4–U6 complex joins the spliceosome, resulting in the looping of the intron so that the branch point and 5′ splice site of the intron are now adjacent to each other. U1 and U4 now disassociate from the spliceosome, and the spliceosome is activated. The pre-mRNA is then cleaved at the 5′ splice site, producing an exon with a 3′–OH. The 5′ end of the intron folds back and forms 5′–2′ phosphodiester linkage through the first transesterification reaction with the adenine nucleotide at the branch point of the intron. This looped structure is called the lariat. Next, the 3′ splice site is cleaved and

then immediately ligated to the 3'–OH of the first exon through the second transesterification reaction. Thus, the exons are now joined and the intron has been excised.

10. Describe two types of alternative processing pathways. How do they lead to the production of multiple proteins from a single gene?
Alternative processing of pre-mRNA can take the form of either alternative splicing of pre-mRNA introns or the alternative cleavage of 3' cleavage sites in a pre-mRNA molecule containing two or more cleavage sites for polyadenylation. Alternative splicing results in different exons of the pre-mRNA being ligated to form mature mRNA. Each mRNA formed by an alternative splicing process will yield a different protein. In pre-mRNA molecules with multiple 3' cleavage sites, cleavage at the different sites will generate mRNA molecules that differ in size. Each alternatively cleaved RNA will code for a different protein. A single pre-mRNA transcript can undergo both alternative processing steps, thus potentially producing multiple proteins.

*11. What is RNA editing? Explain the role of guide RNAs in RNA editing.
RNA editing alters the sequence of a RNA molecule after transcription either by the insertion, deletion, or modification of nucleotides within the transcript. The guide RNAs (gRNAs) provide templates for the alteration of nucleotides in RNA molecules undergoing editing and are complementary to regions within the pre-edited RNA molecule. At these complementary regions, the gRNAs base pair to the pre-edited RNA molecule. This binding determines the location of the alteration of nucleotides.

*12. Summarize the different types of processing that can take place in pre-mRNA.
Several modifications to pre-mRNA take place to produce mature mRNA.
(1) Addition of the 5' cap to the 5' end of the pre-mRNA.
(2) Cleavage of the 3' end of a site downstream of the AAUAAA consensus sequence of the last exon.
(3) Addition of the poly(A) tail to the 3' end of the mRNA immediately following cleavage.
(4) Removal of the introns (splicing).

Section 14.3

*13. What are some of the modifications in tRNA that occur through processing?
(1) The precursor RNA may be cleaved into smaller molecules.
(2) Nucleotides at both the 5' and 3' ends of the tRNAs may be removed or trimmed.
(3) Standard bases can be altered by base-modifying enzymes.

Section 14.4

14. Describe the basic structure of ribosomes in bacterial and eukaryotic cells.
Ribosomes in both eukaryotes and bacteria consist of a complex of protein and RNA molecules. A functional ribosome is composed of a large and a small subunit. The bacterial 70S ribosome consists of a 30S small subunit and a 50S large subunit. Within

the small subunit are a single 16S RNA molecule and 21 proteins. The 23S and the 5S RNA molecules, along with 31 proteins, are found in the large bacterial subunit. The eukaryotic 80S ribosome is comprised of a 60S large subunit and a 40S small subunit. Three RNA molecules, the 28S RNA, the 5.8S RNA, and the 5S RNA, are located in the large subunit as well as 49 proteins. The eukaryotic small subunit contains only a single 18S RNA molecule and 33 proteins.

*15. Explain how rRNA is processed.
Most rRNAs are synthesized as large precursor RNAs that are processed by methylation, cleavage, and trimming to produce the mature mRNA molecules. In E. coli, *methylation occurs to specific bases and the 2'–OH of the ribose sugars of the 30S rRNA precursor. The 30S precursor is cleaved and trimmed to produce the 16S rRNA, 23S rRNA, and the 5S rRNA. In eukaryotes, a similar process occurs. However, small nucleolar RNAs help to cleave and modify the precursor rRNAs.*

Section 14.5

16. What is the origin of siRNAs and microRNAs? What do these RNA molecules do in the cell?
The siRNAs originate from the cleavage of mRNAs, RNA transposons, and RNA viruses by the enzyme Dicer. Dicer may produce multiple siRNAs from a single double-stranded RNA molecule. The double-stranded RNA molecule may occur due to the formation of hairpins or by duplexes between different RNA molecules. The miRNAs arise from the cleavage of individual RNA molecules that are distinct from other genes. The enzyme Dicer cleaves these RNA molecules that have formed small hairpins. A single miRNA is produced from a single RNA molecule.

Both siRNAs and microRNAs silence gene expression through a process called RNA interference. Both function by shutting off gene expression of a cell's own genes or to shut off expression of genes from the invading foreign genes of viruses or tranposons. The microRNAs typically silence genes that are different from those from which the microRNAs are transcribed. However, the siRNAs usually silence genes from which they are transcribed.

17. What are some similarities and differences between siRNAs and miRNAs?

<u>Some similarities:</u>

- *Both silence gene expression of foreign genes or a cell's own genes after combining with proteins to form a RNA-Induced-Silencing Complex (RISC).*
- *Both are dsRNA molecules 21 and 22 nucleotides in length.*
- *Both are produced by the action of the enzyme Dicer on larger RNA molecules.*

<u>Some differences:</u>

- *The miRNAs originate from transcription products of distinct genes, while siRNAs originate from mRNA, RNA transposons, or RNA viruses.*

- *For miRNAs Dicer cleaves single-stranded RNAs that form short hairpins, while for siRNAs Dicer cleaves single-stranded RNAs that form long hairpins or RNA duplexes.*
- *The miRNAs silence genes that are different from the ones from which they were transcribed, while siRNAs silence genes from which they were transcribed.*
- *The miRNAs typically act by inhibiting translation although some do trigger the degradation of the mRNA, while some siRNAs work by stimulating degradation of the mRNA target molecule. Other siRNAs inhibit transcription.*

18. How are miRNAs processed?
The miRNAs are initially transcribed as a longer precursor molecule called a primary miRNA (pri-miRNA) that folds into a double-stranded structure containing at least one hairpin. The pri-miRNA is cleaved into one or more smaller molecules each containing a hairpin. Dicer binds to the hairpin and removes the loop. One of the strands is incorporated into the RNA-induced silencing complex (RISC), while the other is released and degraded.

APPLICATION QUESTIONS AND PROBLEMS

Section 14.1

*19. Duchenne muscular dystrophy is caused by a mutation in a gene that encompasses more than 2 million nucleotides and specifies a protein called dystrophin. However, less than 1% of the gene actually encodes the amino acids in the dystrophin protein. On the basis of what you now know about gene structure and RNA processing in eukaryotic cells, provide a possible explanation for the large size of the dystrophin gene.
The large size of the dystrophin gene is likely due to the presence of many intervening sequences or introns within the coding region of the gene. Excision of the introns through RNA splicing yields the mature mRNA that encodes the dystrophin protein.

Section 14.2

20. How do the mRNA of bacterial cells and the pre-mRNA of eukaryotic cells differ? How do the mature mRNAs of bacterial and eukaryotic cells differ?
Bacterial mRNA is translated immediately upon being transcribed. Eukaryotic pre-mRNA must be processed. Bacterial mRNA and eukaryotic pre-mRNA have similarities in structure. Each has a 5′ untranslated region as well as a 3′ untranslated region. Both also have protein-coding regions. However, the protein-coding region of the pre-mRNA is disrupted by introns. The eukaryotic pre-mRNA must be processed to produce the mature mRNA. Eukaryotic mRNA has a 5′ cap and a poly(A) tail, unlike bacterial mRNAs. Bacterial mRNA also contains the Shine-Dalgarno consensus sequence. Eukaryotic mRNA does not have the equivalent.

*21. Draw a typical eukaryotic gene and the pre-mRNA and mRNA derived from it. Assume that the gene contains three exons. Identify the following items and, for each item, give a brief description of its function.

a. 5′ untranslated region
The 5′ untranslated region lies upstream of the translation start site. In bacteria, the ribosome binding site or Shine-Dalgarno sequence is found within the 5′ untranslated region. However, eukaryotic mRNA does not have the equivalent sequence, and a eukaryotic ribosome binds at the 5′ cap of the mRNA molecule.

b. Promoter
The promoter is the DNA sequence that the transcription apparatus recognizes and binds to initiate transcription.

c. AAUAAA consensus sequence
The AAUAAA consensus sequence lies downstream of the coding region of the gene. It determines the location of the 3′ cleavage site in the pre-mRNA molecule.

d. Transcription start site
The transcription start site begins the coding region of the gene and is located 25 to 30 nucleotides downstream of the TATA box.

e. 3′ untranslated region
The 3′ untranslated region is a sequence of nucleotides at the 3′ end of the mRNA that is not translated into proteins. However, it does affect the translation of the mRNA molecule as well as the stability of the mRNA.

f. Introns
Introns are noncoding sequences of DNA that intervene within coding regions of a gene.

g. Exons
Exons are transcribed regions that are not removed in intron processing. They include the 5′UTR, coding regions that are translated into amino acid sequences, and the 3′UTR.

h. Poly(A) tail
A poly(A) tail is added to the 3′ end of the pre-mRNA. It affects mRNA stability.

i. 5′ cap
The 5′ cap functions in the initiation of translation and mRNA stability.

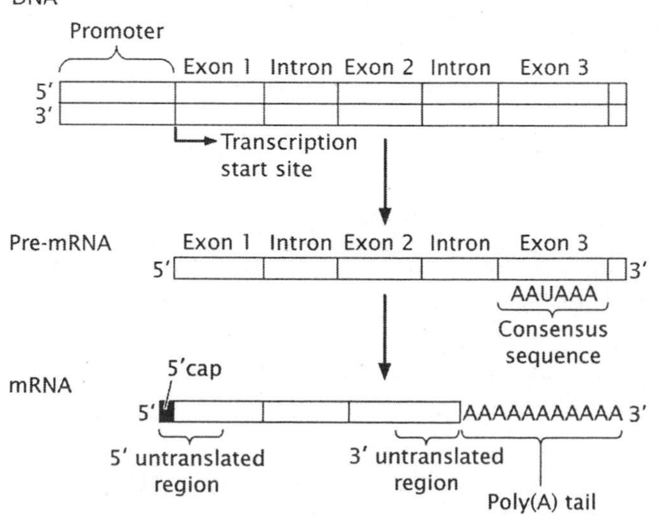

22. How would the deletion of the Shine-Dalgarno sequence affect a bacterial mRNA?
In bacteria, the small ribosomal subunit binds to the Shine-Dalgarno sequence to initiate translation. If the Shine-Dalgarno sequence is deleted, then translation initiation cannot take place, preventing protein synthesis.

*23. How would the deletion of the following sequences or features most likely affect a eukaryotic pre-mRNA?

a. AAUAAA consensus sequence
The deletion of the AAUAA consensus sequence would prevent binding of the cleavage and polydenylation factor (CPSF), thus resulting in no cleavage or polyadenylation of the pre-mRNA. This would affect the stability and translation of the mRNA.

b. 5′ cap
The deletion of the 5′ cap would most likely prevent splicing of the intron that is nearest to the 5′ cap. Ultimately, elimination of the cap will affect the stability of the pre-mRNA as well as its ability to be translated.

c. Poly(A) tail
Polyadenylation increases the stability of the mRNA. If eliminated from the pre-mRNA, then the mRNA would be degraded quickly by nucleases in the cytoplasm.

24. Suppose that a mutation occurs in an intron of a gene encoding a protein. What will the most likely effect of the mutation be on the amino acid sequence of that protein? Explain your answer.
Because introns are removed prior to translation, an intron mutation would have little effect on a protein's amino acid sequence unless the mutation occurred within the 5′ splice site, the 3′ splice site, or the branch point. If mutations within these sequences altered splicing, then the mature mRNA would be altered, thus altering the amino acid sequence of the protein. The result could be a protein with additional amino acid sequence. Or, possibly, the altered splicing could introduce a stop codon that stops translation prematurely.

25. A geneticist induces a mutation in a line of cells growing in the laboratory. The mutation occurs in one of the genes that encodes proteins that are participate in the cleavage and polyadenylation of eukaryotic mRNA. What will the immediate effect of this mutation be on RNA molecules in the cultured cells?
CPSF is needed for cleavage of the 3′UTR and for polyadenylation. Nonfunctional CPSF would result in mRNA lacking a poly(A) tail, and the mRNA would be degraded more quickly in the cytoplasm by nucleases.

*26. A geneticist mutates the gene for proteins that bind to the poly(A) tail in a line of cells growing in the laboratory. What would be the immediate effect of this mutation in the cultured cells?
The stability of the mRNA is dependent on the proteins that bind to the poly(A) tail. If the proteins are unable to bind to the tail, then the mRNA that contain poly(A) tails will be degraded at a much more rapid rate within the cells.

27. A geneticist isolates a gene that contains five exons. He then isolates the mature mRNA produced by this gene. After making the DNA single-stranded, he mixes the single-stranded DNA and RNA. Some of the single-stranded DNA hybridizes (pairs) with the complementary mRNA. Draw a picture of what the DNA-RNA hybrids would look like under the electron microscope.

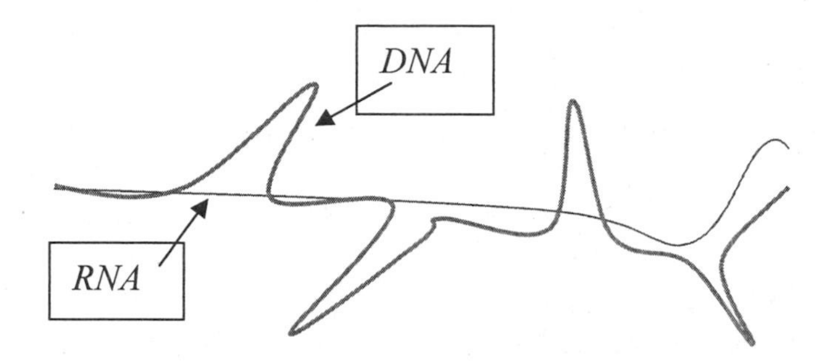

28. A geneticist discovers that two different proteins are encoded by the same gene. One protein has 56 amino acids and the other 82 amino acids. Provide a possible explanation for how the same gene could encode both of these proteins.
The pre-mRNA molecules transcribed from the gene are likely processed by alternative processing pathways. Two possible mechanisms that could have produced the two different proteins from the same pre-mRNA are alternative splicing or multiple 3′ cleavage sites in the pre-mRNA. The cleavage of the pre-mRNA molecule at different 3′ cleavage sites would produce alternatively processed mRNA molecules that differ in size. Translation from each of the alternative mRNAs would produce proteins containing different numbers of amino acids.

Alternative splicing of the pre-mRNA could produce different mature mRNAs, each containing a different number of exons. Again, translation from each alternatively spliced mRNA would generate proteins that differ in the number of amino acids contained.

29. Explain how each of the following processes complicates the concept of colinearity.
a. Trans-splicing
In trans-splicing, exons from different genes are spliced together during RNA processing events. Essentially, the mature mRNA product is not produced by DNA sequences that are contiguous or even necessarily on the same chromosome. This results in an amino acid sequence of the translated protein from trans-spliced mRNA being encoded by two or more different genes. According to the principle of colinearity, we would have expected the DNA sequence of a single gene to correspond to the amino acid sequence of the protein.
b. Alternative splicing
Different mature mRNAs from a single gene can be produce by alternative splicing. Different arrangements of the gene's exons can occur in the mature mRNAs. Thus, different proteins can be encoded within the same gene as opposed to one gene corresponding to one protein as is predicted by the concept of colinearity.

c. RNA editing
In RNA editing, genetic information is added to the pre-mRNA after it is transcribed. In other words, the mature mRNA will contain information that was not part of the DNA from which it was transcribed. The result is that the nucleotide sequence of the gene does not correspond to the amino acid sequence of the protein—a clear violation of the concept of colinearity.

Section 14.4

30. In the early 1990s, Carolyn Napoli and her colleagues were working on petunias, attempting to genetically engineer a variety with dark purple petals by introducing numerous copies of a gene that codes for purple petals (C. Napoli, C. Lemieux, and R. Jorgensen. 1990. *Plant Cell* 2:279–289). Their thinking was that extra copies of the gene would cause more purple pigment to be produced and would result in a petunia with an even darker hue of purple. However, much to their surprise, many of the plants carrying extra copies of the purple gene were completely white or had only patches of color. Molecular analysis revealed that the level of the mRNA produced by the purple gene was reduced 50-fold in the engineered plants compared with levels of mRNA in wild-type plants. Somehow, the introduction of extra copies of the purple gene silenced both the introduced copies and the plant's own purple genes. Explain why the introduction of numerous copies of the purple gene silenced all copies of the purple gene.
The overexpression of the purple gene mRNA led potentially to the formation of double-stranded regions by these RNA molecules because of areas of homology within the mRNAs being produced. These double-stranded molecules stimulated RNA silencing mechanisms or the RNA-Induced-Silencing Complex (RISC) leading to rapid degradation of the mRNA molecules. The result would be a reduction in translation of the protein needed for the production of purple petals and the phenotypic loss of pigmentation.

CHALLENGE QUESTIONS

Section 14.2

31. Alternative splicing occurs in approximately 60% of the human genes that code for proteins. Research has found that how a pre-mRNA is spliced is affected by the pre-mRNA's promoter sequence (D. Auboeuf et al. 2002. *Science* 298:416–419). In addition, factors that affect the rate of elongation of the RNA polymerase during transcription affect the type of splicing that takes place. These findings suggest that the process of transcription affects splicing. Propose one or more mechanisms that would explain how transcription might affect alternative splicing.
Transcription and splicing may be coupled. In other words, splicing reactions may begin before the transcription of the entire mRNA has completed. Components of the transcriptional complex may help recruit proteins needed for splicing to the splice sites, or spliceosomal proteins could participate in the transcriptional complex. The promoter sequence of the gene helps determine which transcription factors bind to the

promoter. The interaction of the transcription complex and its proteins with splicing associated proteins may affect how the spliceosomes are located at splice sites thus affecting splicing. Also, the rate at which RNA polymerase II proceeds could also affect transcription. If transcription occurs at a rapid pace, transcription may be completed before splicing occurs. Posttranscriptional splicing could result in different splice sites being used possibly due to different types of secondary structure in the completed mRNA molecule.

32. Duchene muscular dystrophy (DMD) is an X-linked recessive genetic disease caused by mutations in the gene that encodes dystrophin, a large protein that plays an important role in the development of normal muscle fibers. The dystrophin gene is immense, spanning 2.5 million base pairs, and includes 79 exons and 78 introns. Many of the mutations that cause DMD produce premature stop codons, which bring protein synthesis to a halt, resulting in a greatly shortened and nonfunctional form of dystrophin. Some geneticists have proposed treating DMD patients by introducing small RNA molecules that cause the spliceosome to skip the exon containing the stop codon. The introduction of the small RNAs will produce a protein that is somewhat shortened (because an exon is skipped and some amino acids are missing) but may still result in a protein that has some function (A. Goyenvalle et al. 2004. *Science* 306:1796–1799). The small RNAs used for exon skipping are complementary to bases in the pre-mRNA. If you were designing small RNAs to bring about exon skipping for the treatment of DMD, what sequences should the small RNAs contain?
The proper splicing of the introns depends upon the sequences at the intron 5′ splice site, the branch site located upstream of the 3′ splice site, and the 3′ splice site. To ensure that the exon is removed by splicing, then splice sites of the introns upstream and downstream of the exon need to be affected. One strategy would be to design antisense RNA molecules that can bind to the splice sites or branching points of particular introns. The target would be to block proper splicing of the introns surrounding the exon, but to allow a splicing event that would remove the exon containing the mutation along with its flanking introns. Antisense RNA molecules could be used to block splicing or block the binding of the necessary snRNPs. The use of antisense RNAs complementary to the 3′ splice site and/ or the branch point of the intron upstream of the exon should block the 3′ splice from being cleaved. While an antisense RNA complementary to the 5′ splice site of the intron downstream of the exon should block the 5′ splice site from being cleaved. Ultimately, the upstream and downstream introns along with the exon sandwiched between will be removed resulting in the exon being skipped in the production of the mature mRNA.

33. In eukaryotic cells, a poly(A) tail is normally added to pre-mRNA molecules but not to rRNA or tRNA. Using recombinant DNA techniques, it is possible to connect a protein encoding gene (which is normally transcribed by RNA polymerase II) to a promoter for RNA polymerase I. This hybrid gene is subsequently transcribed by RNA polymerase I and the appropriate pre-mRNA is produced, but this pre-mRNA is not cleaved at the 3′ end and a poly (A) tail is not be added. Propose a mechanism to explain how the type of promoter found at the 5′ end of a gene can affect whether or not a poly (A) tail is added to the 3′ end.

It is likely that the promoter influence on polyadenylation is due to the RNA polymerase II molecule that binds and initiates transcription from that promoter. Because only pre-mRNAs are cleaved and polyadenylated at the 3′ end, RNA polymerase II could play an important role in stimulating 3′ cleavage and polyadenylation. If a protein encoding gene is fused to a RNA polymerase I promoter, then RNA polymerase I will transcribe the protein encoding gene. Because RNA polymerase I will be responsible for the transciption events, RNA polymerase II will not be present and would be unable to stimulate the 3′ cleavage reactions or the addition of the poly (A) tail. Recent research has indicated that RNA polymerase II is a critical component of the 3′ cleavage reaction, which is necessary for the addition of the poly (A) tail.

34. SR proteins are essential to proper spliceosome assembly and are known to be involved in the regulation of alternative splicing. Surprisingly, the role of SR proteins in splice-site selection and alternative splicing is affected by the promoter that is used for transcription of the pre-mRNA. For example, it is possible through genetic engineering to create RNA polymerase II promoters that have somewhat different sequences. When pre-mRNAs with exactly the same sequences are transcribed from two different RNA polymerase II promoters that differ slightly in sequence, which promoter is used can affect how the pre-mRNA is spliced. Propose a mechanism for how the DNA sequence of an RNA polymerase II promoter could affect alternative splicing that takes place in the pre-mRNA.
Different promoters could attract different components of the RNA polymerase II transcriptional complex or even affect how transcription moves from the initiation stage to the elongation stage. Components of the transcription complex may interact with SR proteins and help locate these proteins at sites that affect the splice sites. If different promoters have an affect on the transcription complex and its function, then the changes to the transcriptional complex may also affect their interactions with the SR proteins. The result might be that the SR proteins act at alternative splice sites when different promoters are present. Recent research has indicated that RNA polymerase II does interact with SR proteins.

1,3-12,14,15 18-31 error on 27

Chapter Fifteen: The Genetic Code and Translation

COMPREHENSION QUESTIONS

Section 15.1

1. What is the one gene, one enzyme hypothesis? Why was this hypothesis an important advance in our understanding of genetics?
   The one gene, one enzyme hypothesis proposed by Beadle and Tatum states that each gene encodes a single, separate protein. Now that we know more about the nature of enzymes and genes, it has been modified to the one gene, one polypeptide hypothesis because many enzymes consist of multiple polypeptides. The original hypothesis helped establish a linear link between genes (DNA) and proteins.

Section 15.2

*2. What three different methods were used to help break the genetic code? What did each reveal and what were the advantages and disadvantages of each?
   Marshall Nirenberg and Johann Heinrich Matthaei used the enzyme polynucleotide kinase to create homopolymers of synthetic RNAs. Using a cell-free protein synthesizing system they were able to determine the amino acid coded by each homopolymer. By this method, the meanings for the amino acids specified by the codons UUU, AAA, CCC, and GGG were determined. The disadvantage is that the meanings for only four codons could be determined. The same system was also used to create copolymers that contained random mixtures of two nucleotides in a known ratio. Different amino acids in the protein depended on the ratio of the two nucleotides. To determine or predict the composition of the codons, the frequency of amino acids produced using the copolymer was compared with the theoretical frequencies expected for the codons. A disadvantage of this procedure is that it depended on random incorporation of the nucleotides, which did not always happen. A further problem was that the base sequence of the codon could not be determined—only the bases contained within the codon. The redundancy of the code also provided difficulties because several different codons could specify the same amino acid.

   To solve these problems, Nirenberg and Leder mixed ribosomes bound to short RNAs of known sequences with charged tRNAs. The mixture was passed through a nitrocellulose filter to which the tRNAs paired to ribosome-mRNA stuck. They next determined the amino acids attached to the bound tRNAs. More than 50 codons were identified by this method. The difficulty is that not all tRNAs and codons could be identified with this method.

   Gobind Khorana and his colleagues used a third method. They synthesized RNA molecules of known repeating sequences. Using a cell-free protein synthesizing system they produced proteins of alternating amino acids. However, this procedure could not specify which codon encodes which amino acid.

3. What are isoaccepting tRNAs?
Isoaccepting tRNAs are tRNA molecules that have different anticodon sequences but accept the same amino acids.

*4. What is the significance of the fact that many synonymous codons differ only in the third nucleotide position?
Synonymous codons code for the same amino acid, or, in other words, have the same meaning. A nucleotide at the third position of a codon pairs with a nucleotide in the first position of the anticodon. Unlike the other nucleotide positions involved in the codon-anticodon pairing, this pairing is often weak, or "wobbles," and nonstandard pairings can occur. Because the "wobble," or nonstandard base-pairing with the anticodons, affects the third nucleotide position, the redundancy of codons ensures that the correct amino acid is inserted in the protein when nonstandard pairing occurs.

*5. Define the following terms as they apply to the genetic code:

a. Reading frame
The reading frame refers to how the nucleotides in a nucleic acid molecule are grouped into codons containing three nucleotides. Each sequence of nucleotideshas three possible sets of codons, or reading frames.

b. Overlapping code
If an overlapping code is present, then a single nucleotide is included in more than one codon. The result for a sequence of nucleotides is that more than one type of polypeptide can be encoded within that sequence.

c. Nonoverlapping code
In a nonoverlapping code, a single nucleotide is part of only one codon. This results in the production of a single type of polypeptide from one polynucleotide sequence.

d. Initiation codon
An initiation codon establishes the appropriate reading frame and specifies the first amino acid of the protein chain. Typically, the initiation codon is AUG; however, GUG and UUG can also serve as initiation codons.

e. Termination codon
The termination codon signals the termination or end of translation and the end of the protein molecule. There are three types of termination codons—UAA, UAG, and UGA—which can also be referred to as stop codons or nonsense codons. These codons do not code for amino acids.

f. Sense codon
A sense codon is a group of three nucleotides that code for an amino acid. There are 61 sense codons that code for the 20 amino acids commonly found in proteins.

g. Nonsense codon
Nonsense codons or termination codons signal the end of translation. These codons do not code for amino acids.

h. Universal code
In a universal code, each codon specifies, or codes, for the same amino acid in all organisms. The genetic code is nearly universal, but not completely. Most of the exceptions occur in mitochondrial genes.

i. Nonuniversal codons
Most codons are universal (or nearly universal) in that they specify the same amino acids in almost all organisms. However, there are exceptions where a codon has different meanings in different organisms. Most of the known exceptions are the termination codons, which in some organisms do code for amino acids. Occasionally, a sense codon is substituted for another sense codon.

6. How is the reading frame of a nucleotide sequence set?
The initiation codon on the mRNA sets the reading frame.

Section 15.3

*7. How are tRNAs linked to their corresponding amino acids?
Each of the 20 different amino acids that are commonly found in proteins has a corresponding aminoacyl-tRNA synthetase that covalently links the amino acid to the correct tRNA molecule.

8. What role do the initiation factors play in protein synthesis?
Initiation factors are proteins that are required for the initiation of translation. In bacteria, there are three initiation factors (IF1, IF2, and IF3). Each one has a different role. IF1 promotes the disassociation of the large and small ribosomal subunits. IF3 binds to the small ribosomal subunit and prevents it from associating with the large ribosomal subunit. IF2 is responsible for binding GTP and delivering the fMet-tRNA$_f^{met}$ *to the initiator codon on the mRNA. In eukaryotes, there are more initiation factors, but many have similar roles. Some of the eukaryotic initiation factors are necessary for recognition of the 5' cap on the mRNA. Others possess a RNA helicase activity, which is necessary to resolve secondary structures.*

9. How does the process of initiation differ in bacterial and eukaryotic cells?
Bacterial initiation of translation requires that sequences in the 16S rRNA of the small ribosomal subunit bind to the mRNA at the ribosome binding site or Shine-Dalgarno sequence. The Shine-Dalgarno sequence is essential in placing the ribosome over the start codon (typically AUG). In eukaryotes, there is no Shine-Dalgarno sequence. The small ribosomal subunit recognizes the 5' cap of the eukaryotic mRNA with the assistance of initiation factors. Next, the ribosomal small subunit migrates along the mRNA scanning for the AUG start codon. In eukaryotes, the start codon is located with a consensus sequence called the Kozak sequence (5'–ACCAUGG–3'). Transcription in eukaryotes also requires more initiation factors.

*10. Give the elongation factors used in bacterial translation and explain the role played by each factor in translation.
Three elongation factors have been identified in bacteria: EF-TU, EF-TS, and EF-G. EF-TU joins with GTP and then to a tRNA charged with an amino acid. The charged tRNA is delivered to the ribosome at the "A" site. During the process of delivery, the GTP joined to EF-TU is cleaved to form a EF-TU-GDP complex. EF-TS is necessary to

regenerate EF-TU-GTP. The elongation factor EF-G binds GTP and is necessary for the translocation or movement of the ribosome along the mRNA during translation.

11. What events bring about the termination of translation?
The process of termination begins when a ribosome encounters a termination codon. Because the termination codon would be located at the "A" site, no corresponding tRNA will enter the ribosome. This allows for the release factors (RF_1, RF_2, and RF_3) to bind the ribosome. RF_1 recognizes and interacts with the stop codons UAA and UAG, while RF_2 can interact with UAA and UGA. A RF_3-GTP complex binds to the ribosome. Termination of protein synthesis is complete when the polypeptide chain is cleaved from the tRNA located at the "P" site. During this process, the GTP is hydrolyzed to GDP.

12. Compare and contrast the process of protein synthesis in bacterial and eukaryotic cells, giving similarities and differences in the process of translation in these two types of cells.
Bacterial and eukaryotic cells share several similarities as well as have several differences in protein synthesis. Initially, bacteria and eukaryotes share the universal genetic code. However, the initiation codon, AUG, in eukaryotic cells codes for methionine, whereas in bacteria the AUG codon codes for N-formyl methionine. In eukaryotes, transcription takes place within the nucleus, whereas most translation takes place in the cytoplasm (although some translation does take place within the nucleus). Therefore, transcription and translation in eukaryotes are kept temporally and spatially separate. However, in bacterial cells transcription and translation occur nearly simultaneously.

Stability of mRNA in eukaryotic cells and bacterial cells is also different. Bacterial mRNA is typically short-lived, lasting only a few minutes. Eukaryotic mRNA may last hours or even days. Charging of the tRNAs with amino acids is essentially the same in both bacteria and eukaryotes. The ribosomes of bacteria and eukaryotes are different as well. Bacteria and eukaryotes have large and small ribosomal subunits, but they differ in size and composition. The bacterial large ribosomal consists of two ribosomal RNAs, while the eukaryotic large ribosomal subunit consists of three.

During translation initiation, the bacterial small ribosomal subunit recognizes the Shine-Dalgarno consensus sequence in the 5′ UTR of the mRNA and to regions of the 16S rRNA. In most eukaryotic mRNAs, the small subunit binds the 5′ cap of the mRNA and scans downstream until it encounters the first AUG codon. Finally, elongation and termination in bacterial and eukaryotic cells are functionally similar, although different elongation and termination factors are used.

Section 15.4

13. How do prokaryotic cells overcome the problem of a stalled ribosome on a mRNA that has no termination codon? What about eukaryotic cells?

When a ribosome has stalled on a mRNA, a bacteria cell uses transfer-messenger (tm)RNA to "move" the ribosome. The tmRNA consists of a tRNA component charged with alanine and a mRNA component. At the empty A-site of the stalled ribosome, elongation factor Tu delivers the charged tmRNA. The tmRNA acts as a tRNA and allows for the formation of a peptide bond between the alanine of the tmRNA and the peptide attached to the tRNA at the P site. The now uncharged tRNA from the P site is released and translation continues. However, the ribosome now uses the tmRNA as the template. Ten amino acids are added before a stop codon is reached. The result is the termination of translation and the release of the ribosome. The added amino acids target the peptide for degradation.

Eukaryotic cells use nonstop mRNA decay. When a ribosome is stalled, the codon free A-site of the ribosome is recognized and bound by a protein that recruits other proteins to degrade the mRNA starting at the 3′ end.

14. What are some types of posttranslational modification of proteins?
Several different modifications can occur to a protein following translation. Frequently, the amino terminal methionine may be removed. Sometimes, in bacteria only the formyl group is cleaved from the N-formyl methionine, leaving a methionine at the amino terminal. More extensive modification occurs in some proteins that are originally synthesized as precursor proteins. These precursor proteins are cleaved and trimmed by protease enzymes to produce a functional protein. Glycoproteins are produced by the attachment of carbohydrates to newly synthesized proteins. Molecular chaperones are needed by many proteins to ensure that the proteins are folded correctly. Secreted proteins that are targeted for the membrane or other cellular locations frequently have 15 to 30 amino acids, called the signal sequence, removed from the amino terminal. Finally, acetylation of amino acids in the amino terminal of some eukaryotic proteins also occurs.

*15. Explain how some antibiotics work by affecting the process of protein synthesis.
A number of antibiotics bind the ribosome and inhibit protein synthesis at different steps in translation. Some antibiotics, such as streptomycin, bind to the small subunit and inhibit translation initiation. Other antibiotics, such as chloramphenicol bind to the large subunit and block elongation of the peptide by preventing peptide bond formation. Antibiotics such as tetracycline and neomycin, bind the ribosome near the "A" site yet have different effects. Tetracyclines block entry of charged tRNAs to the "A" site, while neomycin induces translational errors. Finally, some antibiotics such as erythromycin, block the translocation of the ribosome along the mRNA.

APPLICATION QUESTIONS AND PROBLEMS

Section 15.1

*16. Sydney Brenner isolated *Salmonella typhimurium* mutants that were implicated in the biosynthesis of tryptophan and would not grow on minimal medium. When these mutants were tested on minimal medium to which one of four compounds (indole glycerol phosphate, indole, anthranilic acid, and tryptophan) had been added, the growth responses shown in the table on the facing page were obtained.

Give the order of indole glycerol phosphate, indole, anthranilic acid, and tryptophan in a biochemical pathway leading to the synthesis of tryptophan. Indicate which step in the pathway is affected by each of the mutations.

Mutant	Minimal medium	Anthranilic acid	Indole glycerol phosphate	Indole	Tryptophan
trp-1	–	–	–	–	+
trp-2	–	–	+	+	+
trp-3	–	–	–	+	+
trp-4	–	–	+	+	+
trp-6	–	–	–	–	+
trp-7	–	–	–	–	+
trp-8	-	+	+	+	+
trp-9	–	–	–	–	+
trp-10	–	–	–	–	+
trp-11	–	–	–	–	+

Based on the mutant strain's ability to grow on the above substrates, we can group the mutations into four groups that we will call group 1, group 2, group 3, and group 4.

Group 1 mutants can grow only on the minimal medium supplemented with trpytophan. Group 1: trp-*1,* trp-*10,* trp-*11,* trp-*9,* trp-*6, and* trp-*7.*

Group 2 mutants can grow on the minimal medium supplemented with either trpytophan or indole. Group 2: trp-*3.*

Group 3 mutants can grow on the minimal medium supplemented with tryptophan, indole, or indole glycerol phosphate. Group 3: trp-*2 and* trp-*4.*

Group 4 mutants can grow on minimal medium supplemented with the addition of tryptophan, indole, indole glycerol phosphate, or anthranilic acid. Group 4: trp-*8.*

By examining the compounds needed for growth by the different groups of mutants, we can identify the step in the pathway that is blocked by each mutation. For each group, the pathway step blocked will correspond to the step proceeding the last compound on which a mutant strain can grow. Any compound added to the minimal media that proceeds the block will not allow for growth of the mutant strain.

Group 1 mutants can only grow when tryptophan is added to the growth medium. So group 1 mutants are blocked at the last step in the biosynthesis process before tryptophan is synthesized. Because group 2 mutants can grow with either tryptophan or indole added to the growth medium, this suggests that in the pathway indole is the immediate precursor of tryptophan and that group 2 mutants are blocked in the step preceding the synthesis of indole. Using the same type of analysis, we can create the pathway on the following page for synthesis of tryptophan.

Group 4 Group 3 Group 2 Group 1
Precursor → anthranilic acid → Indole glyerol phosphate → Indole →Tryptophan

17. Compounds I, II, and III are in the following biochemical pathway:

precursor → compound I → compound II → compound III
enzyme A enzyme B enzyme C

Mutation *a* inactivates enzyme A, mutation *b* inactivates enzyme B, and mutation *c* inactivates enzyme C. Mutants, each having one of these defects, were tested on minimal medium to which compound I, II, and III was added. Fill in the results expected of these tests by placing a plus sign (+) for growth or a minus sign (–) for no growth in the following table:

Strain with mutation	Minimal medium to which is added: Compound I	Compound II	Compound III
a	+	+	+
b	–	+	+
c	–	–	+

To determine whether growth will occur on the minimal medium with the added compound, the step in the pathway where the mutation occurs must be considered. Because mutation a *affects enzyme A, then any strain with mutation* a *can only grow on minimal media with the addition of compound I, compound II, or compound III. Strains containing mutation* b *can grow only with the addition of either compound II or compound III because enzyme B, which converts compound I to compound II, has been affected. Strains with mutation* c *can only grow with the addition of compound III since the enzyme needed to synthesize compound III from compound II has been mutated.*

Section 15.2

*18. Assume that the number of different types of bases in RNA is four. What would be the minimum codon size (number of nucleotides) required if the number of different types of amino acids in proteins were:
(The number of codons possible must be equal to or greater than the number of different types of amino acids because the codons encode for the different amino acids. To calculate how many possible codons that are possible for a given codon size with

four different types of bases in the RNA, the following formula can be used: 4^n, *where* n *is the number of nucleotides within the codon.)*

a. 2

1, because 4^1 *= 4 codons, which is more than enough to specify 2 different amino acids.*

b. 8

2

c. 17

3

d. 45

3

e. 75

4

19. How many codons would be possible in a triplet code if only three bases (A, C, and U) were used?

To calculate the number of possible codons of a triplet code if only three bases are used, the following equation can be used: 3^n, *where* n *is the number of nucleotides within the codon. So, the number of possible codons is equal to* 3^3, *or 27 possible codons.*

*20. Using the genetic code given in Figure 15.10, give the amino acids specified by the bacterial mRNA sequences and indicate the amino and carboxyl ends of the polypeptide produced.

Each of the mRNA sequences begins with the three nucleotides AUG. This indicates the start point for translation and allows for a reading frame to be set. In bacteria, the AUG initiation codon codes for N-formyl-methionine. Also, for each of these mRNA sequences, a stop codon is present either at the end of the sequence or within the interior of the sequence.

The amino terminal refers to the end of the protein with a free amino group and will be the first peptide in the chain. The carboxyl terminal refers to the end of the protein with a free carboxyl group and is the last amino acid in the chain. For the following peptide chains reading from left to right, the first amino acid is located at the amino end, while the last amino acid is located at the carboxyl end.

a. 5′–AUGUUUAAAUUUAAAUUUUGA–3′

Amino fMet–Phe–Lys–Phe–Lys–Phe Carboxyl

b. 5′–AUGUAUAUAUAUAUAUGA–3′

Amino fMet–Tyr–Ile–Tyr–Ile Carboxyl

c. 5′–AUGGAUGAAAGAUUUCUCGCUUGA–3′

Amino fMet–Asp–Glu–Arg–Phe–Leu–Ala Carboxyl

d. 5′–AUGGGUUAGGGGACAUCAUUUUGA–3′

Amino fMet–Gly Carboxyl (The stop codon UAG occurs after the codon for glycine.)

21. A nontemplate strand on DNA has the following base sequence. What amino acid sequence would be encoded by this sequence?

5′–ATGATACTAAGGCCC–3′

To determine the amino acid sequence, we need to know the mRNA sequence and the codons present. The nontemplate strand of the DNA has the same sequence as the mRNA, except that thymine containing nucleotides are substituted for the uracil containing nucleotides. So the mRNA sequence would be as follows: 5′–AUGAUACUAAGGCCC–3′.

Assuming that the AUG indicates a start codon, then the amino acid sequence would be starting from the amino end of the peptide and ending with the carboxyl end: fMet–Ile–Leu–Arg–Pro.

*22. The following amino acid sequence is found in a tripeptide: Met–Trp–His. Give all possible nucleotide sequences on the mRNA, on the template strand of DNA, and on the nontemplate strand of DNA that could encode this tripeptide.

The potential mRNA nucleotide sequences encoding for the tripeptide Met–Trp–His can be determined by using the codon table found in Figure 15.14. From the table, we can see that the amino acid His has two potential codons, while the amino acids Met and Trp each have only one potential codon. Therefore, there are two different mRNA nucleotide sequences that could encode for the tripeptide. Once the potential mRNA nucleotide sequences have been determined, the template and nontemplate DNA strands can be derived from these potential mRNA sequences.

(1) 5′–AUGUGGCAU–3′

DNA template:	*3′–TACACCGTA–5′*
DNA nontemplate:	*5′–ATGUGGCAT–3′*

(2) 5′–AUGUGGCAC–3′

DNA template:	*3′–TACACCGTG–5′*
DNA nontemplate:	*5′–ATGTGGCAC–3′*

23. How many different mRNA sequences can code for a polypeptide chain with the amino acid sequence Met–Leu–Arg? (Be sure to include the stop codon.)

From Figure 15.14, we can determine that leucine and arginine each have six different potential codons. There are also three potential stop codons. As for methionine, only one codon, AUG, is typically found as the initiation codon. (However, UUG and GUG have been shown to serve as start codons on occasion. For this problem, we will ignore these rare cases.) Therefore, the number of potential sequences is the product of the number of different potential codons for this tripeptide, which gives us a total of $(1 \times 6 \times 6 \times 3) = 108$ different mRNA sequences that can code for the tripeptide Met–Leu–Arg.

*24. A series of tRNAs have the following anticodons. Consider the wobble rules given in Table 15.2 and give all possible codons with which each tRNA can pair.

From the wobble rules outlined in Table 15.2, we can see that when "A" occurs at the 5′ of the anticodon it can pair only with "U" in the 3′ end of the codon. When "C" is present at the 5′ of the anticodon, it can only pair with "G" at the 3′ of the codon. However, both "U" and "G," when present at the 5′ end of the anticodon, can pair with two different nucleotides at the 3′ end of the codon (U with A or G; and G with U or C). The rare base iosine (I) is also found at the 5′ of the anticodon of tRNA on occasion. Iosine can pair with "A," "U," or "C" at the 3′ end of the codon.

a. 5′–GGC–3′
Codons: 3′–CCG–5′ or 3′–UCG–5′

b. 5′–AAG–3′
Codon: 3′–UUC–5′

c. 5′–IAA–3′
Codons: 3′–AUU–5′ or 3′–UUU–5′ or 3′–CUU–5′

d. 5′–UGG–3′
Codons: 3′–ACC–5′ or 3′–GCC–5′

e. 5′–CAG–3′
Codon: 3′–GUC–5′

25. An anticodon on a tRNA has the sequence 5′–GCA–3′.

a. What amino acid is carried by this tRNA?
The anticodon 5′–GCA–3′ would pair with the codon 5′–CGU–3′. Based on the codon table in Figure 15.14, the amino acid encoded by this codon is cysteine. So, this tRNA is most likely carrying cysteine.

b. What would be the effect if the G in the anticodon were mutated to a U?
The anticodon would now be 3′–ACU–5′ and could pair to the codon 5′–UGA–3′, a stop codon. The result would be that amino acid cysteine would be placed where the stop codon 5′–UGA–3′ was located in the mRNA. Essentially, the stop codon would be suppressed and translation could continue.

26. Which of the following amino acid changes could result from a mutation that changed a single base? For each change that could result from the alteration of a single base, determine which position of the codon (first, second, or third nucleotide) in the mRNA must be altered for the change to occur.

a. Leu → Gln
Of the six codons that encode for Leu, only two could be mutated by the alteration of a single base to produce the codons for Gln:
CUA (Leu)—Change the second position to A to produce CAA (Gln).
CUG (Leu)—Change the second position to A to produce CAG (Gln).

b. Phe → Ser
Both Phe codons (UUU and UUC) could be mutated at the second position to produce Ser codons:
UUU (Phe)—Change the second position to C to produce UCU (Ser).
UUC (Phe)—Change the second postion to C to produce UCC (Ser).

c. Phe → Ile
Both Phe codons (UUU and UUC) could be mutated at the first position to produce Ile codons:
UUU (Phe)—Change the first position to A to produce AUU (Ile).
UUC (Phe)—Change the first position to A to produce AUC (Ile).

d. Pro → Ala
All four codons for Pro can be mutated at the first position to produce Ala codons:

CCU (Pro)—Change the first position to G to produce GCU (Ala).
CCC (Pro)—Change the first position to G to produce GCC (Ala).
CCA (Pro)—Change the first position to G to produce GCA (Ala).
CCG (Pro)—Change the first position to G to produce GCG (Ala).

e. Asn → Lys
Both codons for Asn can be mutated at a single position to produce Lys codons:
AAU (Asn)—Change the third position to A to produce AAA (Lys).
AAU (Asn)—Change the third position to G to produce AAG (Lys).
AAC (Asn)—Change the third postion to A to produce AAA (Lys).
AAC (Asn)—Change the third position to G to produce AAG (Lys).

f. Ile → Asn
Only two of the three Ile codons can be mutated at a single position to produce Asn codons:
AUU (Ile)—Change the second position to A to produce AAU (Asn).
AUC (Ile)—Change the second position to A to produce AAC (Asn).

Section 15.3

27. Arrange the following components of translation in the approximate order in which they would appear or be used during protein synthesis.
The components are in order according to when they are used or play a key role in translation. The potential exception is initiation factor 3. Initiation factor 3 could possibly be listed first because it is necessary to prevent the 30s ribosome from associating with the 50s ribosome. It binds to the 30s subunit prior to the formation of the 30s initiation complex. However, during translation events the release of initiation factor 3, allows the 70s initiation complex to form, a key step in translation.

fMet-tRNA^{fMet}

30S initiation complex

initiation factor 3

70S initiation complex

elongation factor Tu

peptidyl transferase

elongation factor G

release factor 1

28. The following diagram illustrates a step in the process of translation. Sketch the diagram and identify the following elements on it.
   a. 5′ and 3′ ends of the mRNA
   b. A, P, and E sites
   c. Start codon
   d. Stop codon
   e. Amino and carboxyl ends of the newly synthesized polypeptide chain
   f. Approximate location of the next peptide bond that will be formed
   g. Place on the ribosome where release factor 1 will bind

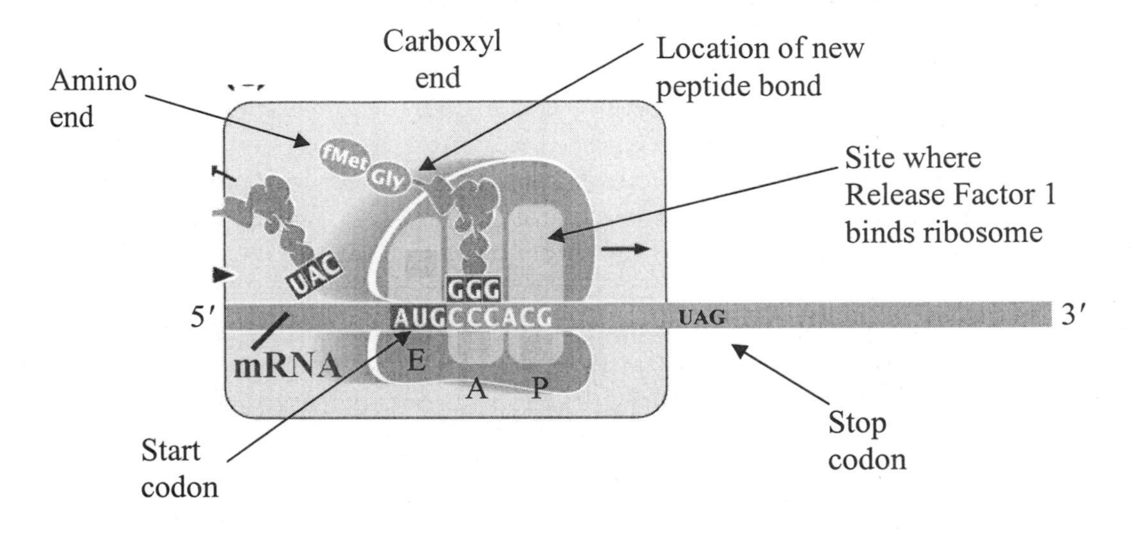

29. Refer to the diagram in Problem 28 to answer the following questions.
   a. What will be the anticodon of the next tRNA added to the A site of the ribosome?
   The anticodon 5′ CGU 3′ is complementary to the codon 5′ ACG 3′, which is located at the A site of the ribosome.

   b. What will be the next amino acid added to the growing polypeptide chain?
   The codon 5′ ACG 3′ encodes the amino acid threonine.

*30. A synthetic mRNA added to a cell-free, protein-synthesizing system produces a peptide with the following amino acid sequence: Met–Pro–Ile–Ser–Ala. What would be the effect on translation if the following components were omitted from the cell-free, protein-synthesizing system? What, if any, type of protein would be produced? Explain your reasoning.
   a. Initiation factor 1
   The lack of IF-1 would decrease the amount of protein synthesized. IF-1 promotes the disassociation of the large and small ribosomal subunits. Since translation initiation requires a free small subunit, initiation would occur at a slower rate because more of the small ribosomal subunits would remain bound to the large ribosomal subunits.

b. Initiation factor 2
No translation would occur. IF-2 is necessary for translation initiation. The lack of IF-2 would prevent fMet-$tRNA_f^{met}$ *from being delivered to the small ribosomal subunit, thus blocking translation.*

c. Elongation factor Tu
Although translation initiation or delivery of the Met to the ribosome-mRNA complex would occur, no further amino acids would be delivered to the ribosome. EF-Tu is necessary for elongation because it binds GTP and the charged tRNA, forming a three-part complex that enters the "A" site of the ribosome. If EF-Tu is not present, then the charged tRNA will not enter the "A" site, thus stopping translation.

d. Elongation factor G
EF-G is necessary for the movement of the ribosome along the mRNA in a 5′ to 3′ direction. Once the formation of the peptide bond occurs between the Met and Pro, the lack of EF-G would prevent the movement of the ribosome along the mRNA, so no new codons would be read. However, the dipeptide Met-Pro would be formed because its formation would not require EF-G.

e. Release factors RF_1, RF_2, and RF_3
The release factors RF_1 *and* RF_2 *recognize the stop codons and bind to the ribosome at the A site. They then interact with* RF_3 *to promote cleavage of the peptide from the tRNA at the "P" site. The absence of the release factors prevents termination of translation at the stop codon, resulting in a larger peptide.*

f. ATP
ATP is required for the charging of the tRNAs with amino acids by the aminoacyl-tRNA synthetases. Without ATP, the charging would not take place, and no amino acids will be available for protein synthesis. So no protein synthesis will occur.

g. GTP
GTP is required for initiation, elongation, and termination of translation. If GTP is absent, no protein synthesis will occur.

31. For each of the following sequences, place a check mark in the appropriate space to indicate the process <u>most immediately</u> affected by deleting the sequence. You should chose only one process for each sequence (one check mark per sequence).

	Process most immediately affected by deletion			
Sequence deleted	*Replication*	*Transcription*	*RNA processing*	*Translation*
a. ori site	✓			
The ori site or origin of replication is necessary for the initiation of replication.				
b. 3′ splice site *consensus sequence*			✓	
The 3′ splice site is necessary for proper excision of the intron. Therefore, RNA processing events will be affected.				
c. poly(A) tail				✓
The poly(A) tail is involved in mRNA stability. If the tail is missing, then the mRNA will be degraded more rapidly thus affecting translation.				
d. terminator		✓		
The terminator is necessary for transcription termination. Deletion of the terminator will result in the production of an abnormally long RNA transcript.				
e. start codon				✓
The start codon is necessary for translation initiation.				
f. –10 consensus seq		✓		
In bacteria, the –10 sequence is an important component of the promoter. Deletion of the –10 sequence will prevent transcription initiation from occurring.				
g. Shine-Dalgarno				✓
The Shine-Dalgarno sequence or ribosome binding site is bound by the 30s subunit during the initiation of translation. If the sequence is deleted, then the ribosome will not bind to the mRNA molecule and translation will not occur.				

32. MicroRNAs are small RNA molecules that bind to the 3′ end of mRNAs and suppress translation (see Chapter 14). How miRNAs suppress translation is still being investigated. Some eukaryotic mRNAs have internal ribosome-binding sites downstream of the 5′ cap, where ribosomes normally bind. In one investigation, miRNAs did not suppress the translation of ribosomes that attach to internal ribosome-binding sites (R. S. Pillai et al. 2005. *Science* 309:1573–1576). What does this finding suggest about how miRNAs suppress translation?

In eukaryotic cells, the ribosome can bind at the *5′ cap or to the mRNA at an internal site. The observation suggests that the miRNAs in this study affected translation by preventing the binding of ribosomes to 5′ cap region of the mRNA perhaps by interfering with the ability of the ribosomes or cap binding proteins to recognize the 5′ cap.*

Section 15.4

33. Mutations that introduce stop codons cause a number of genetic diseases. For example, from 2% to 5% of the people who have cystic fibrosis possess a mutation that causes a premature stop codon in the cystic fibrosis transmembrane conductance regulator (CFTR) gene. This premature stop codon produces a truncated form of CFTR that is nonfunctional and results in the symptoms of cystic fibrosis. One possible way to treat people with genetic diseases caused by these types of mutations is to trick the ribosome into reading through the stop codon, inserting an amino acid into its place. Although the protein produced may have one altered amino acid, it is more likely to be at least partly functional than is the truncated protein produced when the ribosome stalls at the stop codon. Indeed, geneticists have begun to conduct clinical trials on people with cystic fibrosis with the use of a drug called PTC124, which interferes with the ribosome's ability to correctly read stop codons (C. Ainsworth. 2005. *Nature* 438:726–728). On the basis of what you know about the mechanism of nonsense-mediated mRNA decay (NMD), would you expect NMD to be a problem with this type of treatment? Why or why not?
No, NMD should not be a problem with this type of treatment. NMD is thought to be dependent on exon-junction proteins that are not removed from the mRNA by the movement of the ribosomes during translation. These proteins are thought to interact with enzymes that degrade the mRNA. If the first ribosome to read the mRNA inserts an amino acid for the stop codon due to the action of PTC124, then it should not stall at the stop codon. The ribosome's continued movement along the entire mRNA molecule without stalling should remove any exon-junction proteins before the RNA degrading enzymes are recruited. The result is that the mRNA will be stabilized thus allowing more translation events to occur.

CHALLENGE QUESTIONS

Section 15.2

34. The redundancy of the genetic code means that some amino acids are specified by more than one codon. For example, the amino acid leucine is encoded by six different codons. Within a genome, synonymous codons do not occur equally; some synonymous codons occur much more frequently than others, and the preferred codons differ among different species. For example, in one species the codon UUA might be used most often to code for leucine, while in another species the codon CUU might be used most often. Speculate on a reason for this bias in codon usage and why the preferred codons are not the same in all organisms.
Synonymous codon usage patterns may depend on a variety of factors. Two potential factors that could affect usage patterns are the GC content of the organism and the relative amounts of isoaccepting tRNA molecules. The GC content of an organism reflects the relative proportions of nucleotides found in the DNA. In a given organism, the bias for particular synonymous codons may reflect the overall GC content of that organism. For example, in organisms that have a high GC content, you might expect to find that the synonymous codon usage pattern reflects this bias, resulting in a

preference for codons with more Gs and Cs. Therefore, in two organisms that differ in GC content, the synonymous codon usage bias should reflect their differences in base composition.

Isoaccepting tRNAs are those that carry the same amino acid but have different anticodons. These isoaccepting tRNAs act at synonymous codons. For a given organism, the more prevalent synonymous codons may depend on the frequency of its tRNA and complementary anticodon. In different organisms, the concentrations of the various isoaccepting tRNAs will vary leading to different usage patterns of synonymous codons in these organisms.

Section 15.2

35. In what ways are spliceosomes and ribosomes similar? In what ways are they different? Can you suggest some possible reasons for their similarities?
Spliceosomes and ribosomes are both large complexes that are composed of several different RNA and protein molecules. In essence, the spliceosome and ribosome are RNA-based enzymes or ribozymes. The RNA molecules in both structures are necessary for catalysis. The 23S RNA molecule in the ribosome catalyzes the formation of peptide bonds between amino acids. Other rRNAs are also important for protein synthesis. In spliceosomes, the snRNA molecules catalyze the cutting and splicing of pre-mRNA molecules to produce mature mRNA. The catalytic RNAs of the ribosome and the spliceosome may have originated during the RNA world when RNA molecules served to store information and to catalyze reactions that sustained life.

*36. Several experiments were conducted to obtain information about how the eukaryotic ribosome recognizes the AUG start codon. In one experiment, the gene that codes for methionine initiator tRNA ($tRNA_i^{Met}$) was located and changed. The nucleotides that specify the anticodon on $tRNA_i^{Met}$ were mutated so that the anticodon in the tRNA was 5′–CCA–3′ instead of 5′–CAU–3′. When this mutated gene was placed into a eukaryotic cell, protein synthesis took place, but the proteins produced were abnormal. Some of the proteins produced contained extra amino acids, and others contained fewer amino acids.

a. What do these results indicate about how the ribosome recognizes the starting point for translation in eukaryotic cells? Explain your reasoning.
By mutating the anticodon to 5′–CCA–3′ from 5′–CAU–3′ on $tRNA_i^{Met}$, the initiator tRNA will now recognize the codon 5′–UGG–3′, which normally would code only for Trp. If translation initiation by the ribosome in eukaroytes occurs by binding the 5′ cap of the mRNA followed by scanning, then the first 5′–UGG–3′ codon recognized by the mutated $tRNA_i^{Met}$ will be the start site for translation. If the first 5′–UGG–3′ codon occurs prior to the normal 5′–AUG–3′ codon, then a protein containing extra amino acids could be produced. If the first 5′–UGG–3′ codon occurs after the normal 5′–AUG–3′, then a shorter protein will be produced. Finally, truncated proteins could also be produced by the first 5′–UGG–3′ being out of frame of the normal coding sequence. If this happens, then most likely a stop codon will be encountered before the end of the normal coding sequence and will

terminate translation. The data suggest that translation initiation takes place by scanning of the ribosome for the appropriate start sequence.

b. If the same experiment had been conducted on bacterial cells, what results would you expect?

Very little or no protein synthesis would be expected. Translation initiation in bacteria requires the 16S RNA of the small ribosomal subunit to interact with the Shine-Dalgarno sequence. This interaction serves to line up the ribosome over the start codon. If the anticodon has been changed such that the start codon cannot be recognized, then protein synthesis is not likely to take place.

1-8, 9 (first 2), 10-21a, 22

Chapter Sixteen: Control of Gene Expression

COMPREHENSION QUESTIONS

Section 16.1

1. Why is gene regulation important for bacterial cells?
   Gene regulation allows for biochemical and internal flexibility while maintaining energy efficiency by the bacterial cells.

Section 16.2

*2. Name six different levels at which gene expression might be controlled.
   (1) Alteration or modification of the gene structure at the DNA level
   (2) Transcriptional regulation
   (3) Regulation at the level of mRNA processing
   (4) Regulation of mRNA stability
   (5) Regulation of translation
   (6) Regulation by post-translational modification of the synthesized protein

Section 16.3

*3. Draw a picture illustrating the general structure of an operon and identify its parts.

OPERON

5′ Regulator Protein Gene | Promoter   Structural Genes 3′

Operator

4. What is difference between positive and negative control? What is the difference between inducible and repressible operons?
   Positive transcriptional control requires an activator protein to stimulate transcription at the operon. In negative control, a repressor protein inhibits or turns off transcription at the operon.

   An inducible operon normally is not transcribed. It requires an inducer molecule to stimulate transcription either by inactivating a repressor protein in a negative inducible operon or by stimulating the activator protein in a positive inducible operon.

   Transcription normally occurs in a repressible operon. In a repressible operon, transcription is turned off either by the repressor becoming active in a negative repressible operon or by the activator becoming inactive in a positive repressible operon.

*5. Briefly describe the *lac* operon and how it controls the metabolism of lactose.
The lac *operon consists of three structural genes involved in lactose metabolism, the* lacZ *gene, the* lacY *gene, and the* lacA *gene. Each of these three genes has a different role in the metabolism of lactose. The* lacZ *gene codes for the enzyme β-galactosidase, which breaks the disaccharide lactose into galactose and glucose, and converts lactose into allolactose. The* lacY *gene, located downstream of the* lacZ *gene, codes for lactose permease. Permease is necessary for the passage of lactose through the E. coli cell membrane. The* lacA *gene, located downstream of* lacY, *encodes the enzyme thiogalactoside transacetylase whose function in lactose metabolism has not yet been determined. All of these genes share a common overlapping promoter and operator region. Upstream from the lactose operon is the* lacI *gene that encodes the* lac *operon repressor. The repressor binds at the operator region and inhibits transcription of the* lac *operon by preventing RNA polymerase from successfully initiating transcription.*

When lactose is present in the cell, the enzyme β-galactosidase converts some of it into allolactose. Allolactose binds to the lac *repressor, altering its shape and reducing the repressor's affinity for the operator. Since this allolactose-bound repressor does not occupy the operator, RNA polymerase can initiate transcription of the* lac *structural genes from the* lac *promoter.*

6. What is catabolite repression? How does it allow a bacterial cell to use glucose in preference to other sugars?
In catabolite repression, the presence of glucose inhibits or represses the transcription of genes involved in the metabolism of other sugars. Because the gene expression necessary for utilizing other sugars is turned off, only enzymes involved in the metabolism of glucose will be synthesized. Operons that exhibit catabolite repression are under the positive control of catabolic activator protein (CAP). For CAP to be active, it must form a complex with cAMP. Glucose affects the level of cAMP. The levels of glucose and cAMP are inversely proportional—as glucose levels increase, the level of cAMP decreases. Thus, CAP is not activated.

Section 16.4

*7. What is attenuation? What is the mechanism by which the attenuator forms when tryptophan levels are high and the antiterminator forms when tryptophan levels are low?
Attenuation is the termination of transcription prior to the structural genes of an operon. It is a result of the formation of a termination hairpin structure or attenuator in the mRNA.

Two types of secondary structures can be formed by the mRNA 5′ UTR of the trp *operon. If the 5′ UTR forms two hairpin structures from the base pairing of region 1 with region 2 and the pairing of region 3 with region 4, then transcription of the structural genes will not occur. The hairpin structure formed by the pairing of region 3 with region 4 results in a terminator being formed that stops transcription. When region 2 pairs with region 3, the resulting hairpin acts as an antiterminator allowing for transcription to proceed. Region 1 of the 5′ UTR also encodes a small protein and has two adjacent tryptophan codons (UGG).*

Tryptophan levels affect transcription due to the coupling of translation with transcription in bacterial cells. When tryptophan levels are high, the ribosome quickly moves through region 1 and into region 2, thus preventing region 2 from pairing with region 3. Therefore, region 3 is available to form the attenuator hairpin structure with region 4, stopping transcription. When tryptophan levels are low, the ribosome stalls or stutters at the adjacent tryptophan codons in region 1. Region 2 now becomes available to base pair with region 3, forming the antiterminator hairpin. Transcription can now proceed through the structural genes.

Section 16.5

*8. What is antisense RNA? How does it control gene expression?
Antisense RNA molecules are small RNA molecules that are complementary to other DNA or RNA sequences and that form RNA-protein complexes. In bacterial cells, antisense RNA molecules can bind to a complementary region in the 5′ UTR of a mRNA molecule, blocking the attachment of the ribosome to the mRNA and stopping translation or they pair with specific regions of the mRNA and cleave the mRNA stopping translation.

Section 16.6

9. What are riboswitches? How do they control gene expression? How do riboswitches differ from RNA-mediated repression?
Riboswitches are regulatory sequences in RNA molecules. Most can fold into compact secondary structures consisting of a base stem and several branching hairpins. At riboswitches, regulatory molecules bind and influence gene expression by affecting the formation of secondary structures within the mRNA molecule. The binding of the regulatory molecule to a riboswitch sequence may result in repression or induction. Some regulatory molecules bind the riboswitch sequence and stabilize a terminator structure in the mRNA, which results in premature termination of the mRNA molecule. Other regulatory molecules bind riboswitch sequences resulting in the formation of secondary structures that block the ribosome binding sites of the mRNA molecules, thus preventing translation initiation. In induction, the regulatory molecule acts as an inducer, stimulating the formation of a secondary structure in the mRNA that allows for transciption or translation to occur.

RNA-mediated repression occurs through the action of a ribozyme. In RNA-mediated repression, a RNA sequence within the 5′ untranslated region can act as a ribozyme that when stimulated by the presence of a regulatory molecule can induce self-cleavage of the mRNA molecule, which prevents translation of the molecule. When bound by a regulatory molecule, RNA mediated repression results in the self-cleavage of the mRNA molecule. When bound by a regulatory molecule, riboswitch sequences stimulate changes in the secondary structure of the mRNA molecule that affect gene expression.

APPLICATION QUESTIONS AND PROBLEMS

Section 16.3

*10. For each of the following types of transcriptional control, indicate whether the protein produced by the regulator gene will be synthesized initially as an active repressor, inactive repressor, active activator, or inactive activator.

a. Negative control in a repressible operon
Inactive repressor

b. Positive control in a repressible operon
Active activator

c. Negative control in an inducible operon
Active repressor

d. Positive control in an inducible operon
Inactive activator

*11. A mutation occurs at the operator site that prevents the regulator protein from binding. What effect will this mutation have in the following types of operons?

a. Regulator protein is a repressor in a repressible operon.
The regulator protein-corepressor complex would normally bind to the operator and inhibit transcription. If a mutation prevented the repressor protein from binding at the operator, then the operon would never be turned off and transcription would occur all the time.

b. Regulator protein is a repressor in an inducible operon.
In an inducible operon, a mutation at the operator site that blocks binding of the repressor would result in constitutive expression and transcription would occur all the time.

12. The *blob* operon produces enzymes that convert compound A into compound B. The operon is controlled by a regulatory gene *S*. Normally, the enzymes are synthesized only in the absence of compound B. If gene *S* is mutated, the enzymes are synthesized in the presence *and* in the absence of compound B. Does gene *S* produce a repressor or an activator? Is this operon inducible or repressible?
Because the blob *operon is transcriptionally inactive in the presence of B, gene* S *most likely codes for a repressor protein that requires compound B as a corepressor. The data suggest that the* blob *operon is repressible because it is inactive in the presence of compound B, but active when compound B is absent.*

*13. A mutation prevents the catabolite activator protein (CAP) from binding to the promoter in the *lac* operon. What will be the effect of this mutation on transcription of the operon?
Catabolite activator protein binds the CAP site of the lac *operon and stimulates RNA polymerase to bind the* lac *promoter, thus resulting in increased levels of transcription from the* lac *operon. If a mutation prevents CAP from binding to the site, then RNA polymerase will bind the* lac *promoter poorly. This will result in significantly lower levels of transcription of the* lac *structural genes.*

14. Under which of the following conditions would a *lac* operon produce the greatest amount of β-galactosidase? The least? Explain your reasoning.

	Lactose present	Glucose present
Condition 1	Yes	No
Condition 2	No	Yes
Condition 3	Yes	Yes
Condition 4	No	No

Condition 1 will result in the production of the maximum amount of β-galactosidase. For maximum transcription, the presence of lactose and the absence of glucose are required. Lactose (or allolactose) binds to the lac *repressor reducing the affinity of the* lac *repressor to the operator. This decreased affinity results in the promoter being accessible to RNA polymerase. The lack of glucose allows for increased synthesis of cAMP, which can complex with CAP. The formation of CAP-cAMP complexes improves the efficiency of RNA polymerase binding to the promoter, which results in higher levels of transcription from the* lac *operon.*

Condition 2 will result in the production of the least amount of β-galactosidase. With no lactose present, the lac *repressor is active and binds to the operator, inhibiting transcription. The presence of glucose results in a decrease of cAMP levels. A CAP-cAMP complex does not form, and RNA polymerase will not be stimulated to transcribe the* lac *operon.*

15. A mutant strain of *E. coli* produces β-galactosidase in the presence *and* in the absence of lactose. Where in the operon might the mutation in this strain occur?
Within the operon, the operator region is the most probable location of the mutation. If the mutation prevents the lac *repressor protein from binding to the operator, then transcription of the* lac *structural genes will not be inhibited. Expression will be constitutive.*

*16. For *E. coli* strains with the following *lac* genotypes, use a plus sign (+) to indicate the synthesis of β-galactosidase and permease or a minus sign (−) to indicate no synthesis of the enzymes.
In determining if expression of the β-galactosidase and the permease gene will occur, you should consider several factors. The presence of $lacZ^+$ *and* $lacY^+$ *on the same DNA molecule as a functional promoter (*$lacP^+$*) is required because the promoter is a cis-acting regulatory element. However the* $lacI^+$ *gene product or* lac *repressor is trans-acting and does not have to be located on the same DNA molecule as β-galactosidase and permease genes to inhibit expression. For the repressor to function, it does require that the cis-acting* lac *operator be on the same DNA molecule as the functional β-galactosidase and permease genes. Finally, the dominant* $lacI^s$ *gene product is also trans-acting and can inhibit transcription at any functional* lac *operator region.*

Genotype of strain	Lactose absent		Lactose present	
	β-Galactosidase	Permease	β-Galactosidase	Permease
$lacI^{+}lacP^{+}lacO^{+}lacZ^{+}lacY^{+}$	–	–	+	+
$lacI^{-}lacP^{+}lacO^{+}lacZ^{+}lacY^{+}$	+	+	+	+
$lacI^{+}lacP^{+}lacO^{c}lacZ^{+}lacY^{+}$	+	+	+	+
$lacI^{-}lacP^{+}lacO^{+}lacZ^{+}lacY^{-}$	+	–	+	–
$lacI^{-}lacP^{-}lacO^{+}lacZ^{+}lacY^{+}$	–	–	–	–
$lacI^{+}lacP^{+}lacO^{+}lacZ^{-}lacY^{+}$/ $lacI^{-}lacP^{+}lacO^{+}lacZ^{+}lacY^{-}$	–	–	+	+
$lacI^{-}lacP^{+}lacO^{c}lacZ^{+}lacY^{+}$/ $lacI^{+}lacP^{+}lacO^{+}lacZ^{-}lacY^{-}$	+	+	+	+
$lacI^{-}lacP^{+}lacO^{+}lacZ^{+}lacY^{-}$/ $lacI^{+}lacP^{-}lacO^{+}lacZ^{-}lacY^{+}$	–	–	+	–
$lacI^{+}lacP^{-}lacO^{c}lacZ^{-}lacY^{+}$/ $lacI^{-}lacP^{+}lacO^{+}lacZ^{+}lacY^{-}$	–	–	+	–
$lacI^{+}lacP^{+}lacO^{+}lacZ^{+}lacY^{+}$/ $lacI^{+}lacP^{+}lacO^{+}lacZ^{+}lacY^{+}$	–	–	+	+
$lacI^{S}lacP^{+}lacO^{+}lacZ^{+}lacY^{-}$/ $lacI^{+}lacP^{+}lacO^{+}lacZ^{-}lacY^{+}$	–	–	–	–
$lacI^{S}lacP^{-}lacO^{+}lacZ^{-}lacY^{+}$/ $lacI^{+}lacP^{+}lacO^{+}lacZ^{+}lacY^{+}$	–	–	–	–

17. Give all possible genotypes of a *lac* operon that produces β-galactosidase and permease under the following conditions. Do not give partial diploid genotypes.

	Lactose absent		Lactose present		
	β-Galactosidase	Permease	β-Galactosidase	Permease	Genotype
a.	−	−	+	+	$lacI^+ lacP^+ lacO^+ lacZ^+ lacY^+$
b.	−	−	−	+	$lacI^+ lacP^+ lacO^+ lacZ^- lacY^+$
c.	−	−	+	−	$lacI^+ lacP^+ lacO^+ lacZ^+ lacY^-$
d.	+	+	+	+	$lacI^- lacP^+ lacO^+ lacZ^+ lacY^+$
					or
					$lacI^+ lacP^+ lacO^c lacZ^+ lacY^+$
e.	−	−	−	−	$lacI^s lacP^+ lacO^+ lacZ^+ lacY^+$
					or
					$lacI^+ lacP^- lacO^+ lacZ^+ lacY^+$
f.	+	−	+	−	$lacI^- lacP^+ lacO^+ lacZ^+ lacY^-$
					or
					$lacI^+ lacP^+ lacO^c lacZ^+ lacY^-$
g.	−	+	−	+	$lacI^- lacP^+ lacO^+ lacZ^- lacY^+$
					or
					$lacI^+ lacP^+ lacO^c lacZ^- lacY^+$

*18. Explain why mutations at the *lacI* gene are trans in their effects, but mutations in the *lacO* gene are cis in their effects.
The lacI *gene encodes the* lac *repressor protein, which can diffuse within the cell and attach to any operator. It can therefore affect the expression of genes on the same or different molecules of DNA. The* lacO *gene encodes the operator. The binding of the* lac *repressor to the operator affects the binding of RNA polymerase to the DNA, and therefore affects only the expression of genes on the same molecule of DNA.*

*19. The *mmm* operon, which has sequences *A, B, C,* and *D,* encodes enzymes 1 and 2. Mutations in sequences *A, B, C,* and *D* have the following effects, where a plus sign (+) = enzyme synthesized, and a minus sign (−) = enzyme not synthesized.

	mmm absent		*mmm* present	
Mutation in sequence	Enzyme 1	Enzyme 2	Enzyme 1	Enzyme 2
No mutation	+	+	−	−
A	−	+	−	−
B	+	+	+	+
C	+	−	−	−
D	−	−	−	−

a. Is the *mmm* operon inducible or repressible?
The data from the strain with no mutation indicate that the mmm *operon is repressible. The operon is expressed in the absence of* mmm, *but inactive in the presence of* mmm. *This is typical of a repressible operon.*

b. Indicate which sequence (*A, B, C,* or *D*) is part of the following components of the operon:

Regulator gene __B__

When sequence B *is mutated, gene expression is not repressed by the presence of mmm.*

Promoter __D__

When sequence D *is mutated, no gene expression occurs either in the presence or absence of* mmm.

Structural gene for enzyme 1 __A__

When sequence A *is mutated, enzyme 1 is not produced.*

Structural gene for enzyme 2 __C__

When sequence C *is mutated, enzyme 2 is not produced.*

20. Ellis Engelsberg and his coworkers examined the regulation of genes taking part in the metabolism of arabinose, a sugar (E. Engelsberg et al. 1965. *Journal of Bacteriology* 90:946–957). Four structural genes encode enzymes that help metabolize arabinose (genes *A, B, D,* and *E*). An additional gene *C* is linked to genes *A, B,* and *D.* These genes are in the order *D-A-B-C.* Gene *E* is distant from the other genes. Engelsberg and his colleagues isolated mutations at the *C* gene that affected the expression of structural genes *A, B, D,* and *E.* In one set of experiments, they created various genotypes at the *A* and *C* loci and determined whether arabinose isomerase (the enzyme encoded by gene *A*) was produced in the presence or absence of arabinose (the substrate of arabinose isomerase). Results from this experiment are shown in the following table, where a plus sign (+) indicates that the arabinose isomcrasc was synthesized and a minus sign (–) indicates that the enzyme was not synthesized.

Genotype	Arabinose absent	Arabinose present
1. $C^+ A^+$	–	+
2. $C^- A^+$	–	–
3. $C^- A^+/C^+ A^-$	–	+
4. $C^c A^-/C^- A^+$	+	+

a. On the basis of the results of these experiments, is the *C* gene an operator or a regulator gene? Explain your reasoning.

The C *gene is a regulator gene. The* C *gene is trans-acting, thus it affects the expression of the* A *gene located on a different DNA molecule, which is typical of a gene encoding a regulatory protein. If the* C *gene was an operator, it would be cis-acting and only able to regulate the expression of the* A *gene found on the same DNA molecule, which is not the case as demonstrated from genotype 3.*

b. Do these experiments suggest that the arabinose operon is negatively or positively controlled? Explain your reasoning.
Data from these experiments suggest that the arabinose operon is positively controlled. From the above data, the C *gene appears to be a regulator gene that is needed for the transcription of the* A *gene. For the* A *gene to be expressed, a functional* C *gene needs to be present within the cell. In the absence of a functional* C *gene and arabinose, the* A *gene is not expressed. Both results would be explained if the* C *gene encodes a regulator protein that is required to activate transcription of the* A *gene.*

c. What type of mutation is C^c?
The C^c *mutation results in continuous activation of transcription from the* A *gene. In other words,* C^c *leads to constitutive expression of the* A *gene.*

21. In *E. coli,* three structural genes (*A, D,* and *E*) encode enzymes A, D, and E, respectively. Gene *O* is an operator. The genes are in the order *O-A-D-E* on the chromosome. These enzymes catalyze the biosynthesis of valine. Mutations were isolated at the *A, D, E,* and *O* genes to study the production of enzymes A, D, and E when cellular levels of valine were low (T. Ramakrishnan and E. A. Adelberg. 1965. *Journal of Bacteriology* 89:654–660). Levels of the enzymes produced by partial-diploid *E. coli* with various combinations of mutations are shown in the following table.

		Amount of enzyme produced		
	Genotype	E	D	A
1.	$E^+ D^+ A^+ O^+/$ $E^+ D^+ A^+ O^+$	2.40	2.00	3.50
2.	$E^+ D^+ A^+ O^-/$ $E^+ D^+ A^+ O^+$	35.80	38.60	46.80
3.	$E^+ D^- A^+ O^-/$ $E^+ D^+ A^- O^+$	1.80	1.00	47.00
4.	$E^+ D^+ A^- O^-/$ $E^+ D^- A^+ O^+$	35.30	38.00	1.70
5.	$E^- D^+ A^+ O^-/$ $E^+ D^- A^+ O^+$	2.38	38.00	46.70

a. Is the regulator protein that binds to the operator of this operon a repressor (negative control) or an activator (positive control)? Explain your reasoning.
The regulator protein is a repressor. When the operator is defective or nonfunctioning (O^-*), then the expression of enzymes encoded by the* A, D, *and* E *loci is significantly increased over the wild type operator genotype. This suggests that the regulator protein cannot bind the* O^- *region and repress transcription.*

b. Are genes *A, D,* and *E* all under the control of operator *O?* Explain your reasoning.
Yes, all the genes are under control of the operator. When operator (O^+) *and repressor are both functional, then the levels of expression are low. When the operator is nonfunctional the enzyme levels generally increase.*

c. Propose an explanation for the low level of enzyme E produced in genotype 3.
The low level of enzyme E produced in genotype 3 is likely due to a polarity effect. Because they share the same transcriptional control regions, genes A, D, *and* E *are transcribed together producing a single polycistronic mRNA molecule. Gene* E *is located downstream of gene* D *and is thus transcribed after gene* D. *In genotype 3, it is likely that the defect in gene* D *affects transcription elongation that occurs subsequent to the mutation.*

Section 16.4

*22. Listed in parts **a.** through **g.** are some mutations that were found in the 5′ UTR region of the *trp* operon of *E. coli*. What would the most likely effect of each of these mutations be on transcription of the *trp* structural genes?

a. A mutation that prevented the binding of the ribosome to the 5′ end of the mRNA 5′ UTR
If the ribosome does not bind to the 5′ end of the mRNA, then region 1 of the mRNA 5′ UTR will be free to pair with region 2, thus preventing region 2 from pairing with region 3 of mRNA 5′ UTR. Region 3 will be free to pair with region 4, forming the attenuator or termination hairpin. Transcription of the trp *structural genes will be terminated. Essentially, no gene expression will occur.*

b. A mutation that changed the tryptophan codons in region 1 of the mRNA 5′ UTR into codons for alanine
In the wild-type trp *operon, low levels of tryptophan result in the ribosome pausing in region 1 of the mRNA 5′ UTR. The pause permits regions 2 and 3 of the mRNA 5′ UTR to form the antiterminator hairpin, allowing transcription of the structural genes to continue. If alanine codons have replaced tryptophan codons, then under conditions of low tryptophan, the stalling of the ribosome will not occur. The attenuator will form, stopping transcription. The ribosome will stall when alanine is low, so transcription of the structural genes will occur only when alanine is low.*

c. A mutation that created a stop codon early in region 1 of the mRNA 5′ UTR
If region 1 of the mRNA 5′ UTR is free to pair with region 2, then regions 3 and 4 of the mRNA 5′ UTR can form the attenuator. An early stop codon will result in the ribosome "falling off" region 1, allowing it to form a hairpin structure with region 2. Transcription will not occur because regions 3 and 4 are now free to form the attenuator.

d. Deletions in region 2 of the mRNA 5′ UTR
If region 2 of the mRNA 5′ UTR is deleted, then the antiterminator cannot be formed. The attenuator will form and transcription will not occur.

e. Deletions in region 3 of the mRNA 5′ UTR
The trp operon mRNA 5′ UTR will be unable to form the attenuator if region 3 contains a deletion. Attentuation or termination of transcription will not occur, resulting in continued transcription of the trp *structural genes.*

f. Deletions in region 4 of the mRNA 5′ UTR
Deletions in region 4 will prevent formation of the attenuator by the 5′ UTR mRNA. Transcription will proceed.

g. Deletion of the string of adenine nucleotides that follows region 4 in the 5′ UTR
For the attenuator hairpin to function as a terminator, the presence of a string of uracil nucleotides following region 4 in the mRNA 5′ UTR is required. The deletion of the string of adenine nucleotides in the DNA will result in no string of uracil nucleotides following region 4 of the mRNA 5′ UTR. No termination will occur, and transcription will proceed.

23. Some mutations in the *trp* 5′ UTR region increase termination by the attenuator. Where might these mutations occur and how might they affect the attenuator?
Mutations that disrupt the formation of the antiterminator will increase termination by the attenuator. Such disruptions could be caused by a deletion in region 2 that prevents region 2 from pairing with region 3. Mutations in region 1 could also affect the antiterminator if the mutations prevented the ribosome from stalling at the adjacent trpytophan codons within region 1. For example, any mutation that blocks translation initiation or stops translation early within region 1 would not allow the ribosome to migrate on the trp *operon mRNA. Another type of mutation affecting antiterminator formation in region 1 is one that eliminates or replaces the two adjacent tryptophan codons in the small protein. Elimination of these codons would prevent the ribosome from stalling in region 1, thus increasing the rate of the terminator formation.*

24. Some of the mutations mentioned in the Question 23 have an interesting property. They prevent the formation of the antiterminator that normally takes place when the tryptophan level is low. In one of the mutations, the AUG start codon for the 5′ UTR peptide has been deleted. How might this mutation prevent antitermination from occurring?
The AUG start codon is necessary for the translation initiation of the 5′ UTR peptide. If translation does not initiate, then the mRNA 5′ UTR region 1 will be available to pair with region 2. The resulting hairpin will prevent the formation of the antiterminator.

Section 16.5

25. Several examples of antisense RNA regulating translation in bacterial cells have been discovered. Molecular geneticists have also used antisense RNA to artificially control transcription in both bacterial and eukaryotic genes. If you wanted to inhibit the transcription of a bacterial gene with antisense RNA, what sequences might the antisense RNA contain?

To block transcription, you will need to disrupt the action of RNA polymerase either directly or indirectly. Antisense RNA containing sequences complementary to the gene's promoter should inhibit the binding of RNA polymerase. If transcription initiation by RNA polymerase requires the assistance of an activator protein, then antisense RNA complementary to the activator protein-binding site of the gene could also disrupt transcription. By binding the activator site, the antisense RNA would block access to the site by the activator and prevent RNA polymerase from being assisted by the activator to initiate transcription.

CHALLENGE QUESTION

Section 16.4

26. Would you expect to see attenuation in the *lac* operon and other operons that control the metabolism of sugars? Why or why not?
    No, attenuation of the lac *operon and other operons that control the metabolism of sugars is not likely to occur. Most operons involved in the breakdown of sugar already exhibit two levels of control: a repressor and activation by CAP protein. The presence of the sugar alone is not sufficient to elicit high levels of expression from the operon. If present, glucose prevents CAP from being activated and thus glucose will be the primary sugar to be metabolized. If glucose levels are low or absent, then CAP will be activated by cAMP. Therefore, the activation of the sugar operons will require the presence of the sugar and the absence of glucose. With glucose absent, the cell will quickly express the sugar metabolism operon so that energy production can occur. Finally, attenuation of the* trp *operon works because the ultimate result of gene expression in the operon is the amino tryptophan. The presence of the two codons for tryptophan in the mRNA 5' UTR means that the translational rate will be affected by the levels of tryptophan in the cell.*

1-3, 5-8, 10, 13, 15, 17, 19

Chapter Seventeen: Gene Regulation in Eukaryotes

COMPREHENSION QUESTIONS

Section 17.1

*1. List some important differences between bacterial and eukaryotic cells that affect the way in which the genes are regulated.

a. Bacterial genes are frequently organized into operons with coordinate regulation, and genes with operons can be transcribed as on a single long mRNA. Eukaryotic genes are not organized into operons and are singly transcribed from its own promoter.

b. In eukaryotic cells, DNA must be unwound from histone proteins prior to transcription occurring. Essentially, the chromatin must assume a more open configuration state, allowing for access by transcription associated factors.

c. Activator and repressor molecules function in both eukaryotic and bacterial cells. However, in eukaryotic cells activators appear to be more common than in bacterial cells.

d. In bacteria, transcription and translation can occur concurrently. In eukaryotes, the nuclear membrane separates transcription from translation both physically and temporally. This separation results in a greater diversity of regulatory mechanisms that can occur at different points during gene expression.

Section 17.2

*2. Where are DNase I hypersensitivity sites found and what do they indicate about the nature of chromatin?

DNase I hypersensitivity sites are typically found approximately 1000 nucleotides upstream of a transcription start site. This sensitivity to DNase I suggests a more relaxed or open configuration state for the chromatin, allowing for access to the DNA of transcriptional regulatory proteins as well as access by DNase I. Essentially the DNA is more exposed due to the lack of nuclesome structure.

*3. What changes take place in chromatin structure and what role do these changes play in eukaryotic gene regulation?

Changes in chromatin structure can result in repression or stimulation of gene expression. As genes become more transcriptionally active, chromatin shows increased sensitivity to DNase I digestion, suggesting that the chromatin structure is more open. Acetylation of histone proteins by acteyltransferase proteins results in the destabilization of the nucleosome structure and increases transcription as well as hypersensitivity to DNase I. The reverse reaction by deacetylases stabilizes nucleosome structure and lessens DNase I sensitivity. Other transcriptional factors and regulatory proteins, called chromatin remodeling complexes, bind directly to the DNA-altering chromatin structure without acetylating histone proteins. The chromatin remodeling complexes allow for transcription to be initiated by increasing accessibility to the promoters by transcriptional factors.

DNA methylation is also associated with decreased transcription. Methylated DNA sequences stimulate histone deacetylases to remove acetyl groups from the histone proteins, thus stabilizing the nucleosome and repressing transcription. Demethylation of DNA sequences is often followed by increased transcription, which may be related to the deacetylation of the histone proteins.

4. What is the histone code?
   The histone code refers to modifications to the tails of histone proteins. These modifications include the addition or removal of phosphate groups, acetyl groups, or methyl groups to the tails. Information imparted by these modifications affects how genes are expressed.

Section 17.3

5. Briefly explain how transcriptional activator proteins and repressors affect the level of transcription of eukaryotic genes.
   Transcriptional activator proteins stimulate transcription by binding DNA at specific base sequences such as an enhancer or regulatory promoter and attracting or stabilizing the basal transcriptional factor apparatus. Repressor proteins bind to silencer sequences or promoter regulator sequences. These proteins may inhibit transcription by either blocking access to the enhancer sequence by the activator protein, preventing the activator from interacting with the basal transcription apparatus, or preventing the basal transcription factor from being assembled.

*6. What is an enhancer? How does it affect transcription of distant genes?
   Enhancers are DNA sequences that are the binding sites of transcriptional activator proteins. Transcription at a distant gene is affected when the DNA sequence located between the gene's promoter and the enhancer is looped out, allowing for the interaction of the enhancer-bound proteins with proteins needed at the promoter, which stimulates transcription.

7. What is an insulator?
   An insulator or boundary element is a sequence of DNA that inhibits the action of regulatory elements called enhancers in a position dependent manner.

8. What is a response element? How do response elements bring about coordinated expression of eukaryotic genes?
   Response elements are regulatory DNA sequences consisting of short consensus sequences located at various distances from the genes that they regulate. Under conditions of stress, a transcription activator protein binds to the response element and stimulates transcription. If the same response element sequence is located in the control regions of different genes, then these genes will be activated by the same stimuli, thus producing a coordinated response.

Section 17.4

9. Outline the role of alternative splicing in the control of sex differentiation in *Drosophila*.
*Sex development in fruit flies depends on alternative splicing as well as a cascade of genetic regulation. Early in the development of female fruit flies, a female-specific promoter is activated stimulating transcription at the sex-lethal (*Sxl*) gene. Splicing of the pre-mRNA of the transformer (*tra*) gene is regulated by the* Sxl *protein. The mature mRNA produces the* Tra *protein. In conjunction with another protein, the* Tra *protein stimulates splicing of the pre-mRNA from the doublesex (*Dsx*) gene. The resulting* Dsx *protein is required for the embryo to develop female characteristics. Male fruit flies do not produce the* Sxl *protein, which results in the* Tra *pre-mRNA in male fruit flies being spliced at an alternate location. The alternate* Tra *protein is not functional, resulting in the* Dsx *pre-mRNA splicing at a different location as well. Protein synthesis from this mRNA produces a male-specific doublesex protein, which causes development of male-specific traits.*

*10. What role does RNA stability play in gene regulation? What controls RNA stability in eukaryotic cells?
The total amount of protein synthesized is dependent on how much mRNA is available for translation. The amount of mRNA present is dependent on the rates of mRNA synthesis and degradation. Less-stable mRNAs will be degraded faster so there will be fewer copies available to serve as templates for translation.

The presence of the 5′ cap, 3′ poly(A) tail, the 5′ UTR, 3′ UTR, and the coding region in the mRNA molecule affects stability. Poly(A) binding proteins (PABP) bind at the 3′ poly(A) tail. These proteins contribute to the stability of the tail and protect the 5′ cap through direct interaction. Once a critical number of adenine nucleotides have been removed from the tail, the protection is lost and the 5′ cap is removed. The removal of the 5′ cap allows for 5′ to 3′ nucleases to degrade the mRNA.

Section 17.5

*11. Briefly list some of the ways in which siRNAs and miRNAs regulate genes.

(1) *Through cleavage of mRNA sequences through "slicer activity": The binding of RISCs containing either siRNA or miRNA to complementary sequences in mRNA molecules stimulate cleavage of the mRNA through "slicer activity." This is followed by further degradation of the cleaved mRNA.*
(2) *Through binding of complementary regions with the mRNA molecule by miRNAs to prevent translation: The miRNAs as part of RISC bind to complementary mRNA sequences preventing either translation initiation or elongation, which results in premature termination.*
(3) *Through transcriptional silencing due to methylation of either histone proteins or DNA sequences: The siRNA bind to complementary DNA sequences within the nucleus and stimulate methylation of histone proteins. Methylated histones bind DNA more tightly preventing transcriptional factors from*

binding the DNA. The miRNA molecules bind to complementary DNA sequences and stimulate DNA methylases to directly methylate the DNA sequences, which results in transcriptional silencing.

(4) *Through slicer-independent mRNA degradation stimulated by miRNA binding to complementary regions in the 3' UTR of the mRNA: A miRNA binds to the AU rich element in the 3' UTR of the mRNA stimulating degradation using RISC and dicer.*

Section 17.6

*12. What are some of the characteristics of *Arabidopsis thaliana* that make it a good model genetic organism?

Useful characteristics of Arabidopsis thaliana

- *Is an angiosperm and thus shares characteristics and life cycle similarities with other flowering plants.*
- *Is capable of self-fertilization or cross-fertilization.*
- *Is small in size, which is useful in laboratory environments.*
- *Can grow under low illumination levels, which also is useful in laboratory environments.*
- *Is a prolific reproducer with each plant capable of producing 10,000 to 40,000 seeds.*
- *Has a high germination frequency from its seeds.*
- *Has a small genome (~125 million base pairs) that has been completely sequenced.*
- *Has a number of ecotypes or variants that are available, which differ in genotypes and phenotypic characteristics.*
- *Can uptake genes from other organisms by means of a Ti plasmid.*

*13. How does bacterial gene regulation differ from eukaryotic gene regulation? How are they similar?

Bacterial and eukaryotic gene regulation involves the action of protein repressors and protein activators. Cascades of gene regulation in which the activation of one set of genes affects another set of genes takes place in both eukaryotes and bacteria. Regulation of gene expression at the transcriptional level is also common in both types of cells.

Bacterial genes are often clustered in operons and are coordinately expressed through the synthesis of a single polygenic mRNA. Eukaryotic genes are typically separate, with each containing its own promoter and transcribed on individual mRNAs. Coordinate expression of multiple genes is accomplished through the presence of response elements. Genes sharing the same response element will be regulated by the same regulatory factors.

In eukaryotic cells, gene-coding regions are interrupted by introns, which may be much longer than the coding region. Gene expression requires the proper splicing of the pre-mRNA to remove these noncoding regions. In prokaryotic cells, gene-coding regions are usually not interrupted.

In eukaryotic cells, chromatin structure plays a role in gene regulation. Chromatin that is condensed inhibits transcription. Therefore, for expression to occur, the chromatin must be altered to allow for changes in structure. Acetylation of histone proteins and DNA methylation are important in these changes.

At the level of transcription initiation, the process is more complex in eukaryotic cells. In eukaryotes, initiation requires a complex machine involving RNA polymerase, general transcription factors, and transcriptional activators. Bacterial RNA polymerase is either blocked or stimulated by the actions of regulatory proteins.

Finally, in eukaryotes the action of activator proteins binding to enhancers may take place at a great distance from the promoter and structural gene. These distant enhancers occur much less frequently in bacterial cells.

APPLICATION QUESTIONS AND PROBLEMS

Section 17.2

14. A geneticist is trying to determine how many genes are found in a 300,000-bp region of DNA. Analysis shows that four H3K4me3 modifications are found in this piece of DNA. What might their presence suggest about the number of genes located there?
    The histone 3 methylase H3K4me3 adds 3 methyl groups to the lysine 4 in the tail of histone 3. These modifications typically occur near the transcription start site of genes. If four H3K4me3 modifications have been identified, it suggests that at least four genes are present within this region of DNA.

15. In a line of human cells grown in culture, a geneticist isolates a temperature-sensitive mutation at a locus that encodes an acetyltransferase enzyme; at temperatures above 38°C, the mutant cells produce a nonfunctional form of the enzyme. What would be the most likely effect of this mutation when the cells are raised at 40°C?
    Acetyltransferase enzymes add acetyl groups to histone proteins preventing the proteins from forming the 30-nm chromatin fiber. Essentially, the chromatin structure is destabilized, which allows for transcription to occur. If the cells are raised to 40°C, then the acetyltransferase enzyme would not function and acetyl groups would not be added to the histone proteins that are the target of this enzyme. The result would be that the nucleosomes and the chromatin would remain stabilized and block transcriptional activation.

16. What would be the most likely effect of deleting flowering locus D (*FLD*) in *Arbidopsis thalania?*
It is likely that flowering will not occur if the flowering locus D is deleted. The protein encoded by FLD *is a deacetylase enzyme. This deacetylase enzyme removes acetyl groups from histones surrounding the flowering locus C (*FLC*). Once the acetyl groups are removed, the chromatin structure within this region is restored. The restore chromatin inhibits transcription from the* FLC *locus.* FLC *codes for a transcriptional activator whose expression activates other genes that suppress flowering. If* FLC *transcription is active, then flowering will not occur.*

17. X31b is an experimental compound that is taken up by rapidly dividing cells. Research has shown that X31b stimulates the methylation of DNA. Some cancer researchers are interested in testing X31b as a possible drug for treating prostate cancer. Offer a possible explanation for why X31b might be an effective anticancer drug.
Cancer cells are typically rapidly dividing cells. DNA methylation particularly in regions with many CpG sequences (CpG islands) is associated with transcriptional repression. If the X31b molecules can be uptaken by the rapidly dividing cancers cells and then stimulate methylation of DNA sequences in the cancer cells, transcriptional repression of genes in the cancer cells would be expected. The repression of transcription could affect the growth of the cancer cells and potentially cause a loss of viability of these cells.

Section 17.3

18. How do repressors that bind to silencers in eukaryotes differ from repressors that bind to operators in bacteria?
In bacteria, repressors that bind to the operator block RNA polymerase from binding to the promoter. Repressors that bind to silencers in eukaryotes block transcriptional activator proteins from binding at an activator site, thus eliminating transcriptional activation.

19. An enhancer is surrounded by four genes (*A, B, C,* and *D*), as shown in the adjoining diagram. An insulator lies between gene *C* and gene *D*. On the basis of the positions of the genes, the enhancer, and the insulator, the transcription of which genes is most likely to be stimulated by the enhancer? Explain your reasoning.

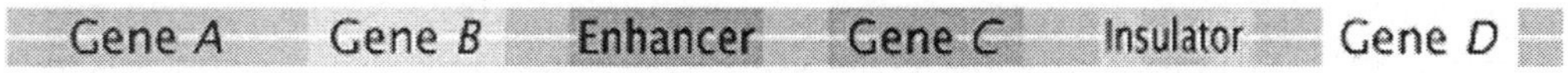

Gene *A* Gene *B* Enhancer Gene *C* Insulator Gene *D*

The action of an enhancer is blocked when the insulator is located between the enhancer and the promoter of the gene. It is likely that genes A, B, *and* C *will be stimulated by the enhancer and that gene* D *will not be stimulated. Insulators block the stimulatory action of enhancers when they lie between the enhancer and the promoter of the gene. In the example from the figure, the insulator is only between gene* D *and the enhancer. The enhancer's effect on genes* A, B, *and* C *is not likely to be affected by the insulator and these genes will be stimulated.*

Section 17.4

*20. What will be the effect on sexual development in newly fertilized *Drosophila* embryos if the following genes are deleted?

*The sex lethal (*Sxl*) gene codes for a protein necessary for the proper splicing of the transformer (*tra*) pre-mRNA and the production of a functional Tra protein. Tra protein in association with the Tra2 protein is needed for female specific splicing of the double-sex (*Dsx*) pre-mRNA. The female-specific Dsx protein is needed for the development of certain female characteristics. A male specific Dsx protein is needed for the development of certain male characteristics.*

a. *sex lethal*

If the Sxl *gene is absent, then female specific splicing of the* tra *pre-mRNA will not occur resulting in the production of a nonfunctional Tra protein. Thus, the fruit flies will develop male specific characteristics.*

b. *transformer*

If the tra *gene is absent, then a male-specific Dsx protein is synthesized allowing for the development of male specific characteristics.*

c. *doublesex*

If the Dsx *gene is deleted, then no doublesex protein will be produced, resulting in flies with intersex characteristics. Female flies lacking the female specific doublesex protein will have both male and female characteristics. Male flies lacking the male-specific doublesex protein will also have both male and female characteristics.*

21. Some eukaryotic mRNAs have an AU-rich element in the 3′ untranslated region. What would be the effect on gene expression if this element were mutated or deleted?

The presence of AU-rich elements is associated with rapid degradation of the mRNA molecules that contain them through a RNA silencing mechanism. If the AU-element was deleted, then the miRNA would not be able to bind to the consensus sequence of the AU-rich element and the RISC degradation would not be initiated. It is likely that this mRNA molecule would be more stable resulting in increased gene expression of the protein coded for by the mRNA.

22. A strain of *Arabidopsis thaliana* possesses a mutation in the *APETALA2* gene, in which much of the 3′ untranslated region of mRNA transcribed from the gene is deleted. What is the most likely effect of this mutation on the expression of the *APETALLA2* gene?

Translation from the APETALA2 *is inhibited by a miRNA that binds within the coding region of the mRNA. Thus, deleting much of the 3′ untranslated region of the* APETALA2 *mRNA will likely not affect the translation regulation by the miRNA molecule. However, the 3′ untranslated region could potentially be needed for mRNA stablility and binding of the ribosome to the mRNA molecule, so the deletion could result in a decrease in expression of the* APETALA2 *gene.*

23. What would be the effect of a mutation that destroyed the ability of poly(A) binding protein (PABP) to attach to a poly(A) tail?
Poly(A) binding protein is necessary for the stability of the mRNA molecules in eukaryotic cells. The protein contributes to the stability of both the poly(A) tail and the 5′ cap. If the PABP protein cannot bind the poly(A) tail, then the 5′ cap will not be protected and thus will be removed, resulting in the mRNA being degraded more rapidly.

CHALLENGE QUESTIONS

Section 17.2

24. In the fungus *Neurospora*, about 2–3% of cytosine bases are methylated. A recent study isolated those DNA sequences in *Neurospora* that contained 5-methylcytosine and found that almost all methylated sequences were located in the relict copies of transposable genetic elements. Based on these observations, propose a possible explanation for why *Neurospora* methylates its DNA and why DNA methylation in this species is associated with transposable genetic elements.
Heavy DNA methylation of cytosine to yield 5-methylcytosine containing nucleotides is associated with transcriptional repression in other organisms such as vertebrate animals and plants. Because almost all of the methylated sequences were associated with relic copies of transposons, this suggests that the methylation may be part of a process for inactivating these transposable elements. The Neurospora *methylation could be a mechanism for inactivating the expression of genes found in these invading nucleic acid sequences and thus preventing the transposon sequences from spreading within the* Neurospora *genome.*

Section 17.3

25. A yeast gene termed *SER3,* which has a role in serine biosynthesis, is repressed during growth in nutrient-rich medium, and so little transcription takes place and little SER3 enzyme is produced. In an investigation of the nature of the repression of the *SER3* gene, a region of DNA upstream of the *SER3* gene was found to be heavily transcribed when the *SER3* gene is repressed (J. A. Martens, L. Laprade, and F. Winston. 2004. *Nature* 429:571–574). Within this upstream region is a promoter that stimulates the transcription of an RNA molecule called *SRG1* (for *SER3* regulatory gene). This RNA molecule has none of the sequences necessary for translation. Mutations in the promoter for *SRG1* result in the disappearance of *SRG1* RNA, and these mutations remove the repression of *SER3*. When RNA polymerase binds to the *SRG1* promoter, the polymerase was found to travel downstream, transcribing the *SGR1* RNA, and to pass through and transcribe the promoter for *SER3*. This activity leads to the repression of *SER3*. Propose a possible explanation for how the transcription of *SGR1* might repress the transcription of *SER3*. (Hint: Remember that the *SGR1* RNA does not encode a protein.)

Potentially, the transcription from the SRG1 *gene interferes with the transcription initiation of the* SER3 *gene. Because part of the* SRG1 *transcript overlaps with the* SER3 *promoter, the transcriptional complex that is transcribing the* SRG1 *gene may prevent transcriptional factors from interacting with the* SER3 *promoter. Therefore, as long as the* SRG1 *gene is being transcribed, then the* SER3 *transcription will not initiate.*

Section 17.5

26. A common feature of many eukaryotic mRNAs is the presence of a rather long 3′ UTR, which often contains consensus sequences. Creatine kinase B (CK-B) is an enzyme important in cellular metabolism. Certain cells—termed U937D cells—have lots of CK-B mRNA, but no CK-B enzyme is present. In these cells, the 5′ end of the CK-B mRNA is bound to ribosomes, but the mRNA is apparently not translated. Something inhibits the translation of the CK-B mRNA in these cells.

    Researchers introduced numerous short segments of RNA containing only 3′ UTR sequences into U937D cells. As a result, the U937D cells began to synthesize the CK-B enzyme, but the total amount of CK-B mRNA did not increase. The introduction of short segments of other RNA sequences did not stimulate the synthesis of CK-B; only the 3′ UTR sequences turned on the translation of the enzyme.

    On the basis of these results, propose a mechanism for how CK-B translation is inhibited in the U937D cells. Explain how the introduction of short segments of RNA containing the 3′ UTR sequences might remove the inhibition.
    From the above experimental data, translation of the CK-B protein is inhibited in the U937D cells—the CK-B mRNA is present and bound to the ribosome, but no protein is synthesized. A possible mechanism for the inhibition of translation could be the binding of translational repressors to the 3′ UTR region of the CK-B mRNA. The action of soluble proteins inhibiting translation seems to be suggested by the response of the U937 cells to the short RNA sequences containing the 3′ UTR. When these sequences are introduced to the U937D cells, the synthesis of CK-B occurs. Possibly, exogenously applied 3′ UTR sequences bind to the translational repressor proteins, making them unavailable to bind to the CK-B mRNA. If these factors are not present on the CK-B mRNA, then synthesis of the CK-B protein can take place.

1–4, 6–11, 14, 16–22, 24–26, 28

Chapter Eighteen: Gene Mutations and DNA Repair

COMPREHENSION QUESTIONS

Section 18.1

*1. What is the difference between somatic mutations and germ-line mutations?
Germ-line mutations are changes in the DNA of germ (reproductive) cells and may be passed to offspring. Somatic mutations are changes in the DNA of an organism's somatic tissue cells and cannot be passed to offspring.

*2. What is the difference between a transition and a transversion? Which type of base substitution is usually more common?
Transition mutations are base substitutions in which one purine (A or G) is changed to the other purine, or a pyrimidine (T or C) is changed to the other pyrimidine. Transversions are base substitutions in which a purine is changed to a pyrimidine or vice versa. Although transversions would seem to be statistically favored because there are eight possible transversions and only four possible transitions, about twice as many transition mutations are actually observed in the human genome.

*3. Briefly describe expanding trinucleotide repeats. How do they account for the phenomenon of anticipation?
Expanding trinucleotide repeats occur when DNA insertion mutations result in an increase in the number of copies of a trinucleotide repeat sequence. Such an increase in the number of copies of a trinucleotide sequence may occur by errors in replication or unequal recombination. Within a given family, a particular type of trinucleotide repeat may increase in number from generation to subsequent generation, increasing the severity of the mutation in a process called anticipation.

4. What is the difference between a missense mutation and a nonsense mutation? A silent mutation and a neutral mutation?
A base substitution that changes the sequence and the meaning of a mRNA codon, resulting in a different amino acid being inserted into a protein, is called a missense mutation. Nonsense mutations occur when a mutation replaces a sense codon with a stop (or nonsense) codon.

A nucleotide substitution that changes the sequence of a mRNA codon, but not the meaning is called a silent mutation. In neutral mutations, the sequence and the meaning of a mRNA codon are changed. However, the amino acid substitution has little or no effect on protein function.

5. Briefly describe two different ways that intragenic suppressors may reverse the effects of mutations.
Intragenic suppression is the result of second mutations within a gene that restore a wild-type phenotype. The suppressor mutations are located at different sites within the gene from the original mutation. One type of suppressor mutation restores the

original phenotype by reverting the meaning of a previously mutated codon to that of the original codon. The suppressor mutation occurs at a different position than the first mutation, which is still present within the codon. Intragenic suppression may also occur at two different locations within the same protein. If two regions of a protein interact, a mutation in one of these regions could disrupt that interaction. The suppressor mutation in the other region would restore the interaction. Finally, a frameshift mutation due to an insertion or deletion could be suppressed by a second insertion or deletion that restores the proper reading frame.

*6. How do intergenic suppressors work?
Intergenic suppressor mutations restore the wild-type phenotype. However, they do not revert the original mutation. The suppression is the result of a mutation in a different gene reducing the phenotypic effect of the original mutation. For example, because many proteins interact with other proteins, the original mutation may have altered one protein in a way that disrupts a protein-protein interaction, while the second mutation alters the second protein to restore the interaction. A second type of intergenic suppression occurs when a mutation within an anticodon region of a tRNA molecule allows the tRNA anticodon to pair with the codon containing the original mutation. This results in the substitution of a functional amino acid in the protein.

Section 18.2

*7. What is the cause of errors in DNA replication?
Two types of events have been proposed that could lead to DNA replication errors: mispairing due to tautomeric shifts in nucleotides and mispairing through wobble or flexibility of the DNA molecule. Current evidence suggests that mispairing through wobble caused by flexibility in the DNA helix is the most likely cause.

8. How do insertions and deletions arise?
Strand slippage that occurs during DNA replication and unequal crossover events due to misalignment at repetitive sequences have been shown to cause deletions and additions of nucleotides to DNA molecules. Strand slippage results from the formation of small loops on either the template or the newly synthesized strand. If the loop forms on the template strand, then a deletion occurs. Loops formed on the newly synthesized strand result in insertions. If, during crossing over, a misalignment of the two strands at repetitive sequence occurs, then the resolution of the cross over will result in one DNA molecule containing an insertion and the other molecule containing a deletion.

*9. How do base analogs lead to mutations?
Base analogs have structures similar to the nucleotides and if present, may be incorporated into the DNA during replication. Many analogs have an increased tendency for mispairing, which can lead to mutations. DNA replication is required for the base analog-induced mutations to be incorporated into the DNA.

10. How do alkylating agents, nitrous acid, and hydroxylamine produce mutations?
Alkylating agents donate alkyl groups (either methyl or ethyl)) to the nucleotide bases. The addition of the alkyl group results in mispairing of the alkylated base and typically leads to transition mutations. Nitrous acid treatment results in the deamination of cytosine, producing uracil, which pairs with adenine. During the next round of replication, a CG to AT transition will occur. The deamination of guanine by nitrous acid produces xanthine. Xanthine can pair with either cytosine or thymine. If paired with thymine, then a CG to TA transition can occur. Hydroxylamine works by adding a hydroxyl group to cytosine, producing hydroxylaminocytosine. The hydroxylaminocytosine has an increased tendency to undergo tautomeric shifts, which allow pairings with adenine, resulting in GC to AT transitions.

11. What types of mutations are produced by ionizing and UV radiation?
Ionizing radiation promotes the formation of radicals and reactive ions that result in the breakage of phosphodiester linkages within the DNA molecule. Single- and double-strand breaks can occur. Double-strand breaks are difficult to repair accurately and may result in the deletion of genetic information. UV radiation promotes the formation of pyrimidine dimers between adjacent pyrimidines in a DNA strand. Inefficient repair of the dimers by error-prone DNA repair systems results in an increased mutation rate.

*12. What is the SOS system and how does it lead to an increase in mutations?
The SOS system is an error-prone DNA repair system consisting of at least 25 genes. Induction of the SOS system results in a bypass of damaged DNA regions, which allows for DNA replication across the damaged regions. However, the bypass of damaged DNA results in a less accurate replication process, and thus more mutations will occur.

Section 18.3

13. What is the purpose of the Ames test? How are *his*$^-$ bacteria used in this test?
The Ames test allows for rapid and inexpensive detection of potentially carcinogenic compounds using bacteria. The majority of carcinogenic compounds result in damage to DNA and are mutagens. The reversion of his$^-$ *bacteria to* his$^+$ *is used to detect the mutagenic potential of the compound being tested.*

Section 18.4

*14. List at least three different types of DNA repair and briefly explain how each is carried out.
(1) Mismatch repair. Replication errors that are the result of base-pair mismatches are repaired. Mismatch-repair enzymes recognize distortions in the DNA structure due to mispairing and detect the newly synthesized strand by the lack of methylation on the new strand. The bulge is excised and DNA polymerase and DNA ligase fill in the gap.

(2) Direct repair. DNA damage is repaired by directly changing the damaged nucleotide back to its original structure.
(3) Base-excision repair. The damaged base is excised, and then the entire nucleotide is replaced.
(4) Nucleotide-excision repair. Repair enzymes recognize distortions of the DNA double-helix. Damaged regions are excised by enzymes, which cut phosphodiester bonds on either side of the damaged region. The gap generated by the excision step is filled in by DNA polymerase.

15. What features do mismatch repair, base-excision repair, and nucleotide-excision repair have in common?
Mismatch repair, base excision repair, and nucleotide-excision repair all result in the removal of nucleotides from DNA. All repair mechanisms that excise nucleotides share a common four-step pathway:
(1) DNA damage is detected.
(2) The damage is excised by DNA repair endonucleases.
(3) Following excision, DNA polymerase adds nucleotides to the free 3′ OH group, using the remaining strand as a template
(4) Ligation of nicks in the sugar phosphate backbone is performed by DNA ligase.

APPLICATION QUESTIONS AND PROBLEMS

Section 18.1

*16. A codon that specifies the amino acid Gly undergoes a single-base substitution to become a nonsense mutation. In accord with the genetic code given in Figure 15.10, is this mutation a transition or a transversion? At which position of the codon does the mutation occur?
By examining the four codons that encode for Gly, GGU, GGC, GGA, and GGG, and the three nonsense codons, UGA, UAA, and UAG, we can determine that only one of the Gly codons, GGA, could be mutated to a nonsense codon by the single substitution of a U for a G at the first position:

GGA → UGA

Because uracil is a pyrimidine and guanine is a purine, the mutation is a transversion.

*17. **a.** If a single transition occurs in a codon that specifies Phe, what amino acids could be specified by the mutated sequence?
Two codons can encode for Phe, UUU, and UUC. A single transition could occur at each of the positions of the codon resulting in different meanings.

Original codon	*Mutated codon (amino acid encoded)*
UUU	*CUU (Leu), UCU (Ser), UUC (Phe)*
UUC	*CUC (Ser), UCU (Ser), UUU (Ser)*

b. If a single transversion occurs in a codon that specifies Phe, what amino acids could be specified by the mutated sequence?

Original codon	*Mutated codon (amino acid encoded)*
UUU	*AUU (Ile), UAU (Tyr), UUA (Leu), GUU (Val), UGU (Cys), UUG (Leu)*
UUC	*AUC (Ile), UAC (Tyr), UUA (Leu), GUC (Val), UGC (Cys), UUG (Leu)*

c. If a single transition occurs in a codon that specifies Leu, what amino acids could be specified by the mutated sequence?

Original codon	*Mutated codon (amino acid encoded)*
CUU	*UUU (Phe), CCU (Pro), CUC (Leu)*
CUC	*UUC (Phe), CCC (Pro), CUG (Leu)*
CUA	*UUA (Leu), CCA (Pro), CUG (Leu)*
CUG	*UUG (Leu), CCG (Pro), CUA (Leu)*
UUG	*CUG (Leu), UCG (Ser), UUA (Ser)*
UUA	*CUA (Leu), UCG (Ser), UUG (Leu)*

d. If a single transversion occurs in a codon that specifies Leu, what amino acids could be specified by the mutated sequence?

Original codon	*Mutated codon (amino acid encoded)*
UUA	*AUA (Met), UAA (Stop), UUU (Phe), GUA (Val), UGA (Stop), UUC (Phe)*
UUG	*AUG (Met), UAG (Stop), UUU (Phe), GUG (Val), UGG (Trp), UUC (Phe)*
CUU	*GUU (Val), CGU (Arg), CUG (Leu), AUU (Ile),*
CUC	*AUC (Ile), CAC (His), CUA (Leu), GUC (Val), CGC (Arg), CUG (Leu)*
CUA	*AUA (Ile), CAA (Gln), CUC (Leu), GUA (Val), CGA (Arg), CUG (Leu)*
CUG	*AUG (Met), CAG (Gln), CUC (Leu), GUG (Val), CGG (Arg), CUU (Leu)*

18. Hemoglobin is a complex protein that contains four polypeptide chains. The normal hemoglobin found in adults—called adult hemoglobin—consists of two α and two β polypeptide chains, which are encoded by different loci. Sickle-cell hemoglobin, which causes sickle-cell anemia, arises from a mutation in the β chain of adult hemoglobin. Adult hemoglobin and sickle-cell hemoglobin differ in a single amino acid: the sixth amino acid from one end in adult hemoglobin is glutamic acid, whereas sickle-cell hemoglobin has valine at this position. After consulting the genetic code provided in Figure 15.10, indicate the type and location of the mutation that gave rise to sickle-cell anemia.

There are two possible codons for glutamic acid, GAA and GAG. Single-base substitutions at the second position in both codons can produce codons that encode valine:

GAA--------> GUA (Val)
GAG--------> GUG (Val)

Both substitutions are transversions. However, in the gene encoding the β chain of hemoglobin, the GAG codon is the wild-type codon and the mutated GUG codon results in the sickle-cell phenotype.

*19. The following nucleotide sequence is found on the template strand of DNA. First, determine the amino acids of the protein encoded by this sequence by using the genetic code provided in Figure 15.10. Then, give the altered amino acid sequence of the protein that will be found in each of the following mutations.

Sequence of DNA template: 3′–TAC TGG CCG TTA GTT GAT ATA ACT–5′
Nucleotide number → 1 24
mRNA sequence: 5′–AUG ACC GGC AAU CAA CUA UAU UGA–3′
amino acid sequence: Amino–Met Thr Gly Asn Gln Leu Tyr Stop–Carboxyl

a. Mutant 1: A transition at nucleotide 11
The transition results in the substitution of Ser for Asn.
original sequence: 3′–TAC TGG CCG TTA GTT GAT ATA ACT–5′
*mutated sequence: 3′–TAC TGG CCG T**C**A GTT GAT ATA ACT–5′*
*mRNA sequence: 5′–AUG ACC GGC A**G**U CAA CUA UAU UGA–3′*
*amino acids: Amino–Met Thr Gly **Ser** Gln Leu Tyr Stop–Carboxyl*

b. Mutant 2: A transition at nucleotide 13
The transition results in the formation of a UAA nonsense codon.
original sequence: 3′–TAC TGG CCG TTA GTT GAT ATA ACT–5′
*mutated sequence: 3′–TAC TGG CCG TTA **A**TT GAT ATA ACT–5′*
*mRNA sequence: 5′–AUG ACC GGC AAU **U**AA CUA UAU UGA–3′*
*amino acid sequence: Amino–Met Thr Gly Asn **STOP**–Carboxyl*

c. Mutant 3: A one-nucleotide deletion at nucleotide 7
The one-nucleotide deletion results in a frameshift mutation.
original sequence: 3′–TAC TGG CCG TTA GTT GAT ATA ACT–5′
*mutated sequence: 3′–TAC TGG **CGT TAG TTG ATA TAA CT**–5′*
*mRNA sequence: 5′–AUG ACC **GCA GUC AAC UAU AUU GA**–3′*
*amino acids: Amino–Met Thr **Ala Ile Asn Tyr Ile** –Carboxyl*

d. Mutant 4: A T→ A transversion at nucleotide 15
The transversion results in the substitution of His for Gln in the protein.
original sequence: 3′–TAC TGG CCG TTA GTT GAT ATA ACT–5′
*mutated sequence: 3′–TAC TGG CCG TTA GT**A** GAT ATA ACT–5′*
*mRNA sequence: 5′–AUG ACC GGC AAU CA**U** CUA UAU UGA–3′*
*amino acids: Amino–Met Thr Gly Asn **His** Leu Tyr Stop–Carboxyl*
or
*mutated sequence: 3′–TAC TGG CCG TTA GT**G** GAT ATA ACT–5′*
*mRNA sequence: 5′–AUG ACC GGC AAU CA**C** CUA UAU UGA–3′*
*amino acids: Amino–Met Thr Gly Asn **His** Leu Tyr Stop–Carboxyl*

e. Mutant 5: An addition of TGG after nucleotide 6
The addition of the three nucleotides results in the addition of Thr to the amino acid sequence of the protein.
original sequence: 3′–TAC TGG CCG TTA GTT GAT ATA ACT–5′
mutated sequence:3′–TAC TGG TGG CCG TTA GTT GAT ATA ACT–5′
mRNA sequence: 5′–AUG ACC ***ACC*** *GGC AAU CAA CUA UAU UGA–3′*
amino acids: Amino–Met Thr ***Thr*** *Gly Asn Gln Leu Tyr Stop–Carboxyl*

f. Mutant 6: A transition at nucleotide 9
The protein retains the original amino acid sequence.
original sequence: 3′–TAC TGG CCG TTA GTT GAT ATA ACT–5′
mutated sequence: 3′–TAC TGG CCA TTA GTT GAT ATA ACT–5′
mRNA Sequence: 5′–AUG ACC GGU AAU CAA CUA UAU UGA –3′
amino acids:Amino–Met Thr ***Gly*** *Asn Gln Leu Tyr Stop–Carboxyl*

20. A polypeptide has the following amino acid sequence:
Met-Ser-Pro-Arg-Leu-Glu-Gly
The amino acid sequence of this polypeptide was determined in series of mutants listed in parts **a**. through **e.** For each mutant, indicate the type of change that occurred in the DNA (single-base substitution, insertion, deletion) and the phenotypic effect of the mutation (nonsense mutation, missense mutation, frameshift, etc.).

a. Mutant 1: Met-Ser-Ser-Arg-Leu-Glu-Gly
A missense mutation has occurred resulting in the substitution of Ser for Pro in the protein. The change is most likely due to a single-base substitution in the Ser codon resulting in the production of a Pro codon. Four of the Ser codons can be changed to Pro codons by a single transition mutation.

Pro	*Ser*
CCU	*UCU*
CCC	*UCC*
CCA	*UCA*
CCG	*UCG*

b. Mutant 2: Met-Ser-Pro
A single-base substitution has occurred in the Arg codon resulting in the formation of a stop codon. Two of the potential codons for Arg can be changed by single substitutions to stop codons. The phenotypic effect is a nonsense mutation.

Arg	*Stop*	
CGA	*UGA*	*transition mutation*
AGA	*UGA*	*transversion mutation*

c. Mutant 3: Met-Ser-Pro-Asp-Trp-Arg-Asp-Lys
The deletion of a single nucleotide at the first position in the Arg codon (most likely CGA) has resulted in a frameshift mutation in which the mRNA is read in a different frame, producing a different amino acid sequence for the protein.

d. Mutant 4: Met-Ser-Pro-Glu-Gly
A six-base pair deletion has occurred, resulting in the elimination of two amino acids (Arg and Leu) from the protein. The result is a truncated polypeptide chain.

e. Mutant 5: Met-Ser-Pro-Arg-Leu-Leu-Glu-Gly
The addition or insertion of three nucleotides into the DNA sequence has resulted in the addition of a Leu codon to the polypeptide chain.

*21. A gene encodes a protein with the following amino acid sequence:

Met-Trp-His-Arg-Ala-Ser-Phe.

A mutation occurs in the gene. The mutant protein has the following amino acid sequence:

Met-Trp-His-Ser-Ala-Ser-Phe.

An intragenic suppressor restores the amino acid sequence to that of the original protein:

Met-Trp-His-Arg-Ala-Ser-Phe.

Give at least one example of base changes that could produce the original mutation and the intragenic suppressor. (Consult the genetic code in Figure 15.10.)

Four of the six Arg codons could be mutated by a single-base substitution to produce a Ser codon. However, only two of the Arg codons mutated to form Ser codons could subsequently be mutated at a second position by a single-base substitution to regenerate the Arg codon. In both events, the mutations are transversions.

Original Arg codon	*Ser codon*	*Restored Arg codon*
CGU	*AGU*	*AGG or AGA*
CGC	*AGC*	*AGG or AGA*

22. A gene encodes a protein with the following amino acid sequence:

Met-Lys-Ser-Pro-Ala-Thr-Pro

A nonsense mutation from a single-base-pair substitution occurs in this gene, resulting in a protein with the amino acid sequence Met-Lys. An intergenic suppressor mutation allows the gene to produce the full-length protein. With the original mutation and the intergenic suppressor present, the gene now produces a protein with the following amino acid sequence:

Met-Lys-Cys-Pro-Ala-Thr-Pro

Give the location and nature of the original mutation and the intergenic suppressor.

The original mutation is located in the Ser codon. Two of the six potential Ser codons (UCA and UCG) can be changed to stop codons by a single-base substitution.

UCA → UGA (tranversion)
UCG → UAG (transversion)

However, only the UCA codon is likely to be suppressed by Cys-tRNA containing a single-base substitution (a transversion mutation) in the anticodon (5'–ACA–3' to 5'–UCA–3'), allowing for pairing with the UGA nonsense codon and suppression of the nonsense phenotype. The suppression is due to the insertion of Cys for the codon UGA.

23. XG syndrome is a rare genetic disease that is due to an autosomal dominant gene. A complete census of a small European country reveals that 77,536 babies were born in 2004, of whom 3 had XG syndrome. In the same year, this country had a population of 5,964,321 people, and there were 35 living persons with XG syndrome. What is the mutation rate of XG syndrome in this country?
The mutation rate for XG syndrome during the year 2000 can be expressed as the number of mutations per gamete, and because each individual is the product of two gametes, then $3/(77{,}536 \times 2) = 1.93 \times 10^{-5}$.

Section 18.2

*24. Can nonsense mutations be reversed by hydroxylamine? Why or why not?
No, hydroxylamine cannot reverse nonsense mutations. Hydroxylamine modifies cytosine-containing nucleotides and can only result in GC to AT transition mutations. In a stop codon, the GC to AT transition will result only in a different stop codon.

For 5′–UGA–3′:	*Template DNA:*	*3′–ACT–5′*
	Coding DNA:	*5′–TGA–3′*
Transition results:	*Template DNA*	*3′–ATT–5′*
GC to AT	*Coding DNA:*	*3′–TAA–5′*
	mRNA codon:	*5′–UAA–3′*

For 5′–UAG–3′:	*Template DNA:*	*3′ AT**C** 5′*
	Coding DNA:	*5′–TA**G**–3′*
Transition results:	*Template DNA:*	*3′–AT**T**–5′*
In GC to AT	*Coding DNA:*	*3′–TA**A**–5′*
	mRNA codon:	*5′–UA**A**–3′*

*25. The following nucleotide sequence is found in a short stretch of DNA:
5′–ATGT–3′
3′–TACA–5′
If this sequence is treated with hydroxylamine, what sequences will result after replication?
Hydroxylamine adds hydroxyl groups to cytosine, enabling the modified cytosine to occasionally pair with adenine, which ultimately can result in a GC to AT transition. Therefore, only one base pair in the sequence will be affected. Ultimately, after replication, one of the dsDNA molecules will have the transition but not the other dsDNA molecule.

Original sequence	*Mutated sequence*
*5′–AT**G**T–3′*	*5′–AT**A**T–3′*
*3′–TA**C**A–5′*	*3′–TA**T**A–5′*

26. The following nucleotide sequence is found in a short stretch of DNA:
5′–AG–3′
3′–TC–5′

a. Give all the mutant sequences that may result from spontaneous depurination occurring in this stretch of DNA.
The strand contains two purines, adenine and guanine. Because repair of depurination typically results in adenine being substituted for the missing purine, only the loss of the guanine by depurination will result in a mutant sequence.

*5′–A**G**–3′* *to* *5′–A**A**–3′*
*3′–T**C**–5′* *3′–T**T**–5′*

b. Give all the mutant sequences that may result from spontaneous deamination occurring in this stretch of DNA.
Deamination of guanine, cytosine, and adenine can occur. However, the deamination of only cytosine and adenine are likely to result in mutant sequences because the deamination products can form improper base pairs. The deamination of guanine does not pair with thymine but can still form two hydrogen bonds with cytosine, thus no change will occur.

*5′–**A**G–3′* *if A is deaminated, then* *5′–**G**G–3′*
*3′–**T**C–5′* *3′–**C**C–5′*

*5′–A**G**–3′* *if C is deaminated, then* *5′–A**A**–3′*
*3′–T**C**–5′* *3′–T**T**–5′*

27. In many eukaryotic organisms, a significant proportion of cytosine bases are naturally methylated to 5-methylcytosine. Through evolutionary time, the proportion of AT base pairs in the DNA of these organisms increases. Can you suggest a possible mechanism by which this increase occurs?
Spontaneous deamination of 5-methylcytosine produces thymine. If the subsequent repair of the GT mispairing is repaired incorrectly or, more likely, not repaired at all because the thymine is a normal base, then a GC to AT transition will result. Over time, the incorrect repairs will lead to an increase in the number of AT base pairs.

Section 18.3

*28. A chemist synthesizes four new chemical compounds in the laboratory and names them PFI1, PFI2, PFI3, and PFI4. He gives the PFI compounds to a geneticist friend and asks her to determine their mutagenic potential. The geneticist finds that all four are highly mutagenic. She also tests the capacity of mutations produced by the PFI compounds to be reversed by other known mutagens and obtains the following results. What conclusions can you make about the nature of the mutations produced by these compounds?

Mutations produced by	Reversed by 2-Aminopurine	Nitrous acid	Hydroxylamine	Acridine orange	
PFI1		Yes	Yes	Some	No
PFI 2	No	No	No	No	
PFI3		Yes	Yes	No	No
PFI4		No	No	No	Yes

First, consider the mutagenic actions of the reversion agents: 2-Aminopurine causes GC to AT and AT to GC transitions; nitrous acid causes GC to AT and AT to GC transitions; hydroxylamine causes GC to AT transitions; acridine orange causes single-base insertions or deletions, resulting in frameshift mutations.

PFI1 causes both types of transitions, GC to AT and AT to GC. PFI1 mutations can be reversed by 2-aminopurine, nitrous acid, and occasionally by hydroxylamine, which suggests that it acts as a transition mutagen. The lack of reversion of all PFI1 mutations by hydroxylamine suggests that some of the mutations were caused by GC to AT substitutions because they are not reverted by hydroxylamine.

PFI2 causes transversions, or large deletions, because mutations caused by PF12 are not reverted by any of the agents.

PFI3 causes GC to AT transitions. Only GC to AT transitions could be reverted by 2-aminopurine and nitrous oxide but not by hydroxylamine.

PFI4 causes single-base insertions or deletions. PF14 is only reverted by acridine orange, an intercalating agent that only reverts single-base insertions or deletions.

29. Mary Alexander studied the effects of radiation on mutation rates in the sperm of *Drosophila melanogaster*. She irradiated *Drosophila* larvae with either 3000 or 3975 roentgens (r), collected the adult males that developed from irradiated larvae, mated them with unirradiated females, and then counted the number of mutant F_1 flies produced by each male. All mutant flies that appeared were used in subsequent crosses to determine if their mutant phenotypes were genetic. She obtained the following results (M. L. Alexander. 1954. *Genetics* 39:409–428):

Group	Number of offspring	Offspring with a genetic mutation
Control (0 r)	45,504	0
Irradiated (3000 r)	49,512	71
Irradiated (3975 r)	50,159	70

a. Calculate the mutation rates of the control group and the two groups of irradiated flies.
Because no mutations were detected in the control group, the mutation rate per gamete must be less than $1/(45{,}504 \times 2)$ or 1.09×10^{-5}.
For flies irradiated with 3000 r, the mutation rate per gamete is $71/(49{,}512 \times 2) = 7.17 \times 10^{-4}$.

For flies irradiated with 3975 r, the mutation rate per gamete is $70/(50{,}159 \times 2) = 6.98 \times 10^{-4}$.

b. On the basis of these data, do you think radiation has any effect on mutation? Explain your answer.
Yes, the radiation significantly increased the mutation rate. Male flies exposed to radiation had higher incidences of offspring with mutations.

30. A genetics instructor designs a laboratory experiment to study the effects of UV radiation on mutation in bacteria. In the experiment, the students expose bacteria plated on petri plates to UV light for different lengths of time, place the plates in an incubator for 48 hours, and then count number of colonies that appear on each plate. The plates that have received more UV radiation should have more pyrimidine dimers, which block replication; thus, fewer colonies should appear on the plates exposed to UV light for longer periods of time. Before the students carry out the experiment, the instructor warns them that, while the bacteria are in the incubator, the students must not open the incubator door unless the room is darkened. Why should the bacteria not be exposed to light?
Exposure of DNA to UV light results in the formation of pyrimidine dimers in the DNA molecule. Often the repair of these dimers leads to mutations. Because the SOS repair system is error-prone and leads to an increased accumulation of mutations, UV light produces more mutations in bacteria when the SOS repair system is activated to repair the damage caused by the UV light. However, many species of bacteria have a direct DNA repair system that can repair pyrimidine dimers by breaking the covalent linkages between the pyrimidines that form the dimer. The enzyme that repairs the DNA is called photolyase, and is activated and energized by light. The photolyase is a very efficient repair enzyme and typically makes accurate repairs of the damage. If the bacteria in the UV radiation experiment are exposed to light, then the photolyase will be activated to repair the damage, resulting in fewer mutations in the irradiated bacteria.

Section 18.4

*31. A plant breeder wants to isolate mutants in tomatoes that are defective in DNA repair. However, this breeder does not have the expertise or equipment to study enzymes in DNA repair systems. How could the breeder identify tomato plants that are deficient in DNA repair? What are the traits to look for?
The plant breeder should look for plants that have increased levels of mutations either in their germ-line or somatic tissues. Potentially mutant plants could be exposed to standard mutagens that damage DNA. If they are defective in DNA repair they should have higher rates of mutation. For example tomato plants with defective DNA repair systems should have an increased mutation rate when exposed to high levels of sunlight.

CHALLENGE QUESTIONS

Section 18.1

32. Robert Bost and Richard Cribbs studied a strain of *E. coli* (*araB14*) that possessed a nonsense mutation in the structural gene that encodes L-ribulokinase, an enzyme that allows the bacteria to metabolize the sugar arabinose (R. Bost and R. Cribbs. 1969. *Genetics* 62:1–8). From the *araB14* strain, they isolated some bacteria that possessed mutations that caused them to revert back to the wild type. Genetic analysis of these revertants showed that they possessed two different suppressor mutations. One suppressor mutation (*R1*) was linked to the original mutation in the L-ribulokinase and probably occurred at the same locus. By itself, this mutation allowed the production of L-ribulokinase, but the enzyme was not as effective in metabolizing arabinose as the enzyme encoded by the wild-type allele. The second suppressor mutation (Su^B) was not linked to the original mutation. In conjunction with the *R1* mutation, Su^B allowed the production of L-ribulokinase, but Su^B by itself was not able to suppress the original mutation.
    a. On the basis of this information, are the *R1* and Su^B mutations intragenic suppressors or intergenic suppressors? Explain your reasoning.
    R1 *is an intragenic suppressor. As the studies indicated, it likely occurred within the same locus as the* araB14 *mutation.* Su^B *is an intergenic suppressor because it was not linked to the original mutation and occurred at a different locus.*
    b. Propose an explanation for how *R1* and Su^B restore the ability of *araB14* to metabolize arabinose and why Su^B is able to more fully restore the ability to metabolize arabinose.
    Potentially, the R1 *mutation changed the nonsense codon found in* araB14 *into a sense codon although not back to the original wild-type codon. The new sense codon in the* R1 *mutation allows for the insertion of an amino acid and a full-length protein is now synthesized that contains a missense mutation.* Su^B *encodes a mutation in a tRNA anticodon that allows for another amino acid to be inserted at the* R1 *mutant codon restoring more of the L-ribulokinase function.*

33. Achondroplasia is an autosomal dominant disorder characterized by disproportionate short stature—the legs and arms are short compared with the head and trunk. The disorder is due to a base substitution in the gene, located on the short arm of chromosome 4, for fibroblast growth factor receptor 3 (FGFR3).

    Although achondroplasia is clearly inherited as an autosomal dominant trait, more than 80% of the people who have achondroplasia are born to parents with normal stature. This high percentage indicates that most cases are caused by newly arising mutations; these cases (not inherited from an affected parent) are referred to as sporadic. Findings from molecular studies have demonstrated that sporadic cases of achondroplasia are almost always caused by mutations inherited from the father (paternal mutations). In addition, the occurrence of achondroplasia is higher among older fathers; indeed, approximately 50% of children with achondroplasia are born

to fathers older than 35 years of age. There is no association with maternal age. The mutation rate for achondroplasia (about 4×10^{-5} mutations per gamete) is high compared with those for other genetic disorders. Explain why most spontaneous mutations for achondroplasia are paternal in origin and why the occurrence of achondroplasia is higher among older fathers.
In men, sperm cells are produced throughout much of their life. The cells responsible for the sperm production are called spermatogonia. These spermatogonia divide by mitosis to produce more spermatogonia and produce spermatocytes, cells that eventually will divide by meiosis to produce sperm cells. These continued cell divisions by the spermatogonia, particularly to produce more spermatogonia, could lead to an increased chance of mutations within the DNA of the spermatogonia cell. Essentially, the more cell divisions the greater the risk for mutation. Also some DNA sequences may be more susceptible to mutations than others. These locations are called hot spots. Potentially, the base substitution occurs at a hot spot in the FGFR3 *gene as more cell divisions occur. In addition, as men age, their exposure to environmental factors over time may increase mutation rates including that of the* FGFR3 *gene.*

However, recent data do not support that the increase in mutations is due only to an increase in mutations as a result of spermatogonia mitosis. A second possibility is that the mutation may confer a positive benefit to the spermatogonia or, ultimately, to the sperm cells produced. Potentially, the mutated sperm have a higher survival rate than normal sperm cells leading to an increased risk of fertilization by the FGFR3 *mutation containing sperm cells.*

34. Tay–Sachs disease is a severe, autosomal recessive genetic disease that produces deafness, blindness, seizures, and eventually death. The disease results from a defect in the *HEXA* gene, which codes for hexosaminidase A. This enzyme normally degrades GM2 gangliosides. In the absence of hexosaminidase A, GM2 gangliosides accumulate in the brain. Recent molecular studies have shown that the most common mutation causing Tay–Sachs disease is a 4 bp insertion that produces a downstream premature stop codon. Further studies reveal that normal transcription of the *HEXA* gene occurs in individuals with Tay–Sachs disease, but the *HEXA* mRNA is unstable. Propose a mechanism to account for how a premature stop codon could cause mRNA instability.
The stability of the mRNA molecule may be dependent on the translational machinery. If the ribosome terminates prematurely, then the mRNA molecule may become a target of ribonucleases. Essentially, the premature stop codon has marked the HEXA *mRNA for destruction in Tay–Sachs individuals.*

Recent data have shown that the premature stop codon within the HEXA *mRNA molecule leads to the rapid destruction of the* HEXA *mRNA within the cell. The destruction is likely accomplished through a process called nonsense-mediated mRNA decay (NMD). In the current model of NMD function, the NMD pathway is stimulated by the premature termination of protein synthesis at a nonsense codon located 50 or more nucleotides upstream of the final exon-exon junction. A exon junction protein complex (EJC) forms approximately 20 to 24 nucleotides upstream of the final exon-exon junction. The EJC stimulates decay of the mRNA molecule unless it is removed by the ribosome during translation. In the normal* HEXA *mRNA translation, the first ribosome removes the EJC because the termination codon is located downstream of the EJC. In individuals with Tay Sachs, the premature stop codon is located upstream of the EJC resulting in termination before the first ribosome can remove the EJC. Because the EJC is not removed, stimulation of NMD occurs and the mutated* HEXA *mRNA is degraded.*

35. Mutations *ochre* and *amber* are two types of nonsense mutations. Before the genetic code was worked out, Sydney Brenner, Anthony O. Stretton, and Samuel Kaplan applied different types of mutagens to bacteriophages in an attempt to determine the bases present in the codons responsible for *amber* and *ochre* mutations. They knew that *ochre* and *amber* mutants were suppressed by different types of mutations, demonstrating that each is a different termination codon. They obtained the following results:
(1) A single-base substitution could convert an *ochre* mutation into an *amber* mutation.
(2) Hydroxylamine induced both *ochre* and *amber* mutations in wild-type phages.
(3) 2-Aminopurine caused *ochre* to mutate to *amber.*
(4) Hydroxylamine did not cause *ochre* to mutate to *amber.*

These data do not allow the complete nucleotide sequence of the *amber* and *ochre* codons to be worked out, but they do provide some information about the bases found in the nonsense mutations.

a. What conclusions about the bases found in the codons of *amber* and *ochre* mutations can be made from these observations?
In considering the data, it is important to remember the mutagenic actions of hydroxylamine and 2-aminopurine. Hydroxylamine produces only GC to AT transition mutations. However, 2-aminopurine can produce both types of transitions, GC to AT and AT to GC.

Because hydroxylamine can be used to produce amber *and* ochre *mutations from wild-type phages, then* amber *and* ochre *codons must contain uracil and/or adenine. The production of* amber *mutations from* ochre *codons by 2-aminopurine but not by hydroxylamine suggests that* ochre *mutations do contain adenine and uracil, while* amber *mutations contain guanine and also contain either adenine and/or uracil.*

b. Of the three nonsense codons (UAA, UAG, UGA), which represents the *ochre* mutation?
Based on the mutagenesis data, only the UAA codon matches the results for the ochre *mutation. It is the only stop codon that does not contain guanine.*

Section 18.3

36. To determine whether radiation associated with the atomic bombings of Hiroshima and Nagasaki produced recessive germ-line mutations, scientists examined the sex ratio of the children of the survivors of the blasts. Can you explain why an increase in germ-line mutations might be expected to alter the sex ratio?

Recessive germ-line mutations on the X chromosome have the potential to alter the sex ratios. Female offspring who have a single recessive mutation on the X chromosome will also possess a nonmutated allele on their other X chromosome. Therefore, females will be heterozygous for the pair of alleles. Male offspring, however, have only one X chromosome, and a mutation within an allele on the X chromosome will result in a phenotypic change. If the allele mutated is essential, then the mutation for males will be lethal. Essentially, recessive mutations on the X chromosome in the germ-line could lead to fewer males being born.

Section 18.4

37. Trichothiodystrophy is an inherited disorder in humans that is characterized by premature aging, including osteoporosis, osteosclerosis, early graying, infertility, and reduced life span. Recent studies have shown that the mutation that causes this disorder occurs in a gene that encodes a DNA helicase. Propose a mechanism for how a mutation in a DNA helicase might cause premature aging. Be sure to relate the symptoms of the disorder to possible functions of the helicase enzyme.
The DNA helicase protein (XPD) is involved in DNA repair process, specifically in nucleotide excision repair to displace damaged nucleotides. If the helicase is not functioning properly, then the damaged nucleotides may not be removed, which may result in the accumulation of DNA damage within the chromosomes of these cells. The accumulation of DNA damage in these cells could lead either to apoptosis (programmed cell death) or to cellular senescence followed by cellular death. In either case, the cells affected will not divide and reproduce through cellular division processes. Cells responsible for bone formation, hair color, and fertility if affected by the DNA damage and subsequent lack of repair would ultimately cease dividing. This lack of new cell production would lead to bone weakening and loss, lack of production of new melanocytes needed for hair color, and a lack of production of reproductive cells.

Because this DNA helicase is also associated with a transcription factor (TFIIH) needed in transcription by RNA polymerase II, a second possibility may involve its role in transcription leading to a decline in levels of transcription. Current data suggest that the DNA helicase may be involved in transcriptionally coupled DNA repair, where its function may be to remove the stalled RNA polymerase from DNA that has been damaged. The removal of the RNA polymerase would allow for repair of the damaged section. If the RNA polymerase is not removed, it may prevent repair. A decline in transcriptional function due to malfunctioning transcription factors or a lack of repair because of a stalled RNA polymerase could lead to a decline in cellular functions and cellular death as well.

Chapter Nineteen: Recombinant DNA Technology

COMPREHENSION QUESTIONS

Section 19.1

1. List some of the effects and practical applications of molecular genetic analyses.
Molecular genetics has had profound effects on all fields of biology. Whole genomes have been sequenced, structures of genes elucidated, the patterns of molecular evolution studied. Recombinant DNA technology is now used to diagnose and screen for genetic diseases, and gene therapy is being explored. Recombinant DNA is used to make pharmaceutical products, such as recombinant insulin and clotting factors. Genetically modified organisms will change the lives of farmers and improve agricultural productivity and the quality of food and fiber.

Section 19.2

2. What common feature is seen in the sequences recognized by type II restriction enzymes?
The recognition sequences are palindromic, and 4–8 base pairs long.

3. What role do restriction enzymes play in bacteria? How do bacteria protect their own DNA from the action of restriction enzymes?
Restriction enzymes cut foreign DNA, such as viral DNA, into fragments. Bacteria protect their own DNA by modifying bases, usually by methylation, at the recognition sites.

*4. Explain how gel electrophoresis is used to separate DNA fragments of different lengths.
Gel electrophoresis uses an electric field to drive DNA molecules through a gel that acts as a molecular sieve. The gel is an aqueous matrix of agarose or polyacrylamide. DNA molecules are loaded into a slot or well at one end of the gel. When an electric field is applied, the negatively charged DNA molecules migrate towards the positive electrode. Shorter DNA molecules are less hindered by the agarose or polyacrylamide matrix and migrate faster than longer DNA molecules, which must wind their way around obstacles and through the pores in the gel matrix.

*5. After DNA fragments are separated by gel electrophoresis, how can they be visualized?
DNA molecules can be visualized by staining with a fluorescent dye. Ethidium bromide intercalates between the stacked bases of the DNA double helix, and the ethidium bromide–DNA complex fluoresces orange when irradiated with an ultraviolet light source. Alternatively, they can be visualized by attaching radioactive or chemical labels to the DNA before it is placed in the gel.

6. What is the purpose of Southern blotting? How is it carried out?
Southern blotting is used to detect and visualize specific DNA fragments that have a sequence complementary to a labeled DNA probe. DNA is first cleaved into fragments with restriction endonucleases. The fragments are separated by size via gel electrophoresis. These fragments are then denatured and transferred by blotting onto the surface of a membrane filter. The membrane filter now has single-stranded DNA fragments bound to its surface, separated by size as in the gel. The filter is then incubated with a solution containing a denatured, labeled probe DNA. The probe DNA hybridizes to its complementary DNA on the filter. After washing away excess unbound probes, the labeled probe hybridized to the DNA on the filter can be detected using the appropriate methods to visualize the label. For radioactively labeled probes, the bound probe is detected by exposure to X-ray film.

*7. Give three important characteristics of cloning vectors.
Cloning vectors should have:
(1) An origin of DNA replication so they can be maintained in a cell
(2) A gene, such as antibiotic resistance, to select for cells that carry the vector
(3) A unique restriction site or series of sites to where a foreign DNA molecule may be inserted

8. Briefly describe two different methods for inserting foreign DNA into plasmids, giving the strengths and weaknesses of each method.
Restriction cloning: Vector and foreign DNA are cut with the same restriction enzyme, then ligated together with DNA ligase. This is the most straightforward, simplest method of cloning. The disadvantage is that matching restriction sites may not be available. Ligation may also produce undesirable products, such as the vector ligating to itself, without the foreign DNA insert.
PCR fragment cloning: DNA fragments generated by PCR may be ligated to plasmid vectors in either of two ways. One way is to synthesize PCR primers that have restriction sites at or near their 5′ ends. The resulting PCR fragments can be digested with the appropriate restriction enzymes to generate sticky ends for restriction cloning as described above. Another way is to use plasmid vectors, called T-vectors, that were cut with a blunt-end generating restriction enzyme, and have single T nucleotides added to the 3′ ends. PCR fragments generated by Taq DNA polymerase have single untemplated A nucleotides at their 3′ ends, so these fragments may be ligated to T-vectors by virtue of the single A-T base pair. The advantage of these methods is that they do not require any specific restriction sites in the DNA fragment to be cloned. The disadvantage is that enough sequence information must be known to design an effective pair of PCR primers.

Ligation of blunt ends with T4 DNA ligase: Any two blunt ends of DNA fragments can be ligated, so the method does not depend on the presence of suitable restriction sites. The disadvantage is that blunt-end ligations are inefficient, and vector self-ligation is difficult to control. A modification of this method treats blunt-ended vectors with alkaline phosphatase to remove 5′ phosphates, to dramatically reduce vector self-ligation.

Addition of linkers: Small synthetic DNA sequences containing restriction sites are ligated to blunt ends of DNA fragments. Restriction digestion then generates sticky ends that can be ligated as in restriction cloning. The advantage is that again, the method does not depend on the availability of suitable restriction sites; any restriction site can be added. The disadvantage is that this involves extra steps and is more difficult than simple restriction cloning.

*9. Briefly explain how an antibiotic-resistance gene and the *lacZ* gene can be used as markers to determine which cells contain a particular plasmid.
Many plasmids designed as cloning vectors carry a gene for antibiotic resistance and the lacZ *gene. The* lacZ *gene on the plasmid has been engineered to contain multiple unique restriction sites. Foreign DNAs are inserted into one of the unique restriction sites in the* lacZ *gene and transformed into* E. coli *cells lacking a functional* lacZ *gene. Transformed cells are plated on a medium containing the appropriate antibiotic to select for cells that carry the plasmid. The medium also contains an inducer for the* lac *operon, so the cells express the* lacZ *gene, and X-gal, a substrate for beta-galactosidase that will turn blue when cleaved by β-galactosidase. The colonies that carry plasmid without foreign DNA inserts will have intact* lacZ *genes, make functional β-galactosidase, cleave X-gal, and turn blue. Colonies that carry plasmid with foreign DNA inserts will not make functional β-galactosidase because the* lacZ *gene is disrupted by the foreign DNA insert. They will remain white. Thus, cells carrying plasmids with inserts will form white colonies.*

*10. Briefly explain how the polymerase chain reaction is used to amplify a specific DNA sequence. What are some of the limitations of PCR?
First, the double-stranded template DNA is denatured by high temperature. Then, synthetic oligonucleotide primers corresponding to the ends of the DNA sequence to be amplified are annealed to the single-stranded DNA template strands. These primers are extended by a thermostable DNA polymerase so that the target DNA sequence is duplicated. These steps are repeated 30 times or more. Each cycle of denaturation, primer annealing, and extension results in doubling the number of copies of the target sequence between the primers.

PCR amplification is limited by several factors. One is that sequence of the gene to be amplified must be known, at least at the ends of the region to be amplified, in order to synthesize the PCR primers. Another is that the extreme sensitivity of the technique renders it susceptible to contamination. A third limitation is that the most common thermostable DNA polymerase used for PCR, Taq DNA polymerase, has a relatively high error rate. A fourth limitation is that PCR amplification is usually limited to DNA fragments of up to a few thousand base pairs; optimized DNA polymerase mixtures and reaction conditions extend the amplifiable length to around 20 kb.

11. What is real-time PCR?
Real-time PCR uses fluorescent probes to detect the formation of specific PCR products, and a sensitive instrument to quantify the amount of PCR product formed

after each cycle of PCR, while the reaction is proceeding. This technique is used to quantify absolute or relative amounts of template DNA in samples.

Section 19.3

*12. How does a genomic library differ from a cDNA library? How is each created?
A genomic library is generated by cloning fragments of chromosomal DNA into a cloning vector. Chromosomal DNA is randomly fragmented by shearing or by partial digestion with a restriction enzyme. A cDNA library is made from mRNA sequences. Cellular mRNAs are isolated and then reverse transcriptase is used to copy the mRNA sequences to cDNA, which are cloned into plasmid or phage vectors.

13. How are probes used to screen DNA libraries? Explain how a synthetic probe can be prepared when the protein product of a gene is known.
The DNA library must first be plated out, either as colonies for plasmid libraries or phage plaques on a bacterial lawn for phage libraries. The colonies or plaques are transferred to membrane filters. A nucleic acid probe can be used to identify colonies or phage plaques that contain identical or similar sequences by hybridization. If no cloned DNA probe is available, but the amino acid sequence of the protein is known, then degenerate synthetic oligonucleotides can be synthesized that represent all possible coding sequences for a sequence of 7 to 10 amino acids. The synthetic oligonucleotides can be end-labeled and used as hybridization probes to screen a DNA library.

*14. Briefly explain in situ hybridization, giving some applications of this technique.
In situ hybridization involves hybridization of radiolabeled or fluorescently labeled DNA or RNA probes to DNA or RNA molecules that are still in the cell. This technique can be used to visualize the expression of specific mRNAs in different cells and tissues and the location of genes on metaphase or polytene chromosomes.

15. Briefly explain how a gene can be isolated through positional cloning.
The approximate location of a gene on a chromosome is identified by recombination or deletion mapping, with markers and deletions with known positions on the chromosome. All genes within this region are characterized to determine which gene has mutations that co-segregate with mutant phenotypes.

16. Explain how chromosome walking can be used to find a gene.
Chromosome walking is used to isolate DNA clones encoding genes defined solely by mutations, for which no DNA or amino acid sequence is known. Mapping experiments locate the gene to a chromosome and to a region of the chromosome. The chromosome walk begins with a neighboring gene that has been previously cloned. Clones that overlap this initial gene are isolated, then the overlapping clones are used to isolate additional overlapping clones, until the set of overlapping clones spans the entire chromosomal region that may contain the target gene. All

genes identified within this region are characterized to determine which gene contains mutations that co-segregate with the mutant phenotype.

Section 19.4

17. What is the purpose of the dideoxynucleoside triphosphate in the dideoxy sequencing reaction?
Dideoxynucleoside triphosphates (ddNTPs) act as a substrate for DNA polymerase but cause termination of DNA synthesis when they are incorporated. Mixed with regular dNTPs, fluorescently labeled ddNTPs generate a series of DNA fragments that have terminated at every nucleotide position along the template DNA molecule being sequenced. These fragments can be separated by gel electrophoresis. Because each of the four ddNTPs carries a different fluorescent label, a laser detector can distinguish which base terminates each fragment. Reading the fragments from shorter to longer, an automated DNA sequencer can determine the sequence of the template DNA molecule.

*18. What is DNA fingerprinting? What types of sequences are examined in DNA fingerprinting?
DNA fingerprinting is the typing of an individual for genetic markers at highly variable loci. This is useful for forensic investigations, to determine whether the suspect could have contributed to the evidentiary DNA obtained from blood or other bodily fluids found at the scene of a crime. Other applications include paternity testing and the identification of bodily remains.

The first loci used for DNA fingerprinting were variable number of tandem repeat (VNTR) loci; these consist of short tandem repeat sequences located in introns or spacer regions between genes. The number of repeat sequences at the locus does not affect the phenotype of the individual in any discernible way, so these loci are highly variable in the population. More recently, loci with smaller repeat sequences of just a few nucleotides, called short tandem repeats (STRs), have been adopted because they can be amplified by PCR. Genotyping at 13 to 15 of these unlinked STR loci can identify one individual among trillions of potential genotypes.

Section 19.5

19. How does a reverse genetics approach differ from a forward genetics approach?
Forward genetics begins with mutant phenotype and proceeds toward cloning and characterization of the DNA encoding the gene. Reverse genetics begins with a DNA sequence and then generates mutants to characterize the functions of the gene.

20. Briefly explain how site-directed mutagenesis is carried out.
In oligonucleotide-directed mutagenesis, an oligonucleotide containing the desired mutation in the sequence is synthesized. This mutant oligonucleotide is annealed to denatured target DNA template and used to direct DNA synthesis. The result is a double-stranded DNA molecule with a mismatch at the site to be mutated. When

transformed into bacterial cells, bacterial repair enzymes will convert the molecule to the mutant form about 50% of the time.

*21. What are knockout mice, how are they produced, and for what are they used?
Knockout mice have a target gene disrupted or deleted ("knocked-out"). First, the target gene is cloned. The middle portion of the gene is replaced with a selectable marker, typically the neo *gene that confers resistance to G418. This construct is then introduced back into mouse embryonic stem cells and cells with G418 resistance are selected. The surviving cells are screened for cells where the chromosomal copy of the target gene has been replaced with the* neo-*containing construct by homologous recombination of the flanking sequences. These embryonic stem cells are then injected into mouse blastocyst-stage embryos and these chimeric embryos are transferred to the uterus of a pseudopregnant female mouse. The knockout cells will participate in the formation of many tissues in the mouse fetus, including germ-line cells. The chimeric offspring are interbred to produce offspring that are homozygous for the knockout allele.*

The phenotypes of the knockout mice provide information about the function of the gene.

22. What are some advantages that mice possess as model genetic organisms?
Mice are small, have a relatively short generation time with multiple progeny per litter, and are easily maintained and bred in the laboratory. Highly inbred strains are available; these provide animals with essentially identical genetic backgrounds. The long history of mouse genetics means that many mutations have been identified and characterized. As a mammal, the mouse is closer to humans in physiology than most other genetic model organisms, and the mouse genome is similar to the human genome in size and content of protein-coding genes. Finally, transgenic techniques allow creation of transgenic mice with either foreign genes added or targeted genes knocked out or replaced.

23. How is RNA interference used in the analysis of gene function?
RNA interference is one potential reverse genetics approach to analyze gene function, by specifically repressing expression of that gene. Double-stranded RNA may be injected directly into a cell or organism or the cell or organism may be genetically modified to express a double-stranded RNA molecule corresponding to the target gene.

Section 19.6

24. What is gene therapy?
Gene therapy is the correction of a defective gene by either gene replacement or the addition of a wild-type copy of the gene. For this to work, enough of the cells of the critically affected tissues or organs must be transformed with the functional copy of the gene to restore normal physiology.

APPLICATION QUESTIONS AND PROBLEMS

Section 19.2

*25. Suppose that a geneticist discovers a new restriction enzyme in the bacterium *Aeromonas ranidae*. This restriction enzyme is the first to be isolated from this bacterial species. Using the standard convention for abbreviating restriction enzymes, give this new restriction enzyme a name (for help, see the footnote to Table 19.1).
The first three letters are taken from the genus and species name, and the Roman numeral indicates the order in which the enzyme was isolated. Therefore, the enzyme should be named AraI.

26. How often, on average, would you expect a type II restriction endonuclease to cut a DNA molecule if the recognition sequence for the enzyme had 5 bp? (Assume that the four types of bases are equally likely to be found in the DNA and that the bases in a recognition sequence are independent.) How often would the endonuclease cut the DNA if the recognition sequence had 8 bp?
Because DNA has four different bases, the frequency of any sequence of n bases is equal to $1/(4^n)$. *A five-bp recognition sequence will occur with a frequency of* $1/(4^5)$, *or once every 1024 bp. An 8-bp recognition sequence will occur with a frequency of 1 per* 4^8, *or 65,536 bp.*

*27. A microbiologist discovers a new type II restriction endonuclease. When DNA is digested by this enzyme, fragments that average 1,048,500 bp in length are produced. What is the most likely number of base pairs in the recognition sequence of this enzyme?
Here, $4^n = 1{,}048{,}500$, *so* $n = 10$. *A 10-bp recognition sequence is most likely.*

28. Will restriction sites for an enzyme that has 4 bp in its restriction site be closer together, farther apart, or similarly spaced, on average, compared with those of an enzyme that has 6 bp in its restriction site? Explain your reasoning.
The restriction sites for an enzyme with a 4-bp recognition sequence should be spaced closer together than the sites for an enzyme with a 6-bp recognition sequence. The 4-bp recognition sequence will occur with an average frequency of once every $4^4 = 256$ *bp, whereas the 6-bp recognition sequence will occur with an average frequency of once every* $4^6 = 4096$ *bp.*

*29. About 60% of the base pairs in a human DNA molecule are AT. If the human genome has 3.2 billion base pairs of DNA, about how many times will the following restriction sites be present?

a. *Bam*HI (restriction site = 5′—GGATCC—3′)
b. *Eco*RI (restriction site = 5′—GAATTC—3′)
c. *Hae*III (restriction site = 5′—GGCC—3′)

We must first calculate the frequency of each base. Given that AT base pairs consist 60% of the DNA, we deduce that the frequency of A is 0.3 and frequency of T is 0.3.

The GC base pairs must consist of 40% of the DNA; therefore, the frequency of G is 0.2 and the frequency of C is 0.2.

a. *Bam*H1 GGATCC *is then (0.2)(0.2)(0.3)(0.3)(0.2)(0.2) = 0.000144*
3,200,000,000(0.000144) = 460,800 times
b. *Eco*RI GAATTC = *(0.2)(0.3)(0.3)(0.3)(0.3)(0.2) = 0.000324*
3,200,000,000(0.000324) = 1,036,800 times
c. *Hae*III GGCC = *(0.2)(0.2)(0.2)(0.2) = 0.0016*
3,200,000,000(0.0016) = 5,120,000 times

*30. Restriction mapping of a linear piece reveals the following *Eco*RI restriction sites.

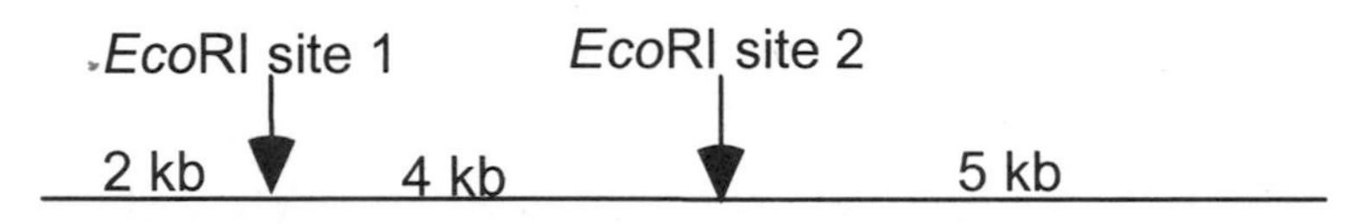

a. This piece of DNA is cut with *Eco*RI, the resulting fragments are separated by gel electrophoresis, and the gel is stained with ethidium bromide. Draw a picture of the bands that will appear on the gel.
b. If a mutation that alters *Eco*RI site 1 occurs in this piece of DNA, how will the banding pattern on the gel differ from the one you drew in part **a.**?
c. If mutations that alter *Eco*RI sites 1 and 2 occur in this piece of DNA, how will the banding pattern on the gel differ from the one you drew in part **a.**?
d. If a 1000-bp insertion occurred between the two restriction sites, how would the banding pattern on the gel differ from the one you drew in part **a.**?
e. If a 500-bp deletion occurred between the two restriction sites, how would the banding pattern on the gel differ from the one you drew in part **a.**?

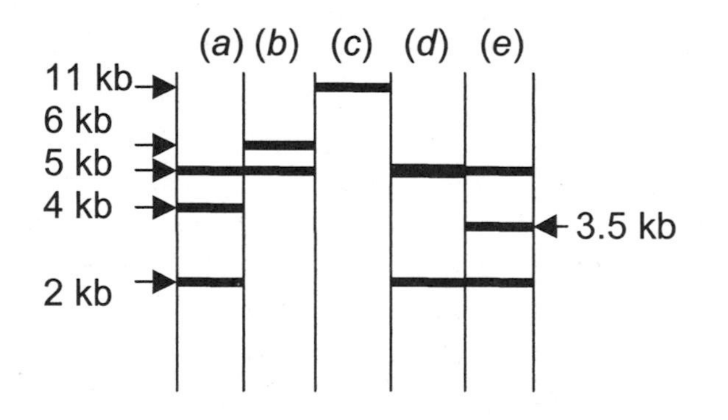

*31. Which vectors (plasmid, λ phage, cosmid, bacterial artificial chromosome) can be used to clone a continuous fragment of DNA with the following lengths?
a. 4 kb – *plasmid*
b. 20 kb – *λphage*
c. 35 kb – *cosmid*
d. 100 kb – *bacterial artificial chromosome*

32. A geneticist uses a plasmid for cloning that has the *lacZ* gene and a gene that confers resistance to penicillin. The geneticist inserts a piece of foreign DNA into a

restriction site that is located within the *lacZ* gene and uses the plasmid to transform bacteria. Explain how the geneticist could identify bacteria that contain a copy of a plasmid with the foreign DNA.

The geneticist should plate the bacteria on agar medium containing penicillin to select for cells that have taken up the plasmid. The medium should also have X-gal and an inducer of the lac *operon, such as IPTG or even lactose. Cells that have taken up a plasmid without foreign DNA will have an intact* lacZ *gene, produce functional β-galactosidase, and cleave X-gal to make a blue dye. These colonies will turn blue. In contrast, cells that have taken up a plasmid containing a foreign DNA inserted into the* lacZ *gene will be unable to make functional β-galactosidase. These colonies will be white.*

Section 19.3

*33. Suppose that you have just graduated from college and have started working at a biotechnology firm. Your first job assignment is to clone the pig gene for the hormone prolactin. Assume that the pig gene for prolactin has not yet been isolated, sequenced, or mapped; however, the mouse gene for prolactin has been cloned and the amino acid sequence of mouse prolactin is known. Briefly explain two different strategies that you might use to find and clone the pig gene for prolactin.

One strategy would be to use the mouse gene for prolactin as a probe to find the homologous pig gene from a pig genomic or cDNA library.

A second strategy would be to use the amino acid sequence of mouse prolactin to design degenerate oligonucleotides as hybridization probes to screen a pig DNA library.

Yet a third strategy would be to use the amino acid sequence of mouse prolactin to design a pair of degenerate oligonucleotide PCR primers to PCR amplify the pig prolactin gene.

34. A genetic engineer wants to isolate a gene from a scorpion that encodes the deadly toxin found in its stinger, with the ultimate purpose of transferring this gene to bacteria and producing the toxin for use as a commercial pesticide. Isolating the gene requires a DNA library. Should the genetic engineer create a genomic library or a cDNA library? Explain your reasoning.

A cDNA library, created from mRNA isolated from the venom gland. Bacteria cannot splice introns. If the engineer wants to express the toxin in bacteria, then he needs a cDNA sequence that has been reverse transcribed from mRNA, and therefore has no intron sequences. The venom gland must be the source of the mRNA for cDNA synthesis, so that the cDNA library will be enriched for toxin cDNAs.

*35. A protein has the following amino acid sequence:
Met-Tyr-Asn-Val-Arg-Val-Tyr-Lys-Ala-Lys-Trp-Leu-Ile-His-Thr-Pro
You wish to make a set of probes to screen a cDNA library for the sequence that encodes this protein. Your probes should be at least 18 nucleotides in length.

a. Which amino acids in the protein should be used to construct the probes so that the least degeneracy results? (Consult the genetic code in Figure 15.12.)
A probe of 18 nucleotides must be based on six amino acids. The six amino acid stretch with the least degeneracy is Val-Tyr-Lys-Ala-Lys-Trp. This sequence avoids the amino acids arg and leu, which have six codons each.

b. How many different probes must be synthesized to be certain that you will find the correct cDNA sequence that specifies the protein?
Val and Ala have four codons each, Tyr and Lys have two codons each, and Trp has one codon. Therefore, there are $4 \times 2 \times 2 \times 4 \times 2 \times 1$ *possible sequences, or 128.*

Section 19.4

36. Suppose that you want to sequence the following DNA fragment:

5′–TCCCGGGAAA-primer site–3′

You first clone the fragment in bacterial cells to produce sufficient DNA for sequencing. You isolate the DNA from the bacterial cells and apply the dideoxy sequencing method. You then separate the products of the polymerization reactions by gel electrophoresis. Draw the bands that should appear on the gel from the four sequencing reactions.

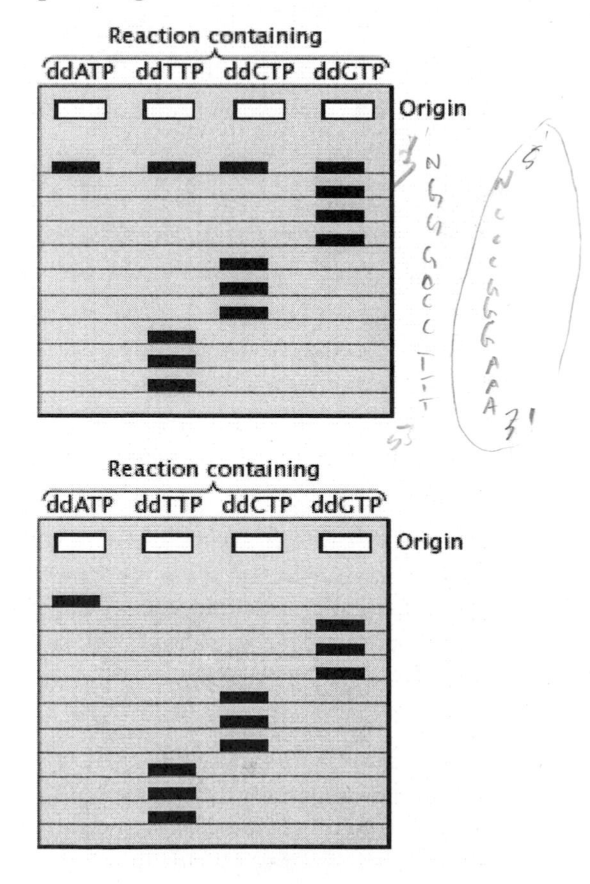

Note that, if the primer is labeled, bands will appear on all four lanes at the 5′ terminus of the DNA template fragment, as shown in the upper figure; the chain will terminate in all four reactions at this position because this is the end of the template. Thus, the 5′ end nucleotide cannot be determined by looking at bands in the sequencing gel. If the dideoxynucleotides are labeled, then the labeled band will appear in only the ddA lane in the uppermost position, as shown in the lower figure.

37. Suppose that you are given a short fragment of DNA to sequence. You clone the fragment, isolate the cloned DNA fragment, and set up a series of four dideoxy reactions. You then separate the products of the reactions by gel electrophoresis and obtain the following banding pattern:

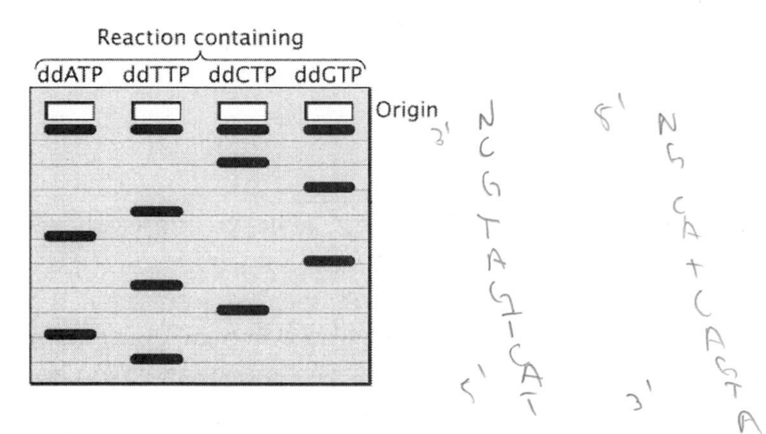

Write out the base sequence of the original fragment that you were given.

Original sequence: 5′– NGCATCAGTA –3′
The base at the 5′ end (N) cannot be determined because the chain stops in all four lanes.

38. The adjoining autoradiograph is from the original study that first sequenced the cystic fibrosis gene (J. R. Riordan et al. 1989. *Science* 245:1066–1073). From the autoradiograph, determine the sequence of the normal copy of the gene and the sequence of the mutated copy of the gene. Identify the location of the mutation that causes cystic fibrosis (CF).

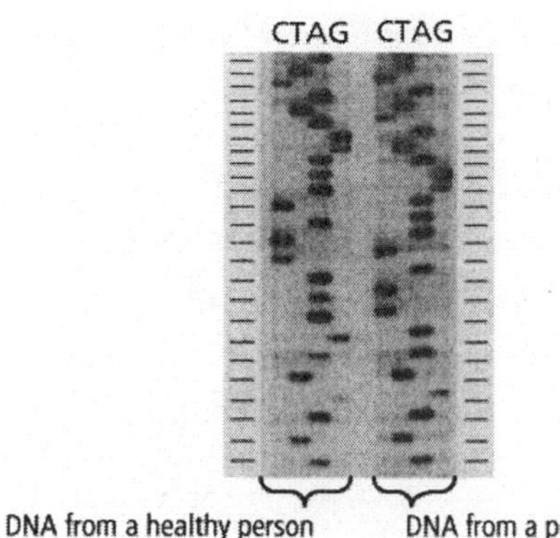

Normal: 5'-ATAGTAGAAACCACAAAGGATACTA...-3'
CF: 5'-ATAGTA[ ]ACCACAAAGGATACTACTT...-3'
The empty bracket in the CF sequence denotes the location of 3 bases deleted in the CF allele.

*39. A hypothetical disorder called G syndrome is an autosomal dominant disease characterized by visual, skeletal, and cardiovascular defects. The disorder appears in middle age. Because the symptoms of the disorder are variable, the disorder is difficult to diagnose. Early diagnosis is important, however, because the cardiovascular symptoms can be treated if the disorder is recognized early. The gene for G syndrome is known to reside on chromosome 7, and it is closely linked to two RFLPs on the same chromosome, one at the *A* locus and one at the *C* locus. The genes at the *G, A,* and *C* loci are very close together, and there is little crossing over between them. The following RFLP alleles are found at the *A* and *C* loci:

A locus: *A1, A2, A3*
C locus: *C1, C2, C3*

Sally, shown in the following pedigree, is concerned that she might have G syndrome. Her deceased mother had G syndrome, and she has a brother with the disorder. A geneticist genotypes Sally and her immediate family for the *A* and *C* loci and obtains the genotypes shown on the pedigree.

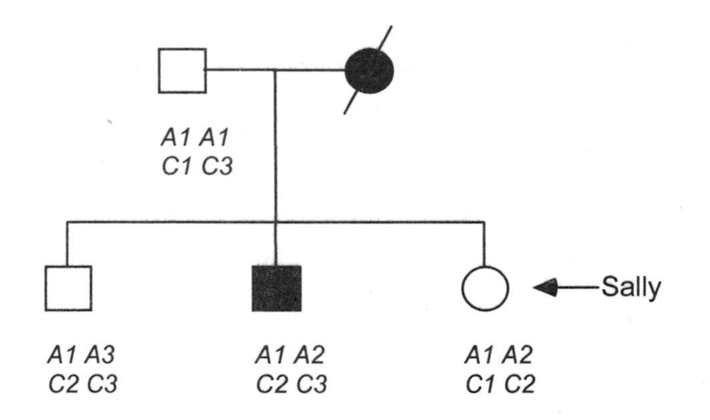

a. Assume that there is no crossing over between the *A, C,* and *G* loci. Does Sally carry the gene that causes G syndrome? Explain why or why not.
First, we must determine the RFLP genotype of Sally's mother. Based on the RFLP genotypes of Sally and her siblings and their father, and given that no recombination takes place among these loci, we deduce that Sally's mother must have had A2A3 *and* C2C2. *For example, Sally's* A2 *and* C2 *could not have come from her father, so they must have come from her mother. Further, we can deduce that* A2 *and* C2 *must be on the same maternal chromosome (because we assume no recombination). That means* A1 *and* C1 *are on the same paternal chromosome. Then Sally's father must have* A1 *and* C3 *on his other chromosome. Now examining Sally's normal brother, we can deduce that he inherited a paternal chromosome with* A1C3, *and a maternal chromosome with* A3C2. *Therefore, the linkage relationships of these chromosomes are* A1C1 *and* A1C3 *from the father, and* A2C2 *and* A3C2 *from the mother. The mother passed on an* A2C2 *to Sally's brother with G syndrome, therefore the G*

syndrome allele must be linked with A2C2. *Because Sally inherited the* A2C2 *chromosome from her mother, she must also have inherited the G syndrome allele, assuming that no crossover occurred between the* A, C, *and* G *loci.*

b. Draw the arrangement of the *A*, *C*, and *G* alleles on the chromosomes for all members of the family.

Father – A1 C1 g, A1 C3 g
Mother – A2 C2 G, A3 C2 g
Sally's unaffected brother – A1 C3 g, A3 C2 g
Sally's affected brother – A1 C3 g, A2 C2 G
Sally – A1 C1 g, A2 C2 G

Section 19.5

*40. You have discovered a gene in mice that is similar to a gene in yeast. How might it be determined whether this gene is essential for development in mice?
This gene must first be cloned, possibly by using the yeast gene as a probe to screen a mouse genomic DNA library. The cloned gene is then engineered to replace a substantial portion of the protein coding sequence with the neo gene. This construct is then introduced into mouse embryonic stem cells. After transfer to the uterus of a pseudopregnant mouse, the progeny are tested for the presence of the knockout allele. Progeny with the knockout allele are interbred. If the gene is essential for embryonic development, no homozygous knockout mice will be born. The arrested or spontaneously aborted fetuses can then be examined to determine how development has gone awry in fetuses that are homozygous for the knockout allele.

41. Andrew Fire, Craig Mello, and their colleagues were among the first to examine the effects of double-stranded RNA on gene expression (A. Fire et al. 1998. *Nature* 391:806–811). In one experiment, they used a transgenic strain of *C. elegans* into which a gene (*gfp*) for a green fluorescent protein had been introduced. They injected some worms with double-stranded RNA complementary to coding sequences of the *gfp* gene and injected other worms with double-stranded RNA complementary to the coding region of a different gene (*unc22C*) that encodes a muscle protein. The adjoining illustration includes photographs of larvae and adult progeny of the injected worms. Green fluorescent protein appears as bright spots in the photographs.

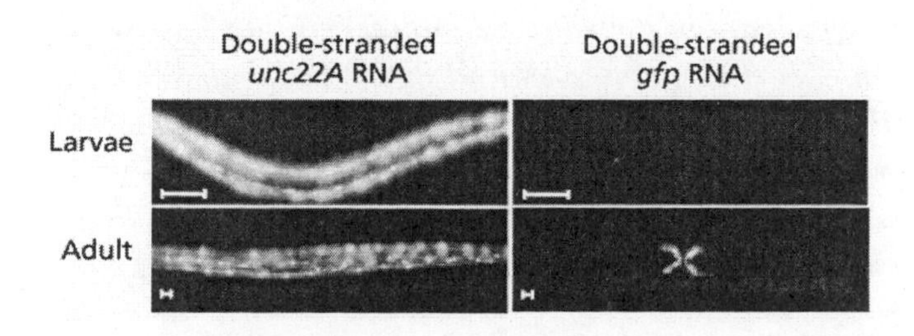

a. Explain these results.
The figure shows that both larvae and adults injected with dsRNA for unc22A *express high levels of gfp protein. However, larvae and adults injected with dsRNA for* gfp *do not express* gfp. *These results indicate that dsRNA specifically inhibits expression of the gene corresponding to the dsRNA, but not unrelated genes.*

b. Fire and Mello conducted another experiment in which they injected double-stranded RNA complementary to the introns and promoter sequences of the *gfp* gene. What results would you expect with this experiment? Explain your answer.
Injection of dsRNA corresponding to introns and promoter sequences would have little effect on gfp *gene expression because RNA interference works by targeting mRNA. Introns and promoter sequences are not present in mRNA.*

CHALLENGE QUESTIONS

Section 19.5

42. Suppose that you are hired by a biotechnology firm to produce a giant strain of fruit flies by using recombinant DNA technology so that genetics students will not be forced to strain their eyes when looking at tiny flies. You go to the library and learn that growth in fruit flies is normally inhibited by a hormone called shorty substance P (SSP). You decide that you can produce giant fruit flies if you can somehow turn off the production of SSP. SSP is synthesized from a compound called XSP in a single-step reaction catalyzed by the enzyme *runtase:*
XSP——————>SSP
runtase
A researcher has already isolated cDNA for runtase and has sequenced it, but the location of the runtase gene in the *Drosophila* genome is unknown.
In attempting to devise a strategy for turning off the production of SSP and producing giant flies by using standard recombinant DNA techniques, you discover that deleting, inactivating, or otherwise mutating this DNA sequence in *Drosophila* turns out to be extremely difficult. Therefore, you must restrict your genetic engineering to gene augmentation (adding new genes to cells). Describe the methods that you will use to turn off SSP and produce giant flies by using recombinant DNA technology.
One possible solution is to create a gene for expression of dsRNA for runtase in the flies. Given the DNA sequence of runtase, one could create a vector with inverted repeats of the cDNA for runtase downstream of a strong promoter, so that hairpin dsRNA molecules of the runtase gene are produced in abundance. These dsRNA molecules will invoke the RNAi pathway to suppress runtase expression. Another route is to design a ribozyme whose targeting sequences are complementary to runtase mRNA. A DNA construct that would express a ribozyme can be transformed into the cell. The expressed ribozyme would then selectively target and cleave runtase mRNA.

Section 19.6

43. Much of the controversy over genetically engineered foods has centered on whether special labeling should be required on all products made from genetically modified crops. Some people have advocated labeling that identifies the product as made from genetically modified plants. Others have argued that only labeling should be required to identify the ingredients, not the process by which they were produced. Take one side in this issue and justify your stand.
    Arguments for labeling products made from genetically modified plants should include risks of allergic reactions to the transgenic protein product, and desires of some consumers who wish to support traditional agricultural practices and oppose genetically modified foods on religious or philosophical grounds, or have concerns about impacts of genetically modified organisms on the diversity of the gene pool and contamination of the gene pool of related species. A relevant precedent is the labeling of organically grown foods, free-range chickens, etc.

    Arguments for labeling to identify only the ingredients center on the fact that many products from genetically modified organisms do not have genetic material and are indistinguishable from products made from non genetically modified organisms. For example, genes for pest resistance will not affect the composition or quality of soybean oil. Another argument is that genetic engineering simply accelerates traditional breeding methods. Modern crops, such as wheat and corn, result from centuries or millenia of breeding; genetic engineering achieves the same result far more quickly and with precision.

Chapter Twenty: Genomics and Proteomics

COMPREHENSION QUESTIONS

Section 20.1

*1. What is the difference between a genetic map and a physical map? Which generally has higher resolution and accuracy and why?
A genetic map locates genes or markers based on genetic recombination frequencies. A physical map locates genes or markers based on the physical length of DNA sequence. Because recombination frequencies vary from one region of the chromosome to another, genetic maps are approximate. Genetic maps also have lower resolution because recombination is difficult to observe between loci that are very close to each other. Physical maps based on DNA sequences or restriction maps have much greater accuracy and resolution, down to a single base pair of DNA sequence.

*2. What is the difference between a map-based approach to sequencing a whole genome and a whole-genome shotgun approach?
The map-based approach first assembles large clones into contigs on the basis of genetic and physical maps and then selects clones for sequencing based on their position in the contig map. The whole genome shotgun approach breaks the genome into short sequence reads, typically 600–700 bps, and then assembles them into contigs on the basis of sequence overlap using powerful computers to search for overlaps.

3. How are DNA fragments ordered into a contig with the use of restriction sites?
Restriction maps are compared among different clones to find overlaps. DNA fragments with overlapping restriction maps are ordered together to form a contig.

*4. Describe the different approaches to sequencing the human genome that were taken by the international collaboration and by Celera Genomics.
The international collaboration took the ordered, map-based approach, beginning with construction of detailed genetic and physical maps. Celera took the whole-genome shotgun approach. Celera did make use of the physical map produced by the international collaboration to order their sequences in the assembly phase.

5. What is a single-nucleotide polymorphism (SNP)? How are SNPs used in genomic studies?
SNPs are single base-pair differences in the sequence of a particular region of DNA from one individual compared to another of the same species or population. SNPs are useful as molecular markers for mapping and pedigree analysis and may themselves be associated with phenotypic differences.

6. What is a haplotype? How do different haplotypes arise?
A haplotype is a particular set of neighboring SNPs or other DNA polymorphisms observed on a single chromosome or chromosome region. They tend to be inherited together as a set because of linkage. Meiotic recombination within the chromosomal region can split the haplotype and create new recombinant haplotypes.

7. What is linkage disequilibrium? How does it result in haplotypes?
Linkage disequilibrium is the association of certain genetic variants (alleles) with each other, so that these combinations occur more frequently in the population than expected based on random assortment. The particular combinations of alleles at different loci within a region of the chromosome is called a haplotype.

8. What is copy number variation? How does it arise?
Copy number variation is the occurrence of more or less than the usual diploid copy number for a DNA sequence of longer than 1000 bp. Such copy number variations could arise through deletion or duplication of a segment of chromosomal DNA.

9. **a.** What is an expressed-sequence tag (EST)?
An *EST is a single-pass sequence read of randomly selected clones from a cDNA library.*
b. How are ESTs created?
First, mRNA is isolated from a whole organism, organ, tissue, or cell line. Then, reverse transcriptase is used to generate cDNAs. The cDNAs are cloned into plasmid or phage vectors. Sequencing primers based on the vector sequence flanking the cloning site are used to sequence the ends of the cDNA inserts.
c. How are ESTs used in genomics studies?
ESTs are used to provide supporting evidence for gene predictions and annotations. ESTs can also show expression patterns for the corresponding gene.

10. How are genes recognized within genomic sequences?
Genes can be recognized by analysis of the following:
(1) The presence of matching cDNA or EST sequences in the database
(2) CpG islands
(3) Open reading frames (ORFs)
(4) Homology (sequence similarity) to previously characterized genes in the same or other species
(5) Computer algorithms that use the preceding criteria to make gene predictions

*11. What are homologous sequences? What is the difference between orthologs and paralogs?
Homologous sequences are derived from a common ancestor. Orthologs are sequences in different species that are descended from a sequence in a common ancestral species. Paralogs are sequences in the same species that originated by

duplication of an ancestral sequence and subsequently diverged. Paralogs may have diverged in function.

12. Describe several different methods for inferring the function of a gene by examining its DNA sequence.
    Homology: For protein-coding genes, the DNA sequence is translated conceptually into the amino acid sequence of the protein. The amino acid sequence of the protein then may yield clues to its function if it is similar to another protein of known function. For example, it is quite easy to recognize histones because their amino acid sequences are highly conserved among eukaryotes. Even if the whole protein is not similar, it may have regions, or domains, that are similar to other domains with known functions or properties. Finally, the amino acid sequence may contain small motifs or signatures that are characteristic of proteins with certain enzymatic activities or properties or subcellular localizations.

    Phylogenetic profile: The coordinated absence or presence of clusters of genes in various species implies that the genes in the cluster have related functions. For example, genes required for nitrogen fixation would all be present in nitrogen-fixing species but absent in other species.

    Gene fusions: In some species, genes of related function have undergone a fusion event to form a single multifunctional polyprotein. Then similar but separate component genes in other species can be presumed to have similar functions.

    Gene clusters or operons: In bacteria, genes in metabolic or functional pathways are often clustered together into an operon. Therefore, all the genes that are co-transcribed into a single polycistronic mRNA should have related functions.

Section 20.2

13. What is a microarray? How can it be used to obtain information about gene function?
    A microarray consists of thousands of DNA fragments spotted onto glass slides in an ordered grid (gene chips) or even proteins or peptides arrayed onto glass slides (protein chips). The identity of the DNA or peptide at each location is known. Gene chips are typically used in hybridization experiments with labeled mRNAs or cDNAs to survey the levels of transcript accumulation for thousands of genes, or even whole genomes, at one time. Peptide or protein chips can be used to identify protein-protein interactions or enzymatic activities or other properties of proteins.

14. Explain how a reporter sequence can be used to provide information about the expression pattern of a gene.
    A reporter sequence is fused to a gene in such a way that the native gene regulatory sequences drive expression of the recombinant gene: reporter fusion product. Typically, genomic DNA sequence including the upstream promoter region and other cis-acting regulatory sequences is ligated to the reporter gene sequence. This construct may then be used to create a transgenic organism expressing the recombinant reporter gene fusion. The reporter may have enzymatic activity (like β-

galactosidase) that is detectable with a substrate that forms a colored product, or with an antibody to the reporter protein, or the reporter may itself be fluorescent (like green fluorescent protein). The gene's own regulatory sequences specify the developmental pattern of expression of the reporter as they would the native gene. If the protein coding region of the gene is also included, in part or in full, the resulting translational fusion product can be used to study the subcellular localization of the protein.

*15. Briefly outline how a mutagenesis screen is carried out.
After random mutagenesis with chemicals or transposons, the mutant progeny population is screened for phenotypes of interest. The mutant gene can be identified by co-segregation with molecular markers or by sequencing the position of transposon insertion. To verify that the mutation identified is truly responsible for the phenotype, a mutation can be introduced into a wild-type copy of the gene and the phenotype observed in the offspring.

Section 20.3

16. What is the relation between genome size and gene number in prokaryotes?
In prokaryotes, the gene number is proportional to the genome size, because most of the genome encodes proteins.

17. Eukaryotic genomes are typically much larger than prokaryotic genomes. What accounts for the increased amount of DNA seen in eukaryotic genomes?
The increased amount of DNA in eukaryotic genomes is due mostly to the increased amount of non-coding DNA, sometimes referred to as "junk" DNA, in introns and intergenic regions, and to transposable elements. A relatively minor contribution to increased genome size is that eukaryotes, especially the complex multicellular species, generally encode more genes, and the average size of eukaryotic proteins may be larger than the average size of prokaryotic proteins.

18. What is one consequence of differences in the G + C content of different genomes?
Since higher G + C content causes greater stability of the DNA duplex, one consequence is that high G + C content is characteristic of cells that are found at high temperatures, such as in hot springs. Increased G + C content also means that proteins will have a higher percentage of amino acids whose codons have G and C residues, such as glycine (GGN) and proline (CCN), alanine (GCN), and arginine (CGN).

*19. What is horizontal gene transfer? How might it take place between different species of bacteria?
Horizontal gene transfer is transmission of genetic material across species boundaries. In bacteria, horizontal gene exchange may occur through uptake of environmental DNA through transformation, by conjugative plasmids with broad host range, or by transfection with bacteriophage with broad host range.

20. DNA content varies considerably among different multicellular organisms. Is this variation closely related to the number of genes and the complexity of the organism? If not, what accounts for the differences?
This question is almost a philosophical one because "complexity" of an organism is not well-defined and thus difficult to quantify. However, we do know that the genomic DNA content can vary widely among related species, so there appears to be little relation between the "complexity" of an organism, the number of genes, and the DNA content. Large differences in DNA content may arise from differences in the frequency and size of introns, the abundance of DNA derived from transposable elements, and duplication of the whole or substantial parts of the genome in the evolutionary history of the species.

*21. More than half of the genome of *Arabidopsis thaliana* consists of duplicated sequences. What mechanisms are thought to have been responsible for these extensive duplications?
The Arabidopsis *genome appears to have undergone at least one round of duplication of the whole genome (tetraploidy) and numerous localized duplications via unequal crossing over.*

22. What is a segmental duplication?
A duplication involving a relatively large (greater than 1000 bp) stretch of chromosomal DNA sequence.

23. What is a gene desert?
An extended chromosomal region with no protein-coding genes. The gene density varies greatly among different regions of eukaryotic chromosomes.

24. The human genome does not encode substantially more protein domains than do invertebrate genomes, and yet it encodes many more proteins. How are more proteins encoded when the number of domains does not differ substantially?
The human genome contains proteins with many more combinations of domains, often featuring multiple domains on a single protein.

25. **a.** What is genomics and how does structural genomics differ from functional genomics?
Genomics is the study of the content, organization, function, operation, and evolution of whole genomes. Structural genomics deals largely with the DNA sequence itself and its physical organization; functional genomics deals largely with the functions of the sequences in the genome, through RNA or protein molecules.

b. What is comparative genomics?
Comparative genomics analyzes the similarities and differences among genomes to achieve insights into genomic or organismal evolution or relationships between groups of genes and physiological functions.

Section 20.4

26. How does proteomics differ from genomics?
Whereas genomics is the analysis of whole genome DNA sequences and their organization, expression, and function, proteomics focuses on the complete set of proteins made by an organism. Unlike the genome, which is constant from cell to cell, the proteome varies within a species from cell type to cell type, with stage of development, and with time in response to signals and other environmental stimuli. Proteins also undergo numerous modifications that affect protein localization and function.

27. How is mass spectrometry used to identify proteins in a cell?
Protein fragments generated by protease digestion are separated by mass and charge. Computer algorithms compare the mass profiles of these fragments with databases of known fragments of proteins or predicted fragments in a genome sequence.

28. What are some of the ethical concerns arising out of the information produced by the Human Genome Project?
The ethical questions concern privacy: Who will have access to a person's genetic profile, and what will be done with that information? Information concerning genetic susceptibility to disease may concern insurance companies and employers and could possibly be used to deny health insurance or employment. Other concerns include the concept of genetic determinacy: How does an individual or society interpret associations between genetic polymorphisms and various phenotypic traits, as susceptibility to various genetic diseases, and correlations with such complex traits such as intelligence and different aspects of behavior? Still other questions are raised about genetic engineering applied to the human germ-line as the ultimate form of eugenics. These are only some of the issues raised by the Human Genome Project and other technologies currently in development.

APPLICATION QUESTIONS AND PROBLEMS

Section 20.1

*29. A 22-kb piece of DNA has the following restriction sites:

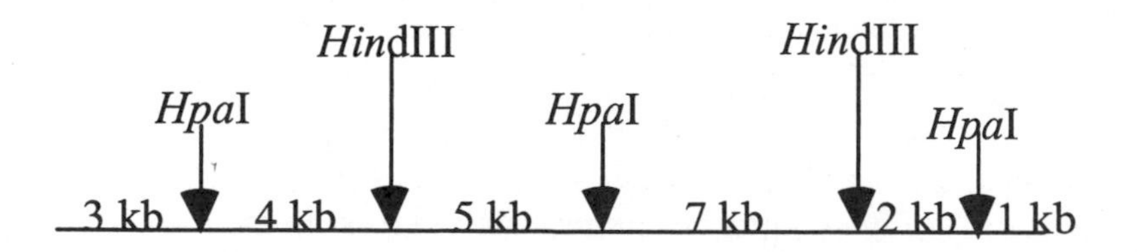

A batch of this DNA is first fully digested by *Hpa*I alone, then another batch is fully digested by *Hind*III alone, and finally a third batch is fully digested by *Hpa*I and

*Hind*III together. The fragments resulting from each of the three digestions are placed in separate wells of an agarose gel, separated by gel electrophoresis, and stained by ethidium bromide. Draw the bands as they would appear on the gel.

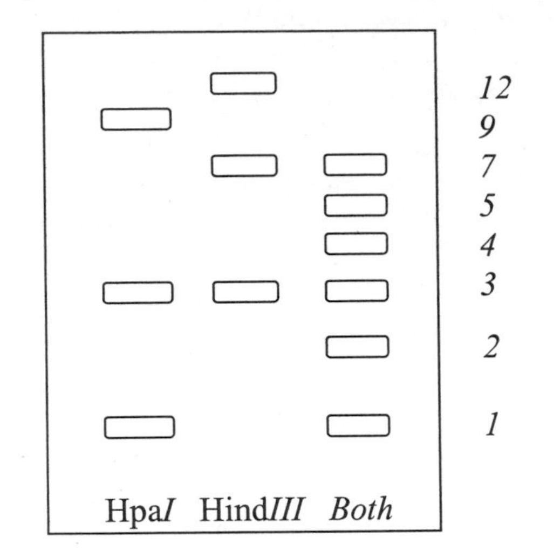

*30. A piece of DNA that is 14 kb long is cut first by *Eco*RI alone, then by *Sma*I alone, and finally by *Eco*RI and *Sma*I together. The following results are obtained:

Digestion by *Eco*RI alone	Digestion by *Sma*I alone	Digestion by *Eco*RI and *Sma*I
3 kb fragment	7 kb fragment	2 kb fragment
5 kb fragment	7 kb fragment	3 kb fragment
6 kb fragment		4 kb fragment
		5 kb fragment

Draw a map of the *Eco*RI and *Sma*I restriction sites on this 14-kb piece of DNA, indicating the relative positions of the restriction sites and the distances between them.
We know that SmaI *cuts only once, in the middle of this piece of DNA, at 7 kb.* Eco*RI cuts twice. Comparing the* Eco*RI digest to the double digest, we see that neither the 3-kb nor the 5-kb fragments is cut by* SmaI; *only the 6-kb* Eco*RI fragment is cut by* SmaI *to 2-kb and 4-kb fragments. Therefore, the 6-kb* Eco*RI fragment is in the middle, and the 3-kb and 5-kb* Eco*RI fragments are at the sides.*

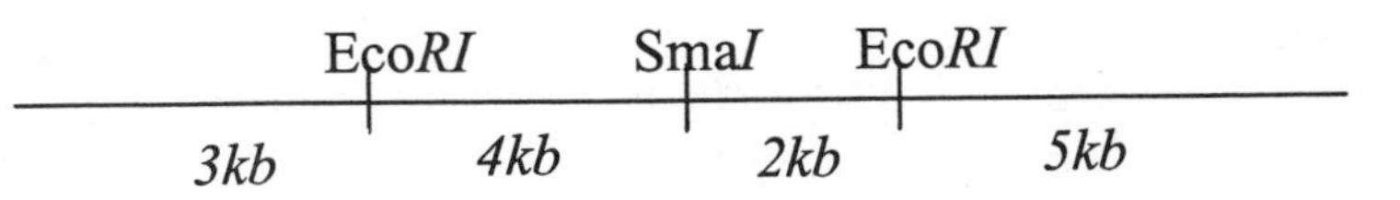

31. The presence (+) or absence (–) of six sequence-tagged sites (STSs) in each of five bacterial artificial chromosome (BAC) clones (A–E) is indicated in the following table. Using these markers, put the BAC clones in their correct order and indicate the locations of the STS sites within them.

BAC clone	STSs 1	2	3	4	5	6
A	+	–	–	–	+	–
B	–	–	–	+	–	+
C	–	+	+	–	–	–
D	–	–	+	–	+	–
E	+	–	–	+	–	–

Solution:

C *3* *2*

D *5* *3*

A *1* *5*

E *4* *1*

B *6* *4*

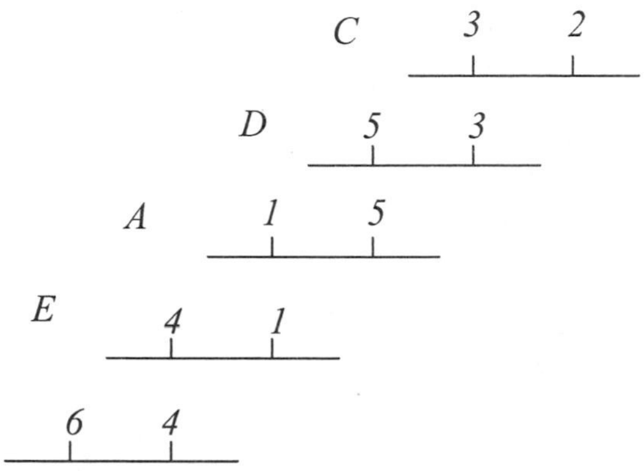

32. How does the density of genes found on chromosome 22 compare with the density of genes found on chromosome 21, two similar-sized chromosomes? How does the number of genes on chromosome 22 compare with the number found on the Y chromosome?

To answer these questions, go to the Ensembl Web site: http://www.ensembl.org/ Under the heading *Species*, click Human. On the left-hand side of the next page are pictures of the human chromosomes. Click on chromosome 22. You will be shown a picture of this chromosome and a histogram illustrating the densities of total genes (uncolored bars) and of known genes (colored bars). The total numbers of genes (gene count) and known genes are given on the upper right-hand side of the page, along with the chromosome length in base pairs.

Now go to chromosome 21 by pulling down the Change Chromosome menu and selecting chromosome 21. Examine the density and total number of genes for chromosome 21. Now do the same for the Y chromosome.

a. Which chromosome has the highest density and greatest number of genes? Which has the fewest?
Chromosome 22 has the highest density and greatest number of genes, with over 500 known and 24 novel genes, whereas the Y chromosome has the lowest density and fewest genes, with fewer than 100 known and 23 novel genes.

b. Examine in more detail the genes at the tip of the short arm of the Y chromosome by clicking on the top bar in the histogram of genes. A more detailed view will be shown. What known genes are found in this region? How many novel genes are there in this region?

The known genes found in this region (0–1,000,000 bp) are PLCXD1, GTPBP6, PPP2R3B, and SHOX. No novel genes are annotated in this region.

Section 20.2

33. Microarrays can be used to determine the levels of gene expression. In one type of microarray, hybridization of the red (experimental) and green (control) cDNAs is proportional to the relative amounts of mRNA in the samples. Red indicates the overexpression of a gene, green indicates the underexpression of a gene in the experimental cells relative to the control cells, yellow indicates equal expression in experimental and control cells, and no color indicates no expression in either experimental or control cells.

    In one experiment, mRNA from a strain of antibiotic-resistant bacteria (experimental cells) is converted into cDNA and labeled with red fluorescent nucleotides; mRNA from a nonresistant strain of the same bacteria (control cells) is converted into cDNA and labeled with green fluorescent nucleotides. The cDNAs from the resistant and nonresistant cells are mixed and hybridized to a chip containing spots of DNA from genes 1 through 25. The results are shown in the illustration on the following page. What conclusions can you make about which genes might be implicated in antibiotic resistance in these bacteria? How might this information be used to design new antibiotics that are less vulnerable to resistance?

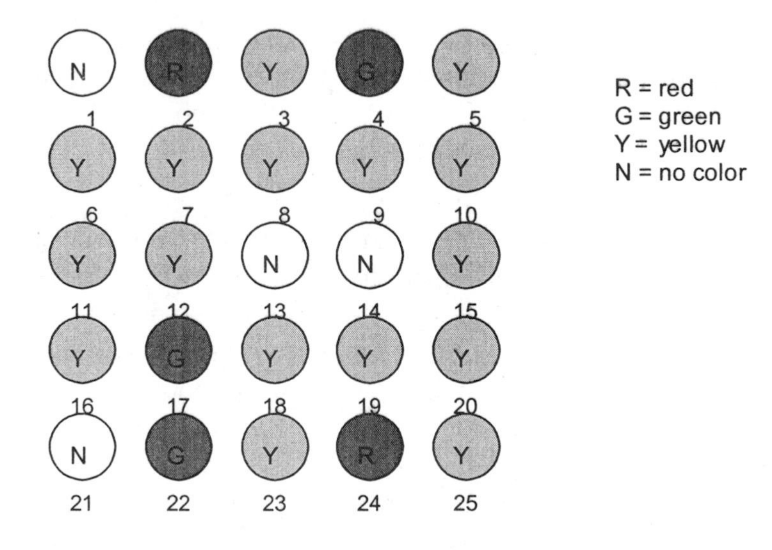

Genes 2 and 24 are expressed at far higher levels in the antibiotic-resistant bacteria than in the nonresistant cells. Conversely, genes 4, 17, and 22 are down-regulated. These genes may be involved in antibiotic resistance: up-regulated genes may be involved in metabolism of the antibiotic or may perform functions that are inhibited by the antibiotic. Down-regulated genes may be involved in import of the antibiotic or represent a cellular mechanism that accentuates the potency of the antibiotic. Characterization of these genes may lead to information regarding the mechanism of antibiotic resistance, and then to the design of new antibiotics that can circumvent this resistance mechanism.

Section 20.3

34. *Dictyostelium discoideum* is a soil-dwelling, social amoeba: much of the time, the organism consists of single, solitary cells, but, during times of starvation, individual amoebae come together to form aggregates that have many characteristics of multicellular organisms. Biologists have long debated whether *D. discoideum* is a unicellular or multicellular organism. In 2005, the genome of *D. discoideum* was completely sequenced. The adjoining table lists some genomic characteristics of *D. discoideum* and other eukaryotes (L. Eichinger et al. 2005. *Nature* 435:43–57).

Feature	***D. discoideum***	***P. falciparium***	***S. cerevisiae***	***A. thaliana***	***D. melanogaster***	***C. elegans***	***H. sapiens***
Organism	amoeba	malaria parasite	yeast	plant	fruit fly	worm	human
Cellularity	?	uni	uni	multi	multi	multi	multi
Genome size (millions bp)	34	23	13	125	180	103	2,852
Number of genes	12,500	5,268	5,538	25,498	13,676	19,893	22,287
Average gene length (BP)	1,756	2,534	1,428	2,036	1,997	2,991	27,000
Genes with introns (%)	69	54	5	79	38	5	85
Mean number of introns	1.9	2.6	1.0	5.4	4.0	5.0	8.1
Mean intron size (bp)	146	179	nd*	170	nd*	270	3,365
Mean G + C (exons)	27%	24%	28%	28%	55%	42%	45%

***nd = not determined**

a. On the basis of the organisms listed in the table other than *D. discoideum*, what are some differences in genome characteristics between unicellular and multicellular organisms?

Unicellular organisms have smaller genomes and fewer genes.

b. On the basis of these data, do you think the genome of *D. discoideum* is more like those of other unicellular eukaryotes or more like those of multicellular eukaryotes? Explain your answer.

The genome size of D. discoideum *is smaller, like unicellular organisms, but the number of genes approaches the number of genes in the* Drosophila *genome. Thus, the* D. discoideum *genome has characteristics of the genomes of both unicellular and multicellular organisms.*

35. What are some of the major differences between the ways in which genetic information is organized in the genomes of prokaryotes versus eukaryotes?
Most prokaryotic genomes consist of a single circular chromosome, whereas eukaryotic genomes have multiple linear chromosomes.

Prokaryotic genomes have none to few introns; complex eukaryotic genomes have numerous introns, with most genes having multiple introns.

Prokaryotic genomes have high gene density, approximately one gene per 1000 base pairs; most eukaryotic genomes have far lower gene density.

Prokaryotic genomes have functionally related genes organized into operons transcribed as polycistronic mRNAs from a single promoter; eukaryotic genomes generally have monocistronic mRNAs, each from its own promoter.

36. How do the following genomic features of prokaryotic organisms compare with those of eukaryotic organisms? How do they compare among eukaryotes?
   a. Genome size: *0.5 million to over 9 million bp in prokaryotes; 13 million (yeast) to over 100 billion in eukaryotes.*
   b. Number of genes: *From 500 to 7000 in prokaryotes; 6000 to 30,000 in eukaryotes.*
   c. Gene density (bp/gene): *Approximately 1000 bp/gene in prokaryotes; varies from 2000 bp/gene to greater than 100,000 bp/gene in eukaryotes.*
   d. G + C content: *From 27% to 72% in prokaryotes; less variable in eukaryotes (about 40% to 50%).*
   e. Number of exons: *With few exceptions, prokaryotic genes have zero or only one exon; multicellular eukaryotic genes typically have multiple exons.*

CHALLENGE QUESTION

Section 20.3

*37. Some researchers have proposed creating an entirely new, free-living organism with a minimal genome, the smallest set of genes that allows for replication of the organism in a particular environment. This organism could be used to design and create, from "scratch," novel organisms that might perform specific tasks, such as the breakdown of toxic materials in the environment.
   a. How might the minimal genome required for life be determined?
   The minimal genome required might be determined by examining simple free-living organisms having small genomes to determine which genes they possess in common. Mutations can then be made systematically to determine which genes are essential for these organisms to survive. The apparently nonessential genes (those genes in which mutations do not affect the viability of the organism) can then be deleted one by one until only the essential genes are left: Elimination of any of these genes will result in loss of viability. Alternatively, essential genes could be assembled through genetic engineering, creating an entirely novel organism.

b. What, if any, social and ethical concerns might be associated with the creation of novel organisms by constructing an entirely new organism with a minimal genome?

This synthetic organism would prove that humans have acquired the ability to create a new species or form of life. Humans would then be able to direct evolution as never before. Social and ethical concerns would revolve around whether human society has the wisdom to temper its power and whether such novel synthetic organisms can or will be used to develop pathogens for biological warfare or terrorism. After all, no person or animal would have been exposed previously or have acquired immunity to such a novel synthetic organism. Also, it would be uncertain how the new organism would affect the ecosystem if it was released or escaped. These are one person's concerns. Other people may have additional concerns.

Chapter Twenty-One: Organelle DNA

COMPREHENSION QUESTIONS

Section 21.1

1. Explain why many traits encoded by mtDNA and cpDNA exhibit considerable variation in their expression, even among members of the same family.
   Within a given eukaryotic cell, there can be thousands of copies of each type of organelle genome. Mutations that occur in one genome can give rise to populations of organelles that differ in DNA sequence within a cell, a phenomenon called heteroplasmy. During cytokinesis or cell division, the process of replicative segregation separates the organelles randomly into the daughter cells. In general, equal proportions of organelles containing mutant sequences and organelles containing wild-type sequences will be distributed to each progeny cell. However, because replicative segregation is random, occasionally one progeny cell may receive all wild-type organelles while the other progeny cell receives the mutant types. When only one type of genome is present within a cell, the condition is known as homoplasmy. Essentially, replicative segregation can lead to cells within a given individual and among family members that differ in organelle phenotype and genotype.

*2. What is the endosymbiotic theory? How does it help to explain some of the characteristics of mitochondria and chloroplasts?
   The endosymbiotic theory proposes that mitochondria and chloroplasts evolved from formerly free-living bacteria that became endosymbiants within a larger eukaryotic cell.

   Chloroplasts and mitochondria contain genomes that encode for proteins, tRNAs, and rRNAs. The chloroplast and mitochondrial genome sizes, circular chromosome structure, and other aspects of genome structure are similar to those of eubacterial cells. Moreover, the chloroplast and mitochondrial ribosomes are similar in size and function to eubacterial ribosomes. DNA sequences in mitochondrial and chloroplast genomes are most similar to those in eubacteria.

3. What evidence supports the endosymbiotic theory?
   Several lines of evidence support the endosymbiotic theory. Perhaps the most convincing lines of evidence are sequence similarities of protein encoding and rRNA genes in mitochondria and chloroplasts to counterpart genes in specific groups of eubacteria. The mtDNA sequences are most similar to DNA from a group of bacteria called the α-proteobacteria, while the cpDNA sequences are most closely linked to the sequences from cyanobacteria. Another line of evidence that supports the endosymbiont theory includes sensitivity of protein synthesis in mitochondria and chloroplasts to antibiotics, such as streptomycin and tetracycline, inhibitors of protein synthesis in eubacteria but not eukaryotes. Conversely, some antibiotics, such as cycloheximide, inhibit eukaryotic protein synthesis but not that of the mitochondria and chloroplasts. A third line of evidence is found in

mitochondrial genomes in which the AUG start codon encodes N-formyl methionine, which is typical of eubacteria translation initiation. Finally, the size and structure of ribosomes in both chloroplasts and mitochondria are similar to that of eubacterial ribosomes.

Section 21.2

4. How are genes organized in the mitochondrial genome? How does this organization differ between ancestral and derived mitochondrial genomes?
The mitochondrial genome consists of circular DNA molecules that encode protein, rRNAs, and tRNAs. How the genes are organized varies among different eukaryotic species with ancestral mitochondrial genomes found in some plants and protists, whereas derived mitochondrial genomes are more common in animals and fungi.

Mitochondrial ancestral genomes typically contain more protein encoding genes and tRNA genes than derived genomes. The rRNA genes of ancestral genomes resemble those of eubacterial rRNA genes. Ancestral genomes include few introns, use universal codons, organize genes into clusters, and contain little noncoding DNA sequences. Finally, DNA sequences of ancestral genomes more closely resemble eubacterial genomes than DNA sequences of derived genomes. Mitochondrial-derived genomes are typically smaller than ancestral genomes. Derived genomes may contain nonuniversal codons, and the rRNA genes differ from those of eubacterial rRNAs.

*5. What are nonuniversal codons? Where are they found?
Nonuniversal codons specify an amino acid in one organism but not in most other organisms. Typically, they are found in mitochondrial genomes, and these exceptions vary in the mitochondria of different organisms.

6. How does replication of mtDNA differ from replication of nuclear DNA in eukaryotic cells?
DNA polymerase gamma is responsible for replication of mitochondrial DNA, whereas the DNA polymerases delta and alpha are necessary for nuclear DNA replication. In mitochondrial DNA, the replication origins for the different strands of the double helix are found at different locations within the genome. Furthermore, replication of the two strands can be asynchronous with both strands exhibiting continuous replication. In nuclear DNA replication, the replication of both strands is synchronous and proceeds from a common origin with continuous replication of one strand and discontinuous replication on the other strand.

*7. The human mitochondrial genome encodes only 22 tRNAs, whereas at least 32 tRNAs are required for cytoplasmic translation. Why are fewer tRNAs needed in mitochondria?
During translation in mitochondria, "wobble" pairing at the third position of the codon occurs more frequently than in translation of nuclear genes. Most anticodons of the mitochondrial tRNAs can pair with more than one codon. Essentially, the first

position of the anticodon can pair with any of the four nucleotides present at the third position of the codon.

8. What are some possible explanations for an accelerated rate of evolution in the sequences of vertebrate mtDNA?
Vertebrate mtDNA has a higher mutation rate than the vertebrate nuclear DNA and the mtDNA of plants, which could be responsible for the accelerated rate of evolution. Reasons for the increased level of mutations include a lack of DNA repair, increased errors in replication, and a higher rate of replication for mtDNA.

9. What are some of the advantages of using yeast for genetic studies?
Some advantages of using yeast for genetic studies:
   - *Yeast are eukaryotic organisms, so they share similarities in genetic and cellular systems with other more complex eukaryotic organisms*
   - *Yeast are unicellular and, like bacterial systems, can be easily and inexpensively grown in the laboratory while requiring little space.*
   - *Yeast can exist either in a haploid or diploid form, so the phenotypes of recessive alleles can easily be identified in the haploid form. The interactions of alleles can be studied in the diploid form.*
   - *Yeast meiotic products can be examined by tetrad analysis. The products of meiosis in yeast are contained with a structure called an ascus, which allows for direct examination of the products of meiosis.*
   - *Yeast can be studied using many molecular biology techniques developed initially for studying bacteria.*
   - *Yeast possess many genes and DNA sequences that are similar to genes in multicellular eukaryotes, including humans.*
   - *The genome of* Saccharomyces cerevisiae, *a yeast, was completely sequenced.*
   - *Yeast naturally contain a plasmid (2μ). This plasmid has been adapted as a vector for transferring genes or DNA sequences of interest into yeast.*
   - *Yeast artificial chromosomes (YACs), which allow for the transfer of large DNA fragments into yeast, have been developed.*

Section 21.3

*10. Briefly describe the organization of genes on the chloroplast genome.
The chloroplast genome is typically a double-stranded circular DNA molecule whose organization and sequences resembles those of eubacterial genomes. Genes are located on both strands of cpDNA and may contain introns. Long noncoding nucleotide sequences occur between genes and there are many duplicated sequences. The chloroplast genome usually encodes ribosomal proteins, 5 rRNAs, 30 to 35 tRNA genes, and proteins involved in photosynthesis, as well as proteins not involved in photosynthesis. A large inverted repeat is also found in the genomes of most chloroplasts.

*11. Briefly describe the general structures of mtDNA and cpDNA. How are they similar? How do they differ? How do their structures compare with the structures of eubacterial and eukaryotic (nuclear) DNA?
Mitochondrial and chloroplast DNAs have many characteristics in common with eubacterial DNAs. Most mitochondrial and chloroplast chromosomes are small, circular, and lack histone proteins—characteristics that are similar to eubacterial, but not eukaryotic, cells. Chloroplasts and some mitochondria produce polycistronic mRNA, another characteristic common to eubacteria. Chloroplast genes, but not most mitochondrial genes, typically possess Shine-Dalgarno sequences. Antibiotics that inhibit eubacterial translation inhibit mitochondrial and chloroplast translation.

Eukaryotic nuclear genomes are typically composed of linear chromosomes in which DNA is bound to histone proteins. Eukaryotic nuclear DNA sequences also contain pre-mRNA introns. Pre-mRNA introns are found in chloroplast but are usually not in mitochondrial genomes. Eukaryotic genomes contain large numbers of repeated sequences and pseudogenes, which are usually not found in mitochondrial genomes.

Section 21.4

12. What is meant by the term "promiscuous DNA"?
DNA sequences that have been exchanged among the nuclear, mitochondrial, or chloroplast genomes are referred to as promiscuous DNA.

APPLICATION QUESTIONS AND PROBLEMS

Section 21.1

13. A wheat plant that is light green in color is found growing in a field. Biochemical analysis reveals that chloroplasts in this plant produce only 50% of the chlorophyll normally found in wheat chloroplasts. Propose a set of crosses to determine whether the light-green phenotype is caused by a mutation in a nuclear gene or a chloroplast gene.
Nuclear and chloroplast genes in wheat will exhibit different inheritance patterns. A nuclear gene is inherited biparentally, whereas a chloroplast gene is inherited uniparentally (or maternally). The differences in inheritance patterns will allow us to determine if the mutation has a nuclear gene or chloroplast gene origin. For the following crosses, the male plant refers to the pollen donor, while the female plant refers to the plant being pollinated and where fertilization will take place.

If the mutation is located within a chloroplast gene, then we would expect the following results, no matter which trait is dominant:
Wild-type wheat male × light-green wheat female → offspring all light-green

Light-green wheat male × wild-type wheat female → offspring all wild-type

Matings between light-green offspring should always produce more light-green progeny, whereas matings between wild-type offspring should always produce wild-type progeny.

If the mutation is in a nuclear gene, then both parents can pass the light-green mutation to their offspring. If we assume that wild type is dominant, then we would expect the following results:
Wild-type wheat male × light-green wheat female → offspring all wild-type
Light-green wheat male × wild-type Wheat → offspring all wild-type

Separate matings between members of the F_1 progeny of each cross should give progeny with a phenotypic ratio of 3:1 wild-type to light-green.

If we assume that the light-green phenotype is dominant, then we would expect the following:
Wild-type wheat male × light-green wheat female → offspring all light-green
Light-green wheat male × wild-type wheat female → offspring all light-green

Separate matings between members of the F_1 progeny of each cross should give progeny with a phenotypic ratio of 3:1 light-green to wild-type.

*14. A rare neurological disease is found in the family illustrated in the following pedigree. What is the most likely mode of inheritance for this disorder? Explain your reasoning.

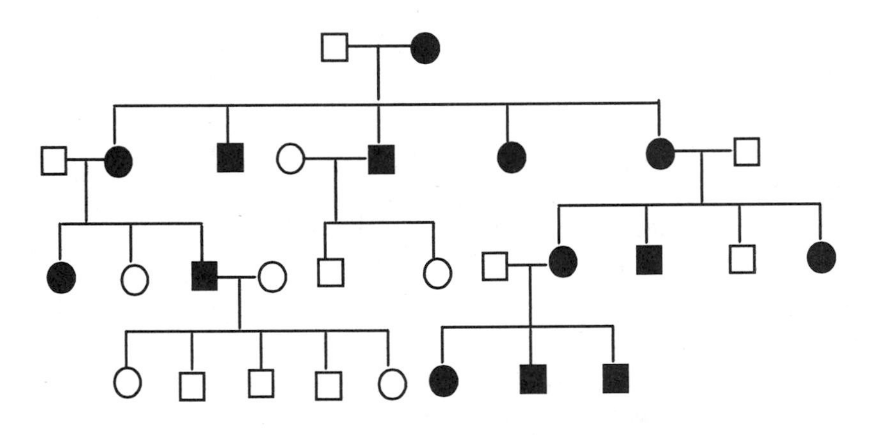

The pedigree indicates that the neurological disorder is a cytoplasmically inherited trait. Only females pass the trait to their offspring. The trait does not appear to be sex-specific in that males as well as females can have the disorder. These characteristics are consistent with cytoplasmic inheritance.

15. Fredrick Wilson and his colleagues studied members of a large family who had low levels of magnesium in their blood (see the adjoining pedigree). They argued that this disorder of magnesium (and associated high blood pressure and high cholesterol) is caused by a mutation in mtDNA (F. H. Wilson et al. 2004. *Science* 306:1190–1194).

a. What evidence suggests that a gene in the mtDNA is causing this disorder?
From the pedigree, it appears that both males and females are affected by the disorder and that it is inherited through the maternal lineage. From the pedigree, we can see that only females pass the trait to their offspring. Maternal inheritance is typical of a genes found in the mitochondria.

b. Could this disorder be caused by an autosomal dominant gene? Why or why not?
No, it is not likely that the disorder is caused by an autosomal dominant gene. From the pedigree, we can see that males with the disorder do not pass it along to their offspring, while females with the disorder pass the trait to their offspring at a relatively high frequency. These results are atypical of an autosomal dominant gene inheritance pattern where both male and female afflicted parents should be equally likely to pass the trait to their offspring.

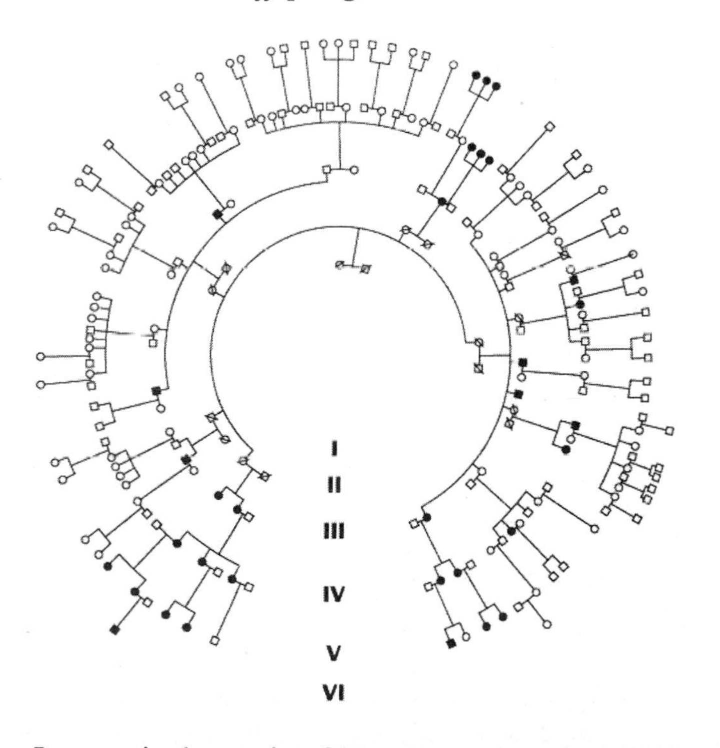

16. In a particular strain of *Neurospora,* a *poky* mutation exhibits biparental inheritance, whereas *poky* mutations in other strains are inherited only from the maternal parent. Explain these results.
In Neurospora, poky *mutations are identified by the mutant organism's slow growth. Maternal parent inheritance indicates that the* poky *mutations have a mitochondrial origin because mitochondrial genomes are inherited maternally in* Neurospora. *Many* poky *mutations have been identified on the mitochondrial genome. For a* Neurospora *strain containing a* poky *mutation with biparental inheritance, the poky mutation probably is of nuclear origin. In* Neurospora,

nuclear genes exhibit biparental inheritance. This particular poky *mutation most likely affects the energy producing pathways in the mitochondria as indicated by the* poky *phenotype, but the mutation is contained within a nuclear gene whose protein product targets mitochondria.*

Section 21.2

*17. A scientist collects cells at various points in the cell cycle and isolates DNA from them. Using density gradient centrifugation, she separates the nuclear and mtDNA. She then measures the amount of mtDNA and nuclear DNA present at different points in the cell cycle. On the following graph, draw a line to represent the relative amounts of nuclear DNA that you expect her to find per cell throughout the cell cycle. Then, draw a dotted line on the same graph to indicate the relative amount of mtDNA that you would expect to see at different points throughout the cell cycle.

Nuclear DNA levels should increase during only the S phase, before declining at cytokinesis. Mitochondrial DNA levels should increase throughout the cell cycle, before declining at cytokinesis.

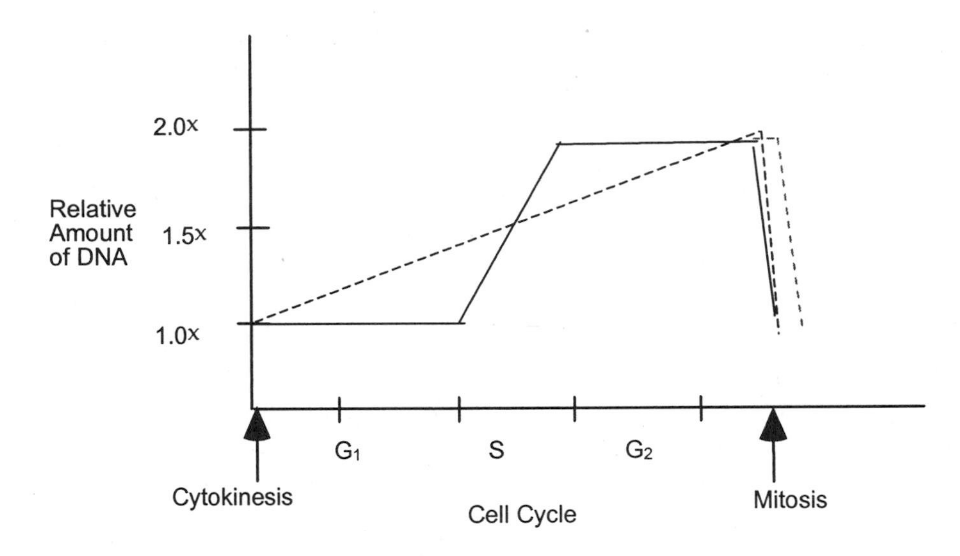

18. In 1979, bones found outside Ekaterinburg, Russia, were shown to be those of Tsar Nicholas and his family, who were executed in 1918 by a Bolshevik firing squad during the Russian Revolution. To prove that the skeletons were those of the royal family, mtDNA was extracted from the bone samples, amplified by PCR, and compared with mtDNA from living relatives of the tsar's family. Why was DNA from the mitochondria analyzed instead of nuclear DNA? What are some of the advantages of using mtDNA for this type of study?
Because mitochondrial DNA in humans is inherited only maternally, the tsar's living relatives of maternal descent should contain very similar mitochondrial DNA sequences. The mitochondrial sequences of the tsar's children and his wife can also be analyzed, but only by comparing their mitochondrial DNA with that of

living maternal relatives to the tsarina. Comparisons of nuclear DNA sequences from the tsar's family with living relatives would prove difficult because of the biparental transfer of nuclear genes and recombination between the nuclear chromosomes. Also, the possibility of similar sequences appearing by chance through matings of the tsar's contemporary relatives could not be ruled out. The nuclear DNA analysis could, however, indicate if the skeletons in the gravesite were related to each other.

An advantage of mitochondrial DNA typing for this type of study is the linear maternal inheritance. Essentially, any maternally related individual should possess very similar mitochondrial DNA sequences to the deceased individual. A second advantage is the number of mitochondrial genomes present. Multiple copies of each chromosome exist within each mitochondrion, and each cell contains multiple mitochondria, thus making it more likely that mitochondrial DNA survived.

19. From Figure 21.7, determine as best you can the percentage of human mtDNA that is coding (transcribed into RNA) and the percentage that is noncoding (not transcribed).
The largest region of noncoding DNA in human mitochondria is located within the D-loop. There are small sections of noncoding sequences (1 to 2 nucleotides in length) at other locations within the genome, but very few. Approximately 93% of the DNA in human mitochondria is coding, whereas only 7% is not.

20. Richard Cordaux and his colleagues analyzed mtDNA sequences to determine whether the recent spread of agriculture into new areas in India was by demic diffusion (traditional farmers replacing hunter–gatherers) or cultural diffusion (the practice of farming spreading among hunter–gatherers; see p. 589). They examined sequences of mtDNA from (1) 229 hunter–gatherers from southern India, (2) 201 recent farmers from southern India, (3) 140 traditional farmers from southern India, and (4) 62 traditional farmers from northern India (R. Cordaux et al. 2004. *Science* 304:1125). On the basis of the mtDNA sequences, they computed *Fst* values, which represent the overall degree of genetic difference between two groups. The *Fst* values for differences between the four groups are listed in the following table.

	Hunter–gatherers from south	**Traditional farmers from south**	**Traditional farmers from north**
Recent farmers from south	0.038	0.028	0.036
Hunter–gatherers from south		0.039	0.048
Traditional farmers from south			0.017

a. Do these data support demic diffusion or cultural diffusion as the mechanism for the spread of farming to the new areas of India? Explain your reasoning.
The data support demic diffusion as the mechanism for the spread of farming into new areas of India. The farmers from the south are genetically more similar to the traditional farmers of the north than to the hunter–gatherers from the south, suggesting a closer genetic relationship between the two groups. This relationship can be more easily explained by a demic diffusion of farmers into the new areas and not by cultural diffusion.

b. These researchers also looked at sequences on the Y chromosome and found the same patterns as those shown by the mtDNA sequences. Why did they use sequences on the Y chromosome?
The Y chromosome is inherited paternally and thus exhibits a direct lineage in males. Examining sequences on the Y chromosome allows for direct comparisons of the paternal lineages of the groups being studied.

Section 21.3

21. Antibiotics, such as chloramphenicol, tetracycline, and erythromycin, inhibit protein synthesis in eubacteria but have no effect on protein synthesis encoded by nuclear genes. Cycloheximide inhibits protein synthesis encoded by nuclear genes but has no effect on eubacterial protein synthesis. How might these compounds be used to determine which proteins are encoded by the mitochondrial and chloroplast genomes?
Proteins that are synthesized by the mitochondrial and chloroplast genomes are typically sensitive to antibiotics that affect eubacterial proteins. The sensitivity is most likely due to the similarities of the translational machinery in eubacteria and in chloroplasts and mitochondria. To determine if a particular protein is coded for on the mitochondrial or chloroplast genomes, we can use antibiotics and determine their effects on protein synthesis. For example, if the antibiotic cycloheximide does not inhibit the synthesis of the proteins in question, the lack of inhibition suggests the proteins are not encoded by the nuclear genome. Subsequently, if the antibiotics chloramphenicol, tetracycline, and erythromycin do inhibit the synthesis of the proteins, then this suggests again that the proteins are not of nuclear origin; they are synthesized either in the mitochondria or the chloroplast. This combination treatment of the different types of antibiotics—those that inhibit nuclear gene protein synthesis and those that inhibit eubacterial as well as organelle protein synthesis—should indicate if proteins are of nuclear or organelle origin.

CHALLENGE QUESTIONS

Section 21.2

22. Steven Frank and Laurence Hurst argued that a cytoplasmically inherited mutation in humans that has severe effects in males but no effect in females will not be eliminated from a population by natural selection because only females pass on mtDNA (S. A. Frank and L. D. Hurst. 1996. *Nature* 383:224). Using this argument, explain why males with Leber hereditary optic neuropathy are more severely affected than females.
    Leber hereditary optic neuropathy (LHON) is the result of mutations in mtDNA. Using the argument of Frank and Hurst, the most severe forms of LHON that appear in males will not be eliminated from the population if they do not also appear in females. If the particular mutation has little or no effect on females, then the females will be able to pass the mutation to their offspring. Male offspring who receive the mutation will be affected, but female offspring will not. The female offspring will then be able to pass the mutation along to their offspring. The selection is for healthy females who can pass the genes to their offspring. For mitochondrial genes, males have a limited role. Therefore, it is possible that mutations with more severe effects in males than females will be present because there is little negative selection pressure against females with deleterious mutations in mitochondrial genes.

23. In a study of a myopathy, several families exhibited vision problems, muscle weakness, and deafness (M. Zeviani et al. 1990. *American Journal of Human Genetics* 47:904–914). Analysis of the mtDNA from affected persons in these families revealed that large numbers of their mtDNA possessed deletions of varying length. Different members of the same family and even different mitochondria from the same person possessed deletions of different sizes, so the underlying defect appeared to be a tendency for the mtDNA of affected persons to have deletions. A pedigree of one of these families studied is shown here. The researchers concluded that this disorder is inherited as an autosomal dominant trait and they mapped the disease-causing gene to a position chromosome 10 in the nucleus.
    a. What characteristics of pedigree rule out inheritance as a trait encoded by a gene in the mtDNA?
    The pedigree indicates that the trait is not just maternally inherited. Both males and females are capable of passing the trait to their offspring. If the trait was encoded by a gene in the mtDNA, we would expect to see only maternal transmission of the trait.
    b. Propose an explanation for how a mutation in a nuclear gene might lead to deletions in mtDNA.
    Genes necessary for mitochondria DNA replication and repair are located on the nuclear genome. If a gene product encoded by chromosome 10 plays a role in either mitochondrial DNA maintenance or stability, then the lack of that functional gene product could result in damage such as deletions to the mitochondrial genome.

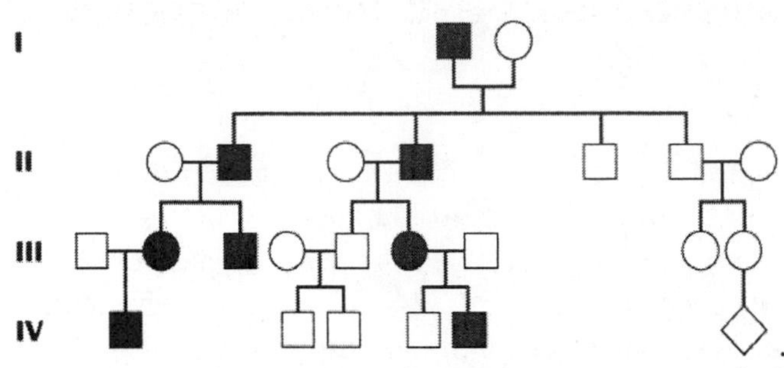

(Adapted from M. Zeniani et al. 1990. *American Journal of Human Genetics* 47:904, 914.)

Section 21.4

24. Mitochondrial DNA sequences have been detected in the nuclear genomes of many organisms, and cpDNA sequences are sometimes found in the mitochondrial genome. Propose a mechanism for how such "promiscuous DNA" might move between nuclear, mitochondrial, and chloroplast genomes.
The transfer of organelle DNA sequences to another location most likely involves multiple steps. Different models for transfer can be proposed. In one model, the nucleus or an organelle may uptake DNA directly from degraded organelles and incorporate the DNA into a genome by double-stranded recombination. A second model involves an RNA intermediate. The organelle gene is initially expressed, and a mature mRNA is produced. The mRNA molecule serves as a template for reverse transcriptase, which can produce a cDNA copy of the mRNA. The cDNA is then incorporated into either a nuclear chromosome or into the genome of a different type of organelle by recombination. In either model, once the DNA is incorporated, replication of this chromosome will propagate the "promiscuous DNA" within either the nuclear or organelle genomes of a cell and the subsequent progeny cells.

Chapter Twenty-Two: Developmental Genetics and Immunogenetics

COMPREHENSION QUESTIONS

Section 22.1

*1. What experiments suggested that genes are not lost or permanently altered in development?
The ability to clone plants and animals from differentiated cells showed that the nuclei of these differentiated cells still retained all of the genetic information required for the development of a whole organism.

Section 22.2

2. Briefly explain how the Dorsal protein is redistributed in the formation of the *Drosophila* embryo and how this redistribution helps to establish the dorsal–ventral axis of the early embryo.
Dorsal protein is distributed uniformly in the cytoplasm of the unfertilized egg. After fertilization, at the stage at which nuclei migrate to the periphery of the egg, Dorsal protein migrates into the nucleus along the future ventral side of the egg, but not along the future dorsal side of the egg. This gradient of nuclear-localized Dorsal protein activates transcription of mesodermal genes along the ventral surface of the embryo. Nuclei that lack Dorsal protein transcribe dorsal-specific genes such as decapentaplegic.

*3. Briefly describe how the *bicoid* and *nanos* genes help to determine the anterior–posterior axis of the fruit fly.
Maternally transcribed bicoid *and* nanos *mRNAs are localized to the anterior and posterior ends of the egg, respectively. After fertilization, these mRNAs are translated, and the proteins diffuse to form opposing gradients: Bicoid protein concentrations are highest at the anterior, whereas Nanos protein concentrations are highest at the posterior. Bicoid protein at the anterior acts as a transcription factor to activate transcription of* hunchback, *a gene required for the formation of head and thoracic structures. Nanos protein at the posterior end inhibits translation of hunchback mRNA, thereby preventing the formation of anterior structures in the posterior regions.*

*4. List the three major classes of segmentation genes and outline the function of each.
(1) Gap genes specify broad regions (multiple adjacent segments) along the anterior–posterior axis of the embryo. Interactions among the gap genes regulate transcription of the pair-rule genes.
(2) Pair-rule genes compartmentalize the embryo into segments and regulate expression of the segment polarity genes. Each pair rule gene is expressed in alternating segments.
(3) Segment polarity genes specify the anterior and posterior compartments within each segment.

5. What role do homeotic genes play in the development of fruit flies?
Homeotic genes specify segment identity—expression of the homeotic genes informs cells of their location or address along the anterior–posterior axis.

Section 22.3

6. How do class A, B, and C genes in plants work together to determine the structures of the flower?
In the first whorl, class A genes determine sepal identity. In the second whorl, expression of class A and class B genes causes petal development. In the third whorl, class B and class C genes together cause stamen development. In the fourth whorl, class C genes act alone to specify carpel development. Class A and class C genes are mutually inhibitory, so that class A and class C genes are not co-expressed in the same whorl.

Section 22.4

*7. What is apoptosis and how is it regulated?
Apoptosis is programmed cell death, characterized by nuclear DNA fragmentation, shrinkage of the cytoplasm and nucleus, and phagocytosis of the remnants of the dead cell. Apoptosis is regulated by internal and external signals that regulate the activation of procaspases, cysteine proteases that are activated by proteolytic cleavage. Once activated, these caspases activate other caspases in a cascade and degrade key cellular proteins.

Section 22.6

*8. Explain how each of the following processes contributes to antibody diversity.
a. Somatic recombination.
Recombination produces many combinations of variable domains with junction segments and diversity segments.
b. Junctional diversity.
During recombination, the V, J, and D joining events are imprecise, resulting in small deletions or insertions and frameshifts.
c. Hypermutation.
The V gene segments are subject to somatic hypermutation—accelerated random mutation—that further diversifies antibodies.

9. What is the function of the MHC antigens? Why are the genes that encode these antigens so variable?
MHC proteins present fragments of foreign or self antigens to T cells. Variability of the MHC proteins has the effect of making individuals virtually unique in terms of their MHC genotype. This variability, or high degree of polymorphism, may have evolved in response to selective pressure to present antigens from many different pathogenic organisms for recognition by T cells.

APPLICATION QUESTIONS AND PROBLEMS

Section 22.1

10. William Jeffrey and his colleagues crossed surface-dwelling Mexican tetra that had fully developed eyes with blind Mexican tetras from caves (see the introduction to this chapter). The progeny from this cross had uniformly small eyes compared with those of surface fish (Y. Yamamoto, D. W. Stock, and W. R. Jeffrey. 2004. *Nature* 431:844–847). What prediction can you make about the expression of *shh* in the embryos of these progeny relative to its expression in the embryos of surface fish?
Widespread expression of shh *in the eye primordium of blind cavefish causes degeneration of lens cells. In these F_1 progeny between blind cavefish and surface fish, the expression of shh is intermediate between blind cavefish and surface fish, resulting in small eyes.*

11. If telomeres are normally shortened after each round of replication in somatic cells (see Chapter 12), what prediction would you make about the length of telomeres in Dolly, the first cloned sheep?
The chromosomes in the mammary cell nucleus that provided the nuclear DNA for Dolly would have had shortened telomeres, compared to chromosomes in germ line cells (sperm and egg). If telomerase expression was not induced in the process of cloning and subsequent embryonic development, we would predict that Dolly's somatic cells will have undergone further loss of telomeric DNA, so that Dolly would have shorter telomeres than other sheep of the same age.

Section 22.2

12. Why do mutations in *bicoid* and *nanos* exhibit genetic maternal effects (a mutation in the maternal parent produces a phenotype that shows up in the offspring, see Chapter 5), but mutations in *runt* and *gooseberry* do not? (Hint: See Tables 22.3 and 22.4.)
Bicoid *and* nanos *mRNAs are produced during oogenesis by the mother and deposited in the egg. Therefore, the zygote's store of these mRNAs depend on the mother's genotype and these genes exhibit genetic maternal effects. In contrast,* runt *and* gooseberry *are transcribed after fertilization from the zygote's own chromosomes. For these latter genes, the zygote's phenotype depends on its own genotype.*

*13. Give examples of genes that affect development in fruit flies by regulating gene expression at the level of (a) transcription and (b) translation.
   a. *The products of* bicoid *and* dorsal *affect embryonic polarity by regulating transcription of target genes.*
   b. *The* nanos *gene regulates translation of* hunchback *mRNA.*

14. What would be the most likely effect on development of puncturing the posterior end of a *Drosophila* egg, allowing a small amount of cytoplasm to leak out, and then injecting that cytoplasm into the anterior end of another egg?
The posterior cytoplasm contains posterior determinants, such as maternal nanos *mRNA. Injecting cytoplasm containing* nanos *mRNA into the anterior end of an egg would interfere with determination of anterior (head and thoracic) structures by* hunchback *and result in an embryo lacking head structures.*

15. Christiane Nüsslein-Volhard and her colleagues carried out several experiments in an attempt to understand what determines the anterior and posterior ends of a *Drosophila* larva (reviewed in C. Nüsslein-Volhard, H. G. Frohnhofer, and R. Lehmann. 1987. *Science* 238:1675–1681). They isolated fruit flies with mutations in the *bicoid* gene (bcd^-). These flies produced embryos that lacked a head and thorax. When they transplanted cytoplasm from the anterior end of an egg from a wild-type female into the anterior end of an egg from a mutant *bicoid* female, normal head and thorax development took place in the embryo. However, transplanting cytoplasm from the posterior end of an egg from a wild-type female into the anterior end of an egg from a *bicoid* female had no effect. Explain these results in terms of what you know about proteins that control the determination of the anterior–posterior axis.
In a wild-type egg, bicoid *mRNA is localized to the anterior pole. Upon fertilization, translation of this mRNA creates a concentration gradient of Bicoid protein diffusing from the anterior towards the posterior. This Bicoid protein gradient promotes the formation of a gradient of* hunchback *mRNA and protein along the anterior–posterior axis. Hunchback protein acts as a transcription factor to turn on genes required for development of anterior (head and thorax) structures. A reverse gradient of nanos protein represses translation of* hunchback *mRNA at the posterior end of the embryo. In a* bcd^- *mutant, Hunchback protein is not synthesized and no head or thorax develops. Such a mutant can be rescued by injecting anterior cytoplasm from a wild-type egg containing* bicoid *mRNA. However, injection of posterior cytoplasm would have no effect because posterior cytoplasm contains no* bicoid *mRNA. Posterior cytoplasm does contain* nanos *mRNA, which would have no effect because its target,* hunchback *mRNA, is lacking in the* bcd^- *mutant embryo.*

*16. What would be the most likely result of injecting *bicoid* mRNA into the posterior end of a *Drosophila* embryo and inhibiting the translation of *nanos* mRNA?
Bicoid *mRNA in the posterior end of the embryo would cause transcription of hunchback in the posterior regions. Without Nanos protein, the* hunchback *mRNA would be translated to create high levels of Hunchback protein in the posterior as well as in the anterior. The result would be an embryo with anterior structures at both ends.*

17. What would be the most likely effect of inhibiting the translation of *hunchback* mRNA throughout the embryo?

If the translation of hunchback *throughout the embryo were inhibited, no anterior structures would form. The embryo would be entirely posteriorized, perhaps forming all abdominal structures.*

*18. Molecular geneticists have performed experiments in which they altered the number of copies of the *bicoid* gene in flies, affecting the amount of Bicoid protein produced.

a. What would be the effect on development of an increased number of copies of the *bicoid* gene?
Females with an increased number of copies of the bicoid *gene would have higher levels of* bicoid *maternal mRNA in the anterior cytoplasm of their eggs, and thus higher levels of Bicoid protein in the embryo after fertilization. The resulting Bicoid protein gradient would extend further to the posterior, resulting in the enlargement of anterior and thoracic structures.*

b. What would be the effect of a decreased number of copies of *bicoid*? Justify your answers.
Conversely, a decreased number of copies of the bicoid *gene would ultimately result in a reduced Bicoid protein gradient in the eggs. Thus, sufficient Bicoid protein concentrations for head structures would be found in a smaller, more anterior portion of the embryo, resulting in an embryo with smaller head structures.*

19. What would be the most likely effect on fruit-fly development of a deletion in the *nanos* gene?
Female flies homozygous for deletions in the nanos *gene would lay eggs lacking* nanos *mRNA. Upon fertilization, the resulting embryos would form all anterior and thoracic structures, with no posterior structures. If the egg does have maternal* nanos *mRNA (because the mother was a heterozygote), and even if the zygote was homozygous for a deletion in the nanos gene, the embryo would develop normally.*

20. Give an example of a gene found in each of the categories of genes (egg-polarity, gap, pair-rule, and so forth) listed in Figure 22.12.
Egg-polarity (maternal effect): *bicoid, nanos*
Gap: *hunchback, Krüppel*
Pair-rule: *even-skipped, fushi tarazu*
Segment-polarity: *gooseberry*

21. In Chapter 1, we considered an early idea about heredity called preformationism, which suggested that inside the egg or sperm is a tiny adult called the homunculus, with all the features of an adult human in miniature. According to this idea, the homunculus simply enlarges during development. What types of evidence presented in this chapter prove that preformationism is false?
Preformationism requires that all or most genes that govern embryonic pattern formation and formation of major body structures and organs act before fertilization. According to the preformation hypothesis, then, all or most developmental mutations would be maternal effect genes. However, many mutations

in zygotically expressed genes have been identified, with phenotypes that arrest development at different stages, or affect development of specific structures and organs. Gap genes, pair-rule genes, segment-polarity genes, and Hox *genes are all transcribed from the zygotic genome after fertilization, and act sequentially during development.*

Section 22.3

22. Explain how: (a) the absence of class B gene expression produces the structures seen in class B mutants (Figure 22.14c); and (b) the absence of class C gene product produces the structures seen in class C mutants (Figure 22.14d).
    a. *Class B genes are normally expressed in the second and third whorls. Class B genes work together with class A genes in the second whorl to specify petals, and class B genes work together with class C genes in the third whorl to specify stamens. In the absence of class B genes, the floral buds express only class A genes in the first and second whorls, which become sepals, and only class C genes in the third and fourth whorls, which become carpels.*
    b. *Class C genes are normally expressed in the third and fourth whorls. In the third whorl, class C genes work together with class B genes to specify stamens, and in the fourth whorl, class C genes alone specify carpels. Class C genes also inhibit expression of class A genes. With no class C gene expression, class A genes will be expressed in all four whorls, with the second and third whorls also expressing class B genes. The resulting pattern of floral development will produce abnormal floral structures that have sepal-petal-petal-sepal patterns.*

23. What would you expect a flower to look like in a plant that lacked class A and class B genes? What about a plant that lacked class B and class C genes?
    A plant that lacked class A and class B genes would express only class C genes in all four whorls, resulting in flowers with only carpels. A plant that lacked both class B and class C genes would express only class A genes in all four whorls, and result in flowers with only sepals.

24. What would be the flower structure of a plant within which expression of the following genes were inhibited:
    a. Expression of class B genes is inhibited in the second whorl but not the third whorl.
    Sepal-sepal-stamen-carpel. The first and second whorls, expressing only class A genes, would both produce sepals. The third whorl, with both class B and class C genes, would produce stamens, and the fourth whorl with only class C genes would produce carpels.
    b. Expression of class C genes is inhibited in the third whorl but not the fourth whorl.
    Sepal-petal-petal-carpel. The first, second, and fourth whorls are unaffected and produce the normal floral parts. The third whorl, without class C gene products, would express class A genes as well as class B genes to specify petals.

c. Expression of class A genes is inhibited in the first whorl but not the second whorl.
Carpel-petal-stamen-carpel. Without class A gene expression, the first whorl would express class C genes and produce carpels. The other whorls have normal gene expression and are unaffected.

d. Expression of class A genes is inhibited in the second whorl but not the first whorl.
Sepal-stamen-stamen-carpel. Lack of class A gene expression in the second whorl would allow class C gene expression and result in stamens because of co-expression of class B and class C genes. The other whorls are unaffected and produce normal floral organs.

Section 22.6

*25. In a particular species, the gene for the kappa light chain has 200 *V* gene segments and 4 *J* segments. In the gene for the lambda light chain, this species has 300 *V* segments and 6 *J* segments. Considering only the variability arising from somatic recombination, how many different types of light chains are possible?
Light chain genes undergo recombination to join one V *gene segment to one* J *segment, in any combination. The number of different possible* V *and* J *combinations is given by the product of the number of* V *segments and the number of* J *segments for each light chain.*
kappa light chain: $200 \times 4 = 800$
lambda light chain: $300 \times 6 = 1800$
Total light chains = kappa + lambda = $800 + 1800 = 2600$

26. In the fictional book *Chromosome 6* by Robin Cook, a biotechnology company genetically engineers individual bonobos (a type of chimpanzee) to serve as future organ donors for clients. The genes of the bonobos are altered so that no tissue rejection takes place when their organs are transplanted into the client. What genes would need to be altered for this scenario to work? Explain your answer.
The MHC genes of the bonobos must be humanized. The T-cells of the immune system recognize antigen complexed with the body's own MHC molecules. Non-self MHC molecules trigger an immune response and tissue rejection. Therefore, the bonobo MHC molecules must be replaced with the human patient's own MHC molecules to prevent rejection.

CHALLENGE QUESTIONS

Section 22.2

27. As we have learned in this chapter, the Nanos protein inhibits the translation of *hunchback* mRNA, thus lowering the concentration of Hunchback protein at the posterior end of a fruit-fly embryo and stimulating the differentiation of posterior characteristics. The results of experiments have demonstrated that the action of

Nanos on *hunchback* mRNA depends on the presence of an 11-base sequence that is located in the 3′ untranslated region of *hunchback* mRNA. This sequence has been termed the Nanos response element (NRE). There are two copies of NREs in the trailer of *hunchback* mRNA. If a copy of NRE is added to the 3′ untranslated region of another mRNA produced by a different gene, the mRNA now becomes repressed by Nanos. The repression is greater if several NREs are added. On the basis of these observations, propose a mechanism for how Nanos inhibits Hunchback translation.
Nanos may bind to the NREs, either directly or indirectly as a complex with other proteins. Multiple NRE binding elements may enhance the binding either by simply providing a higher concentration of binding sites, or through cooperativity (the binding of protein to one NRE enhances the binding of protein to other NREs). The complex of Nanos (and other proteins) bound to the NREs at the 3′ untranslated region may target the mRNA for rapid degradation. Alternatively, the NRE-bound protein complexes may interfere with ribosome binding or translational initiation at the 5′ end of the mRNA, although this postulates that mRNAs assume a circular topology where the 3′ and 5′ ends of the mRNA are in close proximity.

28. Offer a possible explanation for the widespread distribution of *Hox* genes among animals.
One explanation is that the Hox *genes arose in the common evolutionary ancestor of the animals that have* Hox *genes. This ancestor used the ancestral* Hox *gene for body patterning. The complex of* Hox *genes and the overall function was then retained in the descendant lineages.*

Section 22.6

29. Ataxia-telangiectasia (ATM) is a rare genetic neurodegenerative disease. About 20% of people with ATM develop acute lymphocytic leukemia or lymphoma, cancers of the immune cells. Cells in many of these cancers exhibit chromosome rearrangments, with chromosome breaks occurring at antibody and T-cell-receptor genes (A. L. Bredemeyer et al. 2006. *Nature* 442:466–470). Many people with ATM also have a weakened immune system, making them susceptible to respiratory infections. Research has shown that the locus that causes ATM has a role in the repair of double-strand breaks. Explain why people who have a genetic defect in the repair of double-strand breaks might have a high incidence of chromosome rearrangements in their immune cells and why their immune systems might be weakened.
The chromosomal loci that contain the genes for immunoglobulins and T cell receptors undergo DNA rearrangement to create different combinations of V *and* J *segments or* V, D, *and* J *segments. This DNA rearrangement involves generation of double-strand DNA breaks at specific recombination signal sequences in the DNA. If the cell cannot efficiently resolve or repair these double-strand breaks, then the production of mature immunoglobulin and T-cell-receptor genes may be impaired, resulting in a weakened immune system. In addition, the immunoglobulin or T-cell-receptor gene segments may ligate to other chromosomal double-strand break sites that have not been repaired in these cells, and result in a high frequency of chromosomal rearrangements.*

Chapter Twenty-Three: Cancer Genetics

COMPREHENSION QUESTIONS

Section 23.1

*1. What types of evidence indicate that cancer arises from genetic changes?
Higher incidences of many types of cancer are associated with exposure to radiation and other environmental mutagens. In addition, the occurrence of some types of cancer runs in families, and a few cancers are linked to chromosomal abnormalities. Finally, the discovery of oncogenes and specific mutations that cause proto-oncogenes to become oncogenes, or inactivate tumor suppressor genes, proved that cancer has a genetic basis.

2. How is cancer different from most other types of genetic diseases?
Most cancers arise from genetic changes in somatic cells that arise during an individual's lifetime, whereas other types of genetic diseases are inherited through the germline.

Section 23.2

*3. Outline Knudson's multistage theory of cancer and describe how it helps to explain unilateral and bilateral cases of retinoblastoma.
The multistage theory of cancer states that more than one mutation is required for most cancers to develop. Most retinoblastomas are unilateral because the likelihood of any cell acquiring two rare mutations is very low, and thus retinoblastomas develop in only one eye. Bilateral cases of retinoblastoma occur in people born with a predisposing mutation, so that as few as one additional mutational event can result in cancer. Thus, the probability of retinoblastoma is higher in these individuals and likely to occur in both eyes. Because the predisposing mutation is inherited, people with bilateral retinoblastoma have relatives with retinoblastoma.

4. Briefly explain how cancer arises through clonal evolution.
A mutation that relaxes growth control in a cell will cause it to divide and form a clone of cells that are growing or dividing more rapidly than their neighbors. Successive mutations that cause even more rapid growth, or the ability to invade and spread, each produce progeny cells with more aggressive, malignant properties that outgrow their predecessors and take over the original clone.

*5. What is the difference between an oncogene and a tumor–suppressor gene? Give some examples of functions of proto-oncogenes and tumor suppressors in normal cells.
An oncogene stimulates cell division, whereas a tumor–suppressor gene puts the brakes on cell growth. Proto-oncogenes are normal cellular genes that function in cell growth and regulation of the cell cycle: from growth factors such as sis to

receptors like ErbA *and* ErbB, *protein kinases such as* src, *and transcription factors like* myc. *Tumor suppressors inhibit cell cycle progression:* RB *and* p53 *are transcription factors and* NF1 *is a GTPase activator.*

6. What is haploinsufficiency? How might it affect cancer risk?
Haploinsufficiency is a condition where a normally recessive trait affects a heterozygous individual. Haploinsufficiency arises in situations where a single functional copy of a gene is insufficient to produce a wild type phenotype. In most cases, haploinsuffiency reflects the need for a larger quantity of a gene product than is normally produced by a single wild-type allele. In the case of tumor-suppressor genes like RB, *a single functional copy is often enough for the cell to have a normal phenotype, but leaves no "back-up" copy in reserve. In this case, the mutation of the single remaining wild-type allele in any one of the millions of cells in the retina will lead to formation of a cancer cell, hence the "predisposition to cancer" phenotype associated with haploinsuffiency of* RB.

*7. How do cyclins and CDKs differ? How do they interact in controlling the cell cycle?
The CDKs, or cyclin-dependent kinases, have enzymatic activity and phosphorylate multiple substrate molecules when activated by binding the appropriate cyclin. Cyclins are regulators of CDKs and have no enzymatic activity of their own. Each cyclin molecule binds to a single CDK molecule. Whereas CDK levels remain relatively stable, cyclin levels oscillate through the cell cycle.

8. Briefly outline the events that control the progression of cells through the G_1/S checkpoint in the cell cycle.
In G_1*, cyclins D and E accumulate and bind to their respective CDKs. The cyclin D-CDK and cyclin E-CDK phosphorylate RB protein molecules. Phosphorylation of RB inactivates RB and releases active E2F protein. E2F protein transcribes genes required for DNA replication and progression into S phase.*

9. Briefly outline the events that control the progression of cells through the G_2/M checkpoint of the cell cycle.
Cyclin B accumulates through G_2 *and binds to its partner CDK, forming an inactive mitosis-promoting factor (MPF) which is activated by dephosphorylation. When enough MPF activity exceeds a threshold level, the cell commits to mitosis.*

*10. What is a signal-transduction pathway? Why are mutations in components of signal-transduction pathways often associated with cancer?
A signal-transduction pathway is the system that enables a cell to respond appropriately to an external signal. It begins with binding or perception of the external signal molecule, then proceeds through a cascade of intracellular events that relay and amplify the signal to bring about changes in transcription, metabolism, morphology, or other aspects of cell function. Since cell growth and division are regulated by external signals, mutations in signal-transduction components may cause the cell to growth and divide in the absence of external

growth stimuli, or may cause the cell to stop responding to external growth inhibitory signals.

11. How is the Ras protein activated and inactivated?
Ras protein with GDP bound is inactive. Exchanging GDP for GTP activates the Ras protein. This guanine nucleotide exchange is stimulated by adaptor proteins that bind to activated signal receptors.

12. Why do mutations in genes that encode DNA repair enzymes and chromosome segregation often produce a predisposition to cancer?
Mutations that affect DNA repair cause high rates of mutation that may convert proto-oncogenes into oncogenes or inactivate tumor-suppressor genes. Similarly, errors in chromosome segregation cause aneuploidy and chromosomal aberrations that cause loss of tumor-suppressor genes or add extra gene doses of proto-oncogenes.

*13. What role do telomeres and telomerase play in cancer progression?
DNA polymerases are unable to replicate the ends of linear DNA molecules. Therefore, the ends of eukaryotic chromosomes shorten with every round of DNA replication, unless telomerase adds back special non-templated telomeric DNA sequences. Normally, somatic cells do not express telomerase; their telomeres progressively shorten with each cell division until vital genes are lost and the cells undergo apoptosis. Transformed cells (cancerous cells) induce the expression of the telomerase gene, in order to keep proliferating.

Section 23.3

*14. Explain how chromosome deletions, inversions, and translations may cause cancer.
Chromosomal rearrangements may inactivate tumor suppressor genes if the breakpoint occurs within the gene. Alternatively, rearrangements may juxtapose a strong promoter upstream of a proto-oncogene, causing overexpression or unregulated expression of the proto-oncogene. Finally, rearrangements may bring parts of two different genes together, causing the synthesis of a novel protein that is oncogenic.

15. Briefly outline how the Philadelphia chromosome leads to chronic myelogenous cancer.
The Philadelphia chromosome is a shortened chromosome 22 with a translocated tip of chromosome 9. A part of the c-ABL proto-oncogene from chromosome 9 is fused with BCR *gene on chromosome 22. The resulting fusion protein is more active at promoting cell proliferation than the normal c-ABL protein, and causes leukemia.*

16. What is genomic instability? Give some ways in which genomic instability may arise.
Genomic instability is a condition or process that leads to numerous chromosomal rearrangements and aneuploidy, often found in cells of advanced tumors. Mutations that affect the mitotic spindle checkpoint may cause a high frequency of aneuploidy. Other mutations, such as mutations in the APC *gene, may affect the spindle itself or other aspects of the chromosome segregation mechanism. Still other mutations that affect centrosome duplication, such as some* p53 *mutations, could also lead to aneuploidy.*

Section 23.4

*17. How do viruses contribute to cancer?
Retroviruses have strong promoters. Upon integration into the host genome, the retrovirus promoter may drive overexpression of a cellular proto-oncogene. Alternatively, integration of the retrovirus may inactivate a tumor-suppressor gene. A few retroviruses carry oncogenes that are altered versions of host proto-oncogenes. Other viruses, such as human papilloma virus, express gene products (proteins or RNA molecules) that interact with the host cell cycle machinery and inactivate tumor-suppressor proteins.

Section 23.5

18. How is an epigenetic change different from a mutation?
Unlike mutations, epigenetic changes do not alter the sequence of nucleotides in the DNA. Although epigenetic changes are usually transmitted to mitotic progeny cells, epigenetic changes are more readily reversible than mutations.

19. How is DNA methylation related to cancer?
DNA methylation is associated with transcriptional repression. Methylation and silencing of tumor-suppressor genes would increase the risk of cancer; demethylation and activation of proto-oncogenes would also increase the risk of cancer. Hypomethylation (loss of DNA methylation) may also increase the risk of cancer by increasing genomic instability, by mechanisms that are not yet clear.

Section 23.6

20. Briefly outline some of the genetic changes that are commonly associated with the progression of colorectal cancer.
Colorectal cancer begins as benign tumors, called polyps, that enlarge and acquire further mutations that turn them malignant and, finally, invasive and metastatic. These progressive changes are associated with multiple mutations. One common sequence in colorectal cancer is mutation of the APC *gene leads to faster cell division and polyp formation. Oncogenic mutations of the* ras *gene are found in cells from larger polyps. Then mutations in* p53 *and other genes are found in*

malignant tumor cells, which may lead to genomic instability and additional changes that lead to greater malignancy and invasiveness.

APPLICATION QUESTIONS AND PROBLEMS

Section 23.1

*21. The *palladin* gene, which plays a role in pancreatic cancer (see the introduction to this chapter), is said to be an oncogene. Which of its characteristics suggest that it is an oncogene rather than a tumor-suppressor gene?
Because oncogenes promote cell proliferation, they act in a dominant manner. In contrast, mutations in tumor suppressor genes cause loss of function and act in a recessive manner. The mutated palladin *gene caused increased cell migration when introduced into cells that contain wild-type* palladin *genes. Such a dominant effect suggests that* palladin *is an oncogene.*

22. If cancer is fundamentally a genetic disease, how might an environmental factor such as smoking cause cancer?
Environmental factors can cause cancer by acting as mutagens. Higher rates of mutation will lead to higher rates of inactivation of tumor-suppressor genes or conversion of proto-oncogenes to oncogenes.

23. Both genes and environmental factors contribute to cancer. Table 23.2 shows that prostate cancer is 30 times as common among Caucasians from Utah as among Chinese from Shanghai. Briefly outline how you might go about determining if these differences in the incidence of prostate cancer are due to differences in the genetic makeup of two populations or to differences in their environments.
If the differences in cancer rates are due to genetic differences in the two populations, then people who migrated from Utah or Shanghai to other locations would have similar rates of cancer incidence as people who stayed in Utah or Shanghai. Moreover, different ethnic groups in Utah or Shanghai would have different rates of cancer. If the cancer rates are due to environmental factors, then people who migrated from Utah or Shanghai would have rates of cancer determined by their location and not by their place of origin, and different ethnic groups in the same location would have similar rates of cancer.

Section 23.2

*24. A couple has one child with bilateral retinoblastoma. The mother is free from cancer, but the father had unilateral retinoblastoma and he has a brother who has bilateral retinoblastoma.
   a. If the couple has another child, what is the probability that this next child will have retinoblastoma?
   First, we summarize the information with a pedigree. The shaded boxes represent bilateral retinoblastoma; the striped box represents unilateral retinoblastoma.

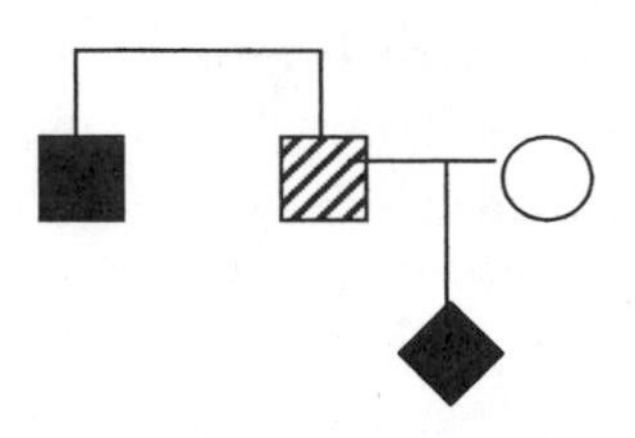

Familial retinoblastoma is caused by mutation of the RB *tumor-suppressor gene. Because the loss of a functional* RB *allele means that only one additional mutation event will completely eliminate* RB *function and lead to retinoblastoma, loss-of-function* RB *mutations have dominant effects with regard to retinoblastoma. If the father with unilateral retinoblastoma is heterozygous for an* RB *mutation, then the chance of another child inheriting the mutant* RB *allele is ½. Note that the father is almost certainly not homozygous for the* RB *mutation because he has only unilateral retinoblastoma and because individuals homozygous for* RB *mutations would be extremely susceptible to multiple types of cancer at early age.*

b. If the next child has retinoblastoma, is it likely to be bilateral or unilateral?
Because retinoblastoma in this family is most likely an inherited disorder, a child with retinoblastoma will more likely have bilateral retinoblastoma. Unilateral retinoblastomas are usually spontaneous in origin, requiring two independent mutations in a single somatic retinal cell. Familial retinoblastomas occur in family members that inherited one of the two mutations required for retinoblastoma. As only one additional mutation is required in the somatic retinal cells, retinoblastoma occurs in both eyes and at earlier ages than spontaneous unilateral retinoblastomas.

c. Propose an explanation for why the father's case of retinoblastoma was unilateral, while his son's and brother's cases were bilateral.
The father may have unilateral retinoblastoma because of variable expressivity of the mutation in the RB *gene. Alleles at another locus or multiple other loci may have contributed to resistance to retinoblastoma in the father, so that he suffered retinoblastoma in only one eye. Alternatively, it may have been just good fortune (random chance) that one of his eyes was spared the second mutation event that led to retinoblastoma in his other eye.*

25. Mutations in the *RB* gene are often associated with cancer. Explain how a mutation that results in a nonfunctional RB protein contributes to cancer.
RB protein is a tumor suppressor, acting at the G_1/S checkpoint to prevent cells from beginning DNA replication. Without functional RB protein, cells are more prone to begin a round of cell division.

26. Cells in a tumor contain mutated copies of a particular gene that promotes tumor growth. Gene therapy can be used to introduce a normal copy of this gene into the tumor cells. Would you expect this therapy to be effective if the mutated gene were an oncogene? A tumor-suppressor gene? Explain your reasoning.

Gene therapy to introduce a normal copy of the gene into tumor cells will not work for oncogenes because oncogenes are dominant, activating mutations of proto-oncogenes. Gene therapy may work if the tumor arises from a mutation that inactivates a tumor-suppressor gene. Loss-of-function mutations are recessive; therefore, a normal copy of the gene will be dominant and restore regulation of cell proliferation in the tumor cells. However, one would have to insert and express the tumor suppressor gene in all tumor cells, which is not possible at this time.

*27. Genes in cancer cells are frequently amplified, meaning that the gene exists in many copies. Would you expect to see gene amplification in oncogenes, proto-oncogenes, or both? Explain your answer.
Gene amplification in cancer cells would most likely affect proto-oncogenes. Oncogenes are altered forms proto-oncogenes with greater activity or lack of regulation. Amplification of oncogenes may have little additional effect for a cancer cell. However, amplification of proto-oncogenes would mimic the effect of an oncogene, and therefore may be frequent in cancer cells.

28. David Seligson and his colleagues examined levels of histone protein modification in prostate tumors and their association with clinical outcomes (D. B. Seligson et al. 2005. *Nature* 435:1262–1266). They used antibodies to stain for acetylation at three different sites and for methylation at two different sites on histone proteins. They found that the degree of histone acetylation and methylation helped predict whether prostate cancer would return within 10 years in the patients who had a prostate tumor removed. Explain how acetylation and methylation might be associated with tumor recurrence in prostate cancer. (Hint: See Chapter 17)
Histone acetylation and methylation patterns are associated with varying transcriptional states of chromatin. Histone modifications may determine whether particular tumor suppressor genes or proto-oncogenes important for prostate cancer are transcribed. Modifications that shut down transcription of tumor suppressors or activate transcription of proto-oncogenes would promote the recurrence of tumors, whereas modifications that activate transcription of tumor suppressor genes or suppress transcription of proto-oncogenes would inhibit the recurrence of tumors.

29. Radiation is known to cause cancer, yet radiation is often used as treatment for some types of cancer. How can radiation be a contributor to both the cause and the treatment of cancer?
Radiation can cause mutations that lead to cancer, such as inactivating a tumor-suppressor gene or causing an oncogenic mutation in a proto-oncogene. On the other hand, radiation will preferentially kill rapidly proliferating cells, such as cancer cells, that are actively replicating their DNA and lack tumor-suppressor functions that ensure DNA damage is repaired before DNA is replicated or before the cell divides.

Section 23.3

30. Some cancers are consistently associated with the deletion of a particular part of a chromosome. Does the deleted region contain an oncogene or a tumor-suppressor gene? Explain.
The deleted region contains a tumor-suppressor gene. Tumor suppressors act as brakes on cell proliferation. The deletion of tumor-suppressor genes will therefore permit the uncontrolled cell proliferation that is characteristic of cancer. Oncogenes, on the other hand, function as stimulators of cell division. Deletion of oncogenes will therefore prevent cell proliferation, and usually cannot cause cancer.

Section 23.5

31. Some cancers have been treated with drugs that demethylate DNA. Propose an explanation for how these drugs might work. Do you think the genes causing cancers that respond to the demethylation are likely to be oncogenes or tumor-suppressor genes? Explain your reasoning.
Drugs that demethylate DNA would presumably activate expression of demethylated genes. Cancer growth and progression may be inhibited if these drugs are able to turn on expression of tumor suppressor genes that had been silenced by DNA methylation. If DNA demethylation turned on expression of oncogenes, cancer growth and progression would be accelerated.

CHALLENGE QUESTIONS

Section 23.2

32. Many cancer cells are immortal (will divide indefinitely) because they have mutations that allow telomerase to be expressed. How might this knowledge be used to design anticancer drugs?
Because cancer cells depend on telomerase activity to preserve their telomeres, drugs that target telomerase enzymatic activity may limit the ability of cancer cells to divide indefinitely.

33. Bloom syndrome is an autosomal recessive disease that exhibits haploinsufficiency. As described on p. 631 a recent survey showed that people heterozygous for mutations at the *BLM* locus are at increased risk of colon cancer. Suppose you are a genetic counselor. A young woman whose mother has Bloom syndrome is referred to you; the young woman's father has no family history of Bloom syndrome. The young woman asks whether she is likely to experience any other health problems associated with her family history of Bloom syndrome. What advice would you give her?
The young woman must be heterozygous for the mutation at the BLM *locus because her mother was homozygous for the mutation. Although the young woman does not have Bloom syndrome, haploinsufficiency at this locus will result in some increased*

risk of colon cancer. Her cells will have a reduced amount of the BLM *helicase involved in DNA double-strand break repair and will be more susceptible to mutations that may lead to cancer.*

34. Imagine that you discover a large family in which bladder cancer is inherited as an autosomal dominant trait. Briefly outline a series of studies that you might conduct to identify the gene that causes bladder cancer in this family.
Because this cancer is inherited as a dominant trait, the cause is most likely an oncogenic mutation, rather than inactivation of a tumor-suppressor gene. One possible approach to identify the bladder cancer oncogene would be to isolate DNA from bladder cancer cells and transfect cultured normal cells from a distinct genetic background. Isolate any colonies of transformed cells that arise and determine which common chromosomal region has been taken up by the transformed cell lines. The region of the gene could be futher refined by transfection with subframents of that region. Genes within this transforming region would be sequenced to identify mutations. Finally, candidate mutations would be tested by genetically engineering cells to carry that precise mutation to see if such cells become cancerous.

Chapter Twenty-Four: Quantitative Genetics

COMPREHENSION QUESTIONS

Section 24.1

*1. How does a quantitative characteristic differ from a discontinuous characteristic?
Discontinuous characteristics have only a few distinct phenotypes. In contrast, a quantitative characteristic shows a continuous variation in phenotype.

2. Briefly explain why the relation between genotype and phenotype is frequently complex for quantitative characteristics.
Quantitative characteristics are polygenic, so many genotypes are possible. Moreover, most quantitative characteristics are also influenced by environmental factors. Therefore, the phenotype is determined by complex interactions of many possible genotypes and environmental factors.

*3. Why do polygenic characteristics have many phenotypes?
Many genotypes are possible with multiple genes. Even for the simplest two-allele loci, the number of possible genotypes is equal to 3^n*, where* n *is the number of loci or genes. Thus, for 3 genes, we have 27 genotypes, 4 genes yields 81, and so forth. If each genotype corresponds to a unique phenotype, then we have the same numbers of phenotypes: 27 possible phenotypes for 3 genes and 81 possible phenotypes for 4 genes. Finally, the phenotype for a given genotype may be influenced by environmental factors, leading to an even greater array of phenotypes.*

Section 24.2

*4. Explain the relation between a population and a sample. What characteristics should a sample have to be representative of the population?
A sample is a subset of the population. To be representative of the population, a sample should be randomly selected and sufficiently large to minimize random differences between members of the sample and the population.

5. What information do the mean and variance provide about a distribution?
The mean is the center of the distribution. The variance is how broad the distribution is around the mean.

6. How is the standard deviation related to the variance?
The standard deviation is the square root of the variance.

*7. What information does the correlation coefficient provide about the association between two variables?
The magnitude or absolute value of the correlation coefficient reports how strongly the two variables are associated. A value close to +1 or –1 indicates a strong association; values close to zero indicate weak association.

8. What is regression? How is it used?
Regression is a mathematical relationship between correlated variables. Regression is used to predict the value of a variable from the value of a correlated variable.

Section 24.3

*9. List all the components that contribute to the phenotypic variance and define each component.

V_G – *Component of variance due to variation in genotype*
V_A – *Component of variance due to additive genetic variance*
V_D – *Component of variance due to dominance genetic variance*
V_I – *Component of variance due to genic interaction variance*
V_E – *Component of variance due to environmental differences*
V_{GE} – *Component of variance due to interaction between genes and environment*

*10. How do the broad-sense and narrow-sense heritabilities differ?
The broad-sense heritability is the portion of phenotypic variance that is due to all types of genetic variance, including additive, dominance, and genic interaction variances. The narrow-sense heritability is just that portion of the phenotypic variance due to additive genetic variance.

11. Briefly outline some of the ways that heritability can be calculated.
Elimination of variance components from the equation: $V_P = V_G + V_E + V_{GE}$. *By either eliminating genetic variance (*$V_G = 0$*) with genetically identical individuals, or by eliminating* V_E *with individuals raised in identical environments, we can determine the values of* V_E *or* V_G*, respectively. If* V_P *can be determined under conditions of genotypic variance, then the missing term* V_G *or* V_E *can be calculated by simple subtraction.*

Parent-offspring regression: The mean phenotypic values of the parents are plotted against the mean phenotypic values of the offspring for a series of families. The narrow-sense heritability equals the regression coefficient.

Comparison of phenotypes for different degrees of relatedness: Compare phenotypes of monozygotic and dizygotic twins. Twice the difference in correlation coefficients of monozygotic and dizygotic twins yields an estimate of the broad-sense heritability.

Response to selection: The response to selection is equal to the product of the narrow-sense heritability and the selection differential.

12. Briefly discuss common misunderstandings or misapplications of the concept of heritability.
    (1) Heritability is the portion of phenotypic variance due to genetic variance; it does not indicate to what extent the phenotype itself is determined by genotype.
    (2) Heritability applies to populations; it does not apply to individuals.
    (3) Heritability is determined for a particular population in a particular environment at a particular time. Heritability determined for one population does not apply to other populations, or even the same population facing different environmental conditions at a different period.
    (4) A trait with high heritability may still be strongly influenced by environmental factors.
    (5) High heritability does not mean that differences between populations are due to differences in genotype.

13. Briefly explain how genes affecting a polygenic characteristic are located with the use of QTL mapping.
    Two homozygous, highly inbred strains that differ at many loci are crossed and the F_1 are interbred. Quantitative traits are measured and correlated with the inheritance of molecular markers throughout the genome. The correlations are used to infer the presence of a linked QTL.

Section 24.4

*14. How is the response to selection related to the narrow-sense heritability and the selection differential? What information does the response to selection provide?
    The response to selection (R) = *narrow-sense heritability* (h^2) × *selection differential* (S). *The value of* R *predicts how much the mean quantitative phenotype will change with different selection in a single generation.*

15. Why does the response to selection often level off after many generations of selection?
    After many generations, the response to selection plateaus because of two factors. First, the genetic variation may be depleted—all the individuals in the population now have the alleles that maximize the quantitative trait; with no genetic variation, there can be no selection or response to selection. Second, even if genetic variation persists, artificial selection may be limited by an opposing natural selection.

APPLICATION QUESTIONS AND PROBLEMS

Section 24.1

*16. For each of the following characteristics, indicate whether it would be considered a discontinuous characteristic or a quantitative characteristic. Briefly justify your answer.

a. Kernel color in a strain of wheat, in which two codominant alleles segregating at a single locus determine the color. Thus, there are three phenotypes present in this strain: white, light red, and medium red.
This is a discontinuous characteristic because only a few distinct phenotypes are present and it is determined by alleles at a single locus.

b. Body weight in a family of Labrador retrievers. An autosomal recessive allele that causes dwarfism is present in this family. Two phenotypes are recognized: dwarf (less than 13 kg) and normal (greater than 13 kg).
This is a discontinuous characteristic because there are only two phenotypes (dwarf and normal) and a single locus determines characteristic.

c. Presence or absence of leprosy. Susceptibility to leprosy is determined by multiple genes and numerous environmental factors.
This is a quantitative characteristic because susceptibility is a continuous trait that is determined by multiple genes and environmental factors. It is an example of a quantitative phenotype with a threshold effect.

d. Number of toes in guinea pigs, which is influenced by genes at many loci.
A quantitative characteristic because it is determined by many loci. The number of toes is an example of a meristic characteristic.

e. Number of fingers in humans. Extra (more than five) fingers are caused by the presence of an autosomal dominant allele.
A discontinuous characteristic because there are only a few distinct phenotypes determined by alleles at a single locus.

*17. Assume that plant weight is determined by a pair of alleles at each of two independently assorting loci (*A* and *a*, *B* and *b*) that are additive in their effects. Further assume that each allele represented by an uppercase letter contributes 4 g to weight and each allele represented by a lowercase letter contributes 1 g to weight.

a. If a plant with genotype *AA BB* is crossed with a plant with genotype *aa bb*, what weights are expected in the F_1 progeny?
All the F_1 progreny will have genotype Aa Bb, *so they should all have 4 + 1 + 4 + 1 = 10 grams of weight.*

b. What is the distribution of weight expected in the F_2 progeny?
We can group the 16 expected genotypes by the number of uppercase and lowercase alleles:
4 uppercase: AA BB = *1/16 with 16 grams*
3 uppercase: 2 Aa BB, 2 AA Bb = *4/16 with 13 grams*
2 uppercase: 4 Aa Bb, aa BB, AA bb = *6/16 with 10 grams*
1 uppercase: 2 Aa bb, 2 aa Bb = *4/16 with 7 grams*
0 uppercase: aa bb = *1/16 with 4 grams*

*18. Assume that three loci, each with two alleles (*A* and *a*, *B* and *b*, *C* and *c*), determine the differences in height between two homozygous strains of a plant. These genes are additive and equal in their effects on plant height. One strain (*aabbcc*) is 10 cm in height. The other strain (*AABBCC*) is 22 cm in height. The two strains are crossed, and the resulting F_1 are interbred to produce F_2 progeny. Give the phenotypes and the expected proportions of the F_2 progeny.

The AABBCC *strain is 12 cm taller than the* aabbcc *strain. We therefore calculate that each dominant allele adds 2 cm of height above the baseline 10 cm of the all-recessive strain. The F_1, with genotype* AaBbCc, *therefore will be 10 + 6 = 16 cm tall. The seven different possible phenotypes with respect to plant height and the expected frequencies in the F_2 are listed in the following table:*

Number of dominant alleles	*Height (cm)*	*Proportion of F_2 progeny*
6	22	1/64
5	20	6/64
4	18	15/64
3	16	20/64
2	14	15/64
1	12	6/64
0	10	1/64
		Total = 64/64

The proportions can be determined by counting the numbers of boxes with one dominant allele, two dominant alleles, and so on from an 8 × 8 Punnett square.

*19. A farmer has two homozygous varieties of tomatoes. One variety, called *Little Pete*, has fruits that average only 2 cm in diameter. The other variety, *Big Boy*, has fruits that average a whopping 14 cm in diameter. The farmer crosses *Little Pete* and *Big Boy*; he then intercrosses the F_1 to produce F_2 progeny. He grows 2000 F_2 tomato plants and doesn't find any F_2 offspring that produce fruits as small as *Little Pete* or as large as *Big Boy*. If we assume that the differences in fruit size of these varieties are produced by genes with equal and additive effects, what conclusion can we make about the minimum number of loci with pairs of alleles determining the differences in fruit size of the two varieties?
That six or more loci are involved. Generally, $(1/4)^n$ of the F_2 progeny should resemble one of the homozygous parents, where n *is the number of loci with pairs of alleles that determine the differences in the trait. If 5 genes were involved, the farmer should have found approximately 1/1000 of the F_2 that resembled either* Little Pete *or* Big Boy. *Because he did not, we can conclude that at least 6 loci are involved in the difference in fruit size between the two varieties: $(1/4)^6$ = 1/4096 would be expected to resemble one of the parents if six loci were involved.*

20. Seed size in a plant is a polygenic characteristic. A grower crosses two pure-breeding varieties of the plant and measures seed size in the F_1 progeny. He then backcrosses the F_1 plants to one of the parental varieties and measures seed size in the backcross progeny. The grower finds that seed size in the backcross progeny has a higher variance than seed size in the F_1 progeny. Explain why the backcross progeny are more variable.
The F_1 progeny all have the same genetic makeup: they are all heterozygotes for the loci that differ between the two pure-breeding strains. The backcross progeny will

have much greater genetic diversity as a result of the genetic diversity from meiosis of the F_1 heterozygotes.

Section 24.2

*21. Listed below are the numbers of digits per foot in 25 guinea pigs. Construct a frequency distribution for these data.
4, 4, 4, 5, 3, 4, 3, 4, 4, 5, 4, 4, 3, 2, 4, 4, 5, 6, 4, 4, 3, 4, 4, 4, 5

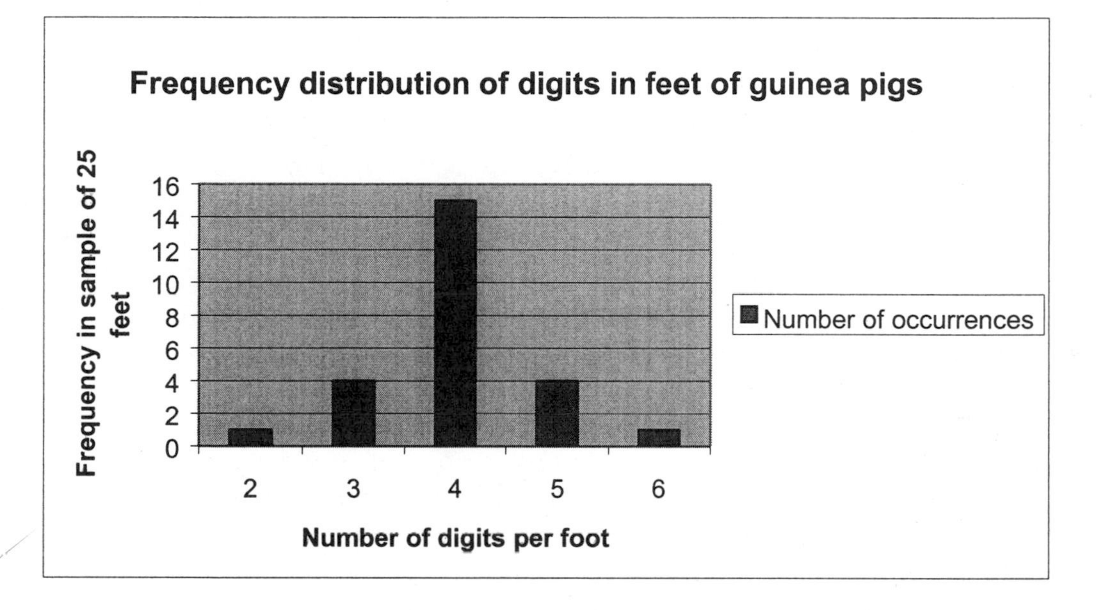

22. Ten male Harvard students were weighed in 1916. Their weights are given in the following table. Calculate the mean, variance, and standard deviation for these weights.

Weight (kg) of Harvard students (class of 1920)
51
69
69
57
61
57
75
105
69
63

x_i	$\lvert x_i - \text{mean}\rvert$	$(x_i - \text{mean})^2$
51	16.6	275.56
69	1.4	1.96
69	1.4	1.96
57	10.6	112.36
61	6.6	43.56
57	10.6	112.36
75	7.4	54.76
105	37.4	1398.76
69	1.4	1.96
63	4.6	21.16
Sum = 676		Σ = 2024.4
Mean = 67.6		

The sum of the weights is 676, divided by 10 students, yields a mean of 67.6 kg.

The variance is $s^2 = \frac{\sum(x_i - \bar{x})^2}{n-1}$ *= 2024.4/9 = 224.9*

The standard deviation = $s = \sqrt{s^2}$ *= 15*

23. Among a population of tadpoles, the correlation coefficient for size at metamorphosis and time required for metamorphosis is –.74. On the basis of this correlation, what conclusions can you make about the relative sizes of tadpoles that metamorphose quickly and those that metamorphose more slowly?
Size at metamorphosis and time required for metamorphosis are inversely correlated (negative correlation coefficient). The greater the time required for metamorphosis, the smaller the size, and vice versa. Therefore, tadpoles that metamorphose quickly are larger than tadpoles that metamorphose slowly.

*24. A researcher studying alcohol consumption in North American cities finds a significant, positive correlation between the number of Baptist preachers and alcohol consumption. Is it reasonable for the researcher to conclude that the Baptist preachers are consuming most of the alcohol? Why or why not?
No, correlation does not mean causation. A number of other factors may be responsible for the observed correlation. For example, Baptist preachers may feel their services are more needed in areas with high alcohol consumption. The reader should think of other plausible scenarios.

25. Body weight and length were measured on six mosquito fish; these measurements are given in the following table. Calculate the correlation coefficient for weight and length in these fish.

Wet weight (g)	Length (mm)
115	18
130	19
210	22
110	17
140	20
185	21

The correlation coefficient r *is calculated from the formula* $r = \frac{cov_{xy}}{s_x s_y}$.

First, we calculate the covariance of the two traits and their standard deviations.

x_i	y_i	$x_i - \bar{x}$	$y_i - \bar{y}$	$(x_i - \bar{x})(y_i - \bar{y})$	$(x_i - \bar{x})^2$	$(y_i - \bar{y})^2$
115	*18*	*−33.33*	*−1.5*	*50*	*1111.11*	*2.25*
130	*19*	*−18.33*	*−0.5*	*9.17*	*336.11*	*0.25*
210	*22*	*61.67*	*2.5*	*154.17*	*3802.78*	*6.25*
110	*17*	*−38.33*	*−2.5*	*95.83*	*1469.44*	*6.25*
140	*20*	*−8.33*	*0.5*	*−4.17*	*69.44*	*0.25*
185	*21*	*36.67*	*1.5*	*55*	*1344.44*	*2.25*
$\bar{x}$ *= 148.3*		$\bar{y}$ *= 19.5*		$\Sigma = 360$	$\Sigma = 8133.33$	$\Sigma = 17.5$

$cov_{xy} = 72$ $s^2_x = 1626.67$ $s^2_y = 3.5$

$$r = \frac{72}{(40.33)(1.87)} = 0.95$$

$s_x = 40.33$ $s_y = 1.87$

*26. The heights of mothers and daughters are given in the following table.

Height of mother (inches)	Height of daughter (inches)
64	66
65	66
66	68
64	65
63	65
63	62
59	62
62	64
61	63
60	62

a. Calculate the correlation coefficient for the heights of the mothers and daughters.

The correlation coefficient r *is calculated from the formula* $r = \frac{cov_{xy}}{s_x s_y}$.

x	y	$x_i - \bar{x}$	$y_i - \bar{y}$	$(x_i - \bar{x})(y_i - \bar{y})$	$(x_i - \bar{x})^2$	$(y_i - \bar{y})^2$
64	66	1.3	1.7	2.21	1.69	2.89
65	66	2.3	1.7	3.91	5.29	2.89
66	68	3.3	3.7	12.21	10.89	13.69
64	65	1.3	0.7	0.91	1.69	0.49
63	65	0.3	0.7	0.21	0.09	0.49
63	62	0.3	−2.3	−0.69	0.09	5.29
59	62	−3.7	−2.3	8.51	13.69	5.29
62	64	−0.7	−0.3	0.21	0.49	0.09
61	63	−1.7	−1.3	2.21	2.89	1.69
60	62	−2.7	−2.3	.21	7.29	5.29
$\bar{x} = 62.7$	$\bar{y} = 64.3$			$\Sigma = 35.9$	$\Sigma = 44.1$	$\Sigma = 38.1$
				$cov(x,y) = 3.99$	$s^2_x = 4.9$	$s^2_y = 4.23$
					$s_x = 2.2$	$s_y = 2.1$
				$b = 0.81$		
				$b(\bar{x}) = 51$		
				$a = 13$		

$$r = \frac{3.99}{(2.2)(2.1)} = .80$$

b. Using regression, predict the expected height of a daughter whose mother is 67 inches tall.

In the regression equation, $y = a + bx$, b *is given by the formula* $b = \frac{cov_{xy}}{s_x^2}$ *and the value of* a *by the equation* $a = \bar{y} - b\bar{x}$. *If the mother is 67 inches tall, the regression equation* $y = a + bx$ *becomes* $y = 13.26 + 0.81(67) = 67.8$ *inches.*

Section 24.3

*27. Phenotypic variation in tail length of mice has the following components:

Additive genetic variance (V_A)	= 0.5
Dominance genetic variance (V_D)	= 0.3
Genic interaction variance (V_I)	= 0.1
Environmental variance (V_E)	= 0.4
Genetic-environmental interaction variance (V_{GE})	= 0.0

a. What is the narrow-sense heritability of tail length?
Narrow-sense heritability is $V_A/V_P = 0.5/1.3 = 0.38$.

b. What is the broad-sense heritability of tail length?
Broad-sense heritability is $V_G/V_P = (V_A + V_D + V_I)/V_P = 0.9/1.3 = 0.69$.

28. The narrow-sense heritability of ear length in Reno rabbits is 0.4. The phenotypic variance (V_P) is 0.8 and the environmental variance (V_E) is 0.2. What is the additive genetic variance (V_A) for ear length in these rabbits?
Narrow-sense heritability $= V_A/V_P = 0.4$
Given that $V_P = 0.8$, $V_A = 0.4(0.8) = 0.32$

*29. Assume that human ear length is influenced by multiple genetic and environmental factors. Suppose you measured ear length on three groups of people, in which group A consists of five unrelated persons, group B consists of five siblings, and group C consists of five first cousins.

a. Assuming that the environment for each group is similar, which group should have the highest phenotypic variance? Explain why.
Group A, because unrelated individuals have the greatest genetic variance.

b. Is it realistic to assume that the environmental variance for each group is similar? Explain your answer.
No. Siblings from the same family and who are raised in the same house should have smaller environmental variance than group A of unrelated individuals.

30. A characteristic has a narrow-sense heritability of 0.6.

a. If the dominance variance (V_D) increases and all other variance components remain the same, what will happen to the narrow-sense heritability? Will it increase, decrease, or remain the same? Explain.
The narrow-sense heritability will decrease. Narrow-sense heritability is V_A/V_P. *Increasing the* V_D *will increase the total phenotypic variance* V_P. *If* V_A *remains unchanged, then the proportion* V_A/V_P *will become smaller.*

b. What will happen to the broad-sense heritability? Explain.
The broad-sense heritability V_G/V_P *will increase.* V_G *is the sum of* $V_A + V_D + V_I$. V_P *is the sum of* $V_A + V_D + V_I + V_E$. *Increasing the numerator and denominator of the fraction by the same arithmetic increment will result in a larger fraction, if the fraction is smaller than 1, as must be the case for* V_G/V_P.

c. If the environmental variance (V_E) increases and all other variance components remain the same, what will happen to the narrow-sense heritability? Explain.
The narrow sense heritability V_A/V_P *will decrease because the total phenotypic variance* V_P *will increase if* V_E *increases.*

d. What will happen to the broad-sense heritability? Explain.
The broad-sense heritability V_G/V_P *will decrease because* $V_P = V_G + V_E$ *will increase.*

31. Flower color in the varieties of pea plants studied by Mendel is controlled by alleles at a single locus. A group of peas homozygous for purple flowers is grown in a garden. Careful study of the plants reveals that all their flowers are purple, but there

is some variability in the intensity of the purple color. If heritability were estimated for this variation in flower color, what would it be? Explain your answer.
The plants are homozygous for the single color locus; therefore, there is no genetic variance: $V_G = 0$. *Because heritability is* V_G/V_P, *if* V_G *is zero, then heritability is zero.*

*32. A graduate student is studying a population of bluebonnets along a roadside. The plants in this population are genetically variable. She counts the seeds produced by 100 plants and measures the mean and variance of seed number. The variance is 20. Selecting one plant, the graduate student takes cuttings from it and cultivates these cuttings in the greenhouse, eventually producing many genetically identical clones of the same plant. She then transplants these clones into the roadside population, allows them to grow for one year, and then counts the number of seeds produced by each of the cloned plants. The graduate student finds that the variance of seed number among these cloned plants is 5. From the phenotypic variance of the genetically variable and genetically identical plants, she calculates the broad-sense heritability.

a. What is the broad-sense heritability of seed number for the roadside population of bluebonnets?
In the genetically identical population, $V_G = 0$, *and* $V_P = V_E = 5$. *In the original population,* $V_G = V_P - V_E = 20 - 5 = 15$. *The broad-sense heritability is then* $V_G/V_P = 15/20 = 0.75$.

b. What might cause this estimate of heritability to be inaccurate?
This estimate may be inaccurate if the environmental variance of the genetically identical population is different from the environmental variance of the genetically diverse population. For example, the weather conditions may be different and the transplanted clones may not be dispersed over as wide an area as the original 100 roadside plants.

33. Many researchers have estimated heritability of human traits by comparing the correlation coefficients of monozygotic and dizygotic twins. One of the assumptions of using this method is that two monozygotic twins experience environments that are no more similar to each other than those experienced by two dizygotic twins. How might this assumption be violated? Give some specific examples of ways that the environments of two monozygotic twins might be more similar than the environments of two dizygotic twins.
One obvious way a monozygotic twins may have a more similar environment is if the dizygotic twins differ in sex. Dizygotic twins also differ more in physical traits than monozygotic twins. Such differences, in hair color, eye color, height, weight, and others lead to different preferences in clothing, whether or when eye glasses or braces are required, and differences in preferred activities such as different aptitudes for sports.

34. A genetics researcher determines that the broad-sense heritability of height among Southwestern University undergraduate students is 0.90. Which of the following conclusions would be reasonable? Explain your answer.

a. Because Sally is a Southwestern University undergraduate student, 10% of her height is determined by nongenetic factors.
b. Ninety percent of variation in height among all undergraduate students in the United States is due to genetic differences.
c. Ninety percent of the height of Southwestern University undergraduate students is determined by genes.
d. Ten percent of the variation in height of Southwestern University undergraduate students is determined by variation in non genetic factors.
e. Because the heritability of height among Southwestern University students is so high, any change in the students environment will have minimal impact on their height.

Heritability is the proportion of total phenotypic variance that is due to genetic variance, and applies only to the particular population. Thus, the only reasonable conclusion is (d). Statement (a) is not justified because the heritability value does not apply to absolute height, but to the variance in height among Southwestern undergraduates. Statement (b) is not justified because the heritability has been determined only for Southwestern University students; students at other universities, with different ethnic backgrounds and from different regions of the country may have different heritability for height. Statement (c) is again not justified because the heritability refers to the variance in height rather than absolute height. Statement (e) is not justified because the heritability has been determined for the range of variation in nongenetic factors experienced by the population under study; environmental variation outside this range (such as severe malnutrition) may have profound effects on height.

*35. The length of the middle joint of the right index finger was measured on 10 sets of parents and their adult offspring. The mean parental lengths and the mean offspring lengths for each family are listed in the following table. Calculate the regression coefficient for regression of mean offspring length against mean parental length and estimate the narrow sense heritability for this characteristic.

Mean parental length (mm)	Mean offspring length (mm)
30	31
35	36
28	31
33	35
26	27
32	30
31	34
29	28
40	38
33	34

The narrow-sense heritability is equal to the regression coefficient b *of a regression of the means of the parents and the means of the offspring.*

x	y	$x_i - \bar{x}$	$y_i - \bar{y}$	$(x_i - \bar{x})(y_i - \bar{y})$	$(x_i - \bar{x})^2$	$(y_i - \bar{y})^2$
30	31	−1.7	−1.4	2.38	2.89	1.96
35	36	3.3	3.6	11.88	10.89	12.96
28	31	−3.7	−1.4	5.18	13.69	1.96
33	35	1.3	2.6	3.38	1.69	6.76
26	27	−5.7	−5.4	30.78	32.49	29.16
32	30	0.3	−2.4	−0.72	0.09	5.76
31	34	−0.7	1.6	−1.12	0.49	2.56
29	28	−2.7	−4.4	11.88	7.29	19.36
40	38	8.3	5.6	46.48	68.89	31.36
33	34	1.3	1.6	2.08	1.69	2.56
$\bar{x} = 31.7$	$\bar{y} = 32.4$			$\Sigma = 112.2$	$\Sigma = 140.1$	$\Sigma = 114.4$
				cov(x,y) = 12.47	$s^2_x = 15.6$	12.7

b = 0.80

$b(\bar{x}) = 25.4$

a = 7.0

From the above table, the narrow-sense heritability = b = *0.8.*

36. *Drosophila buzzati* is a fruit fly that feeds on the rotting fruits of cacti in Australia. Timothy Prout and Stuart Barker calculated the heritabilities of body size, as measured by thorax length, for a natural population of *D. buzzati* raised in the wild and for a population of *D. buzzati* collected in the wild but raised in the laboratory (T. Prout and J. S. F. Barker. 1989. *Genetics* 123:803–813). They found the following heritabilities.

Population	**Heritability of body size (± standard error)**
Wild population	0.0595 ± 0.0123
Laboratory-reared population	0.3770 ± 0.0203

Why do you think the heritability measured in the laboratory-reared population is higher than that measured in the natural population raised in the wild?
Heritability is the proportion of total phenotypic variance that is due to genetic variance: $H^2 = V_G/V_P$. *The difference in heritability between the wild population and the laboratory-reared population could be due to either the wild population having less genetic variance, or the wild population having greater phenotypic variance due to greater environmental or genetic-environmental interaction variance. Because the laboratory-reared population was collected in the wild, it is unlikely that the laboratory-reared population has greater genetic variance than the wild population. Therefore, the more likely explanation is that the wild population has greater phenotypic variance due to environmental factors. Variance due to differences in availability of food, parasitism, and ambient temperature may be some environmental factors.*

*37. Mr. Jones is a pig farmer. For many years, he has fed his pigs the food left over from the local university cafeteria, which is known to be low in protein, deficient in vitamins, and downright untasty. However, the food is free and his pigs do not complain. One day a salesman from a feed company visits Mr. Jones. The salesman claims that his company sells a new, high-protein, vitamin-enriched feed that enhances weight gain in pigs. Although the food is expensive, the salesman claims that the increased weight gain of the pigs will more than pay for the cost of the feed, increasing Mr. Jones' profit. Mr. Jones responds that he took a genetics class when he went to the university and that he has conducted some genetic experiments on his pigs; specifically, he has calculated the narrow-sense heritability of weight gain for his pigs and found it to be .98. Mr. Jones says that this heritability value indicates that 98% of the variance in weight gain among his pigs is determined by genetic differences, and therefore the new pig feed can have little effect on the growth of his pigs. He concludes that the feed would be a waste of his money. The salesman does not dispute Mr. Jones' heritability estimate, but he still claims that the new feed can significantly increase weight gain in Mr. Jones' pigs. Who is correct and why?

The salesman is correct because Mr. Jones' determination of heritability was conducted for a population of pigs under one environmental condition: low nutrition. His findings do not apply to any other population or even to the same population under different environmental conditions. High heritability for a trait does not mean that environmental changes will have little effect.

Section 24.4

38. Joe is breeding cockroaches in his dorm room. He finds that the average wing length in his population of cockroaches is 4 cm. He picks six cockroaches that have the largest wings; the average wing length among these selected cockroaches is 10 cm. Joe interbreeds these selected cockroaches. From previous studies, he knows that the narrow-sense heritability for wing length in his population of cockroaches is 0.6.

a. Calculate the selection differential and expected response to selection for wing length in these cockroaches.

Use the equation: $R = h^2 \times S$, *where* S *is the selection differential. In this case,* S = *10 cm – 4 cm = 6 cm, and we are given that the narrow-sense heritability* h^2 *is 0.6. Therefore, the response to selection* R = *0.6(6 cm) = 3.6 cm.*

b. What should be the average wing length of the progeny of the selected cockroaches?

The average wing length of the progeny should be the mean wing length of the population plus R: *4 cm + 3.6 cm = 7.6 cm.*

39. Three characteristics in beef cattle—body weight, fat content, and tenderness—are measured and the following variance components are estimated:

	Body weight	Fat content	Tenderness
V_A	22	45	12
V_D	10	25	5
V_I	3	8	2
V_E	42	64	8
V_{GE}	0	0	1

In this population, which characteristic would respond best to selection? Explain your reasoning.
Tenderness would respond best because it has the highest narrow-sense heritability. The response to selection is given by the equation $R = h^2 \times S$, *where the narrow-sense heritability* h^2 *is equal to* V_A/V_P.

*40. A rancher determines that the average amount of wool produced by a sheep in his flock is 22 kg per year. In an attempt to increase the wool production of his flock, the rancher picks five male and five female sheep with the greatest wool production; the average amount of wool produced per sheep by those selected is 30 kg. He interbreeds these selected sheep and finds that the average wool production among the progeny of the selected sheep is 28 kg. What is the narrow-sense heritability for wool production among the sheep in the rancher's flock?
We use the equation $R = h^2 \times S$. *The value of* R *is given by the difference in the average wool production of the progeny of the selected sheep compared to the rest of the flock: 28 kg – 22 kg = 6 kg. The value of* S *is the difference between the selected sheep and the flock: 30 kg – 22 kg = 8 kg. Then,* $h^2 = R/S = 6/8 = 0.75$.

41. A strawberry farmer determines that the average weight of individual strawberries produced by plants in his garden is 2 g. He selects the 10 plants that produce the largest strawberries; the average weight of strawberries among these selected plants is 6 g. He interbreeds these selected strawberry plants. The progeny of these selected plants produce strawberries that weigh 5 grams. If the farmer were to select plants that produce an average strawberry weight of 4 grams, what would be the predicted weight of strawberries produced by the progeny of these selected plants?
Here we can use the equation $R = h^2 \times S$. *R, the response to selection, is the difference between the mean of the starting population and the mean of the progeny of the selected parents. In this case,* $R = 5\text{ g} - 2\text{ g} = 3\text{ g}$. *S, the selection differential, is the difference between the mean of the starting population and the mean of the selected parents; in this case* $S = 6\text{ g} - 2\text{ g} = 4\text{ g}$. *Substituting in the equation, we get* $3\text{ g} = h^2(4\text{ g})$; $h^2 = 0.75$. *If the selected plants averaged 4 g, then S would be 2 g and* $R = 0.75(2\text{ g}) = 1.5\text{ g}$. *Therefore, the predicted average weight of strawberries from the progeny plants would be 2 g + 1.5 g = 3.5 g.*

42. The narrow-sense heritability of wing length in a population of *Drosophila melanogaster* is 0.8. The narrow-sense heritability of head width in the same population is 0.9. The genetic correlation between wing length and head width is –0.86. If a geneticist selects for increased wing length in these flies, what will happen to head width?
The head width will decrease. These two traits have high negative genetic correlation. Therefore, selection for one trait will affect the other trait inversely.

43. Pigs have been domesticated from wild boars. Would you expect to find higher heritability for weight among domestic pigs or wild boars? Explain your answer.
Wild boars will probably have higher heritability than domestic pigs. Domestic pigs, because of many generations of breeding and selection, are likely to have less variance, and more homozygosity, for genes that affect commercial traits such as weight.

CHALLENGE QUESTIONS

Section 24.1

44. Manic-depressive illness is a psychiatric disorder that has a strong hereditary basis, but the exact mode of inheritance is not known. Previous research has shown that siblings of patients with manic-depressive illness are more likely also to develop the disorder than are siblings of unaffected individuals. A recent study demonstrated that the ratio of manic-depressive brothers to manic-depressive sisters is higher when the patient is male than when the patient is female. In other words, relatively more brothers of manic-depressive patients also have the disease when the patient is male than when the patient is female. What does this new observation suggest about the inheritance of manic-depressive illness?
These observations suggest that an X-linked locus or loci may influence manic-depressive illness. Males inherit their X-chromosome genes only from their mother. Females inherit X-chromosome genes from both parents. Therefore, the brothers of an affected male inherited their X chromosome alleles from the same parent, the mother. On the other hand, an affected female may have inherited a contributory X-linked QTL locus allele from either parent; if this allele came from her father, there is no chance that her brothers inherited the same X-linked allele.

Section 24.3

45. We have explored some of the difficulties in separating genetic and environmental components of human behavioral characteristics. Considering these difficulties and what you know about calculating heritability, propose an experimental design for accurately measuring the heritability of musical ability.
For the purpose of this discussion, let us assume that we have a reliable and accurate method of quantifying musical ability. I propose a study comparing musical abilities in individuals with different degrees of relatedness. I would compare two groups: one group would consist of monozygotic (identical) twins

raised apart; the second group would consist of dizygotic (fraternal, or nonidentical) twins raised apart. Both groups should have comparable environmental variance, but the monozygotic twins share 100% of their genes, whereas the dizygotic twins share only 50% of their genes. By correlating the musical abilities of the two groups, we can estimate the broad-sense heritability with equation 22.20:

$$H^2 = 2(r_{MZ} - r_{DZ})$$

where r_{MZ} is the correlation coefficient of musical ability in the monozygotic group and r_{DZ} is the correlation coefficient in the dizygotic group.

46. A student who has just learned about quantitative genetics says, "Heritability estimates are worthless! They do not tell you anything about the genes that affect a characteristic. They do not provide any information about the types of offspring to expect from a cross. Heritability estimates measured in one population cannot be used for other populations, so they do not even give you any general information about how much of a characteristic is genetically determined. I cannot see that heritabilities do anything other than make undergraduate students sweat during tests." How would you respond to this statement? Is the student correct? What good are heritabilities, and why do geneticists bother to calculate them?
One of the most valuable aspects of heritability is that it allows geneticists to predict the response to selection, either natural or artificial. In breeding plants and animals for desired QTL traits, knowing the heritability of the trait in the breeding population allows the geneticist (breeder) to make better predictions about the effectiveness of any artificial selection program.

Another important application concerns susceptibility to disease in humans (or livestock or cultivated plants). Knowing to what extent susceptibility to a particular disease is influenced by genes or by environment is essential in making public health policy decisions. If a large part of the variance is due to environmental factors, then the overall health of the population may be improved by addressing environmental improvements (e.g., reducing cigarette smoke to combat the incidence of lung cancer).

Section 24.4

47. Eugene Eisen selected for increased 12-day litter weight (total weight of a litter of offspring 12 days after birth) in a population of mice (E. J. Eisen. 1972. *Genetics* 72:129–142). The 12-day litter weight of the population steadily increased, but then leveled off after about 17 generations. At generation 17, Eisen took one family of mice from the selected population and reversed the selection procedure: in this group, he selected for decreased 12-day litter size. This group immediately responded to decreased selection; the 12-day litter weight dropped 4.8 g within one generation and dropped 7.3 g after 5 generations. Based on the results of the reverse selection, what is the most likely explanation for the leveling off of 12-day litter weight in the original population?
The leveling off in the response to selection for increased litter weight may be due to either of two causes: elimination or reduction of genetic variance because

maximum litter weight had been achieved, or opposing selection countering further increase in litter weight. The results of the reverse selection, showing immediate response, indicate that genetic variance was still significant. Therefore, the most likely explanation is opposing selection. Further increase in litter weight may cause problems for the mother during pregnancy, for example.

Chapter Twenty-Five: Population Genetics

COMPREHENSION QUESTIONS

Section 25.1

1. What is a Mendelian population? How is the gene pool of a Mendelian population usually described?
A Mendelian population is a group of sexually reproducing individuals mating with each other and sharing a common gene pool. The gene pool is usually described by genotype frequencies and allele frequencies.

Section 25.2

2. What are the predictions given by the Hardy–Weinberg law?
The Hardy–Weinberg law states that a large population mating randomly with no effects from selection, migration, or mutation will have the following relationship between the genotype frequencies and allele frequencies:

$f(AA) = p^2$; $f(Aa) = 2pq$; $f(aa) = q^2$, *where* p *and* q *equal the allelic frequencies Moreover, the allele frequencies do not change from generation to generation, as long as the above conditions hold.*

*3. What assumptions must be met for a population to be in Hardy-Weinberg equilibrium?
Large population, random mating, and not affected by migration, selection, or mutation

4. What is random mating?
Random mating takes place when each genotypes mate with each other according to their respective frequencies, so the frequency of an AA *individual mating with another* AA *individual will be* $f(AA)^2$, *the frequency of* AA *mating with* Aa *will be* $2f(AA)f(Aa)$, *the mating of* Aa *with* Aa *will be* $f(Aa)^2$, *and so forth.*

*5. Give the Hardy–Weinberg expected genotypic frequencies for (a) an autosomal locus with three alleles and (b) an X-linked locus with two alleles.
a. *If the frequencies of alleles A1, A2, and A3 are defined as p, q, and r, respectively:*
$f(A1A1) = p^2$
$f(A1A2) = 2pq$
$f(A2A2) = q^2$
$f(A1A3) = 2pr$
$f(A2A3) = 2qr$
$f(A3A3) = r^2$

b. *For an X-linked locus with two alleles:*
$f(X^1X^1) = p^2$ *among females;* $p^2/2$ *for the whole population*
$f(X^1X^2) = 2pq$ *among females;* pq *for the whole population*
$f(X^2X^2) = q^2$ *among females;* $q^2/2$ *for the whole population*
$f(X^1Y) = p$ *among males;* $p/2$ *for the whole population*
$f(X^2Y) = q$ *among males;* $q/2$ *for the whole population*

Section 25.3

6. Define inbreeding and briefly describe its effects on a population.
Inbreeding is preferential mating between genetically related individuals. Inbreeding increases homozygosity and reduces heterozygosity in the population.

Section 25.4

7. What determines the allelic frequencies at mutational equilibrium?
At mutational equilibrium, the allelic frequencies are determined by the forward and reverse mutation rates.

*8. What factors affect the magnitude of change in allelic frequencies due to migration?
The proportion of the population due to migrants (m) and the difference in allelic frequencies between the migrant population and the original resident population.

9. Define genetic drift and give three ways that it can arise. What effect does genetic drift have on a population?
Genetic drift is change in allelic frequencies resulting from sampling error. It may arise through a long-term limitation on population size, founder effect that occurs when the population is founded by a small number of individuals, or a bottleneck effect when the population undergoes a drastic reduction in population size. Genetic drift causes changes in allelic frequencies and loss of genetic variation because some alleles are lost as other alleles become fixed. It also causes genetic divergence between populations because the different populations undergo different changes in allelic frequencies and become fixed for different alleles.

*10. What is effective population size? How does it affect the amount of genetic drift?
The effective population size N_E *differs from the actual population size if the sex ratio is disproportionate, or if a few dominant individuals of either sex contribute disproportionately to the gene pool of the next generation. When the sex ratio differs from 1:1, the effective population size can be calculated as follows:* $N_E = 4 \times$ *number of males* $\times$ *number of females/(number of males + number of females).*

The smaller the effective population size, the greater the magnitude of the genetic drift.

11. Define natural selection and fitness.
Natural selection is the differential reproduction of genotypes. Fitness is the relative reproductive success of a genotype.

12. Briefly discuss the differences between directional selection, overdominance, and underdominance. Describe the effect of each type of selection on the allelic frequencies of a population.
Directional selection is when one allele has greater fitness than another. Overdominance is when the heterozygote has greater fitness than either of the homozygotes. Underdominance is when the heterozygote has less fitness than either of the homozygotes. Directional selection will cause the allele with greater fitness to increase in frequency and eventually reach fixation. Overdominance establishes a balanced equilibrium that maintains both alleles. Underdominance results in an unstable equilibrium that will degenerate once disturbed and move away from equilibrium until one allele is fixed.

13. What factors affect the rate of change in allelic frequency due to natural selection?
The intensity of selection (the difference in fitness between the different genotypes) and the dominance relationships affect the rate of change. The rate of change also depends on the allelic frequency: the rate of decline of a deleterious allele will be proportional to the allelic frequency.

Section 25.4

*14. Compare and contrast the effects of mutation, migration, genetic drift, and natural selection on genetic variation within populations and on genetic divergence between populations.
Mutation increases genetic variation within populations and increases divergence between populations because different mutations arise in each population.

Migration increases genetic variation within a population by introducing new alleles, but decreases divergence between populations.

Genetic drift decreases genetic variation within populations because it causes alleles eventually to become fixed, but it increases divergence between populations because drift takes place differently in each population.

Natural selection may either increase or decrease genetic variation, depending on whether the selection is directional or balanced. It may increase or decrease divergence between populations, depending on whether different populations have similar or different selection pressures.

APPLICATION QUESTIONS AND PROBLEMS

Section 25.1

15. How would you respond to someone who said that models are useless in studying population genetics because they represent oversimplifications of the real world?
While it is important to keep in mind that models do represent simplifications, they nevertheless provide important and valuable information and predictions about the effects of size, mutation, migration, inbreeding, and selection on the gene pool of the population.

*16. Voles (*Microtus ochrogaster*) were trapped in old fields in southern Indiana and were genotyped for a transferrin locus. The following numbers of genotypes were recorded.

T^ET^E	T^ET^F	T^FT^F
407	170	17

Calculate the genotypic and allelic frequencies of the transferrin locus for this population.

The total number of voles is 594.

$f(T^ET^E) = 407/594 = .685$

$f(T^ET^F) = 170/594 = .286$

$f(T^FT^F) = 17/594 = .029$

$f(T^E) = f(T^ET^E) + f(T^ET^F)/2 = .685 + .143 = .828$

$f(T^F) = f(T^FT^F) + f(T^ET^F)/2 = .029 + .143 = .172$

17. Jean Manning, Charles Kerfoot, and Edward Berger studied the frequencies at the phosphoglucose isomerase (GPI) locus in the cladoceran *Bosmina longirostris*. At one location, they collected 176 animals from Union Bay in Seattle, Washington, and determined their GPI genotypes by using electrophoresis (J. Manning, W. C. Kerfoot, and E. M. Berger. 1978. *Evolution* 32:365–374).

Genotype	Number
S1S1	4
S1S2	38
S2S2	134

Determine the genotypic and allelic frequencies for this population.

$f(S1S1) = 4/176 = 0.023$

$f(S1S2) = 38/176 = 0.216$

$f(S2S2) = 134/176 = 0.761$

$f(S1) = (8 + 38)/352 = 0.13$

$f(S2) = (268 + 38)/352 = 0.87$

18. Orange coat color in cats is due to an X-linked allele (X^O) that is codominant to the allele for black (X^+). Genotypes of the orange locus of cats in Minneapolis and St. Paul, Minnesota, were determined and the following data were obtained.

X^OX^O females	11
X^OX^+ females	70
X^+X^+ females	94
X^OY males	36
X^+Y males	112

Calculate the frequencies of the X^O and X^+ alleles for this population.

We add up all the X^O or X^+ alleles and divide by the total number of X^O and X^+ alleles.

The number of X^O alleles = $2(X^O X^O) + (X^OX^+) + (X^OY) = 22 + 70 + 36 = 128$

The number of X^+ alleles = $2(X^+X^+) + (X^oX^+) + (X^+Y) = 188 + 70 + 112 = 370$
$f(X^o) = 128/(128 + 370) = 128/498 = .26$
$f(X^+) = 370/498 = .74$

Section 25.2

19. A total of 6129 North American Caucasians were blood typed for the MN locus, which is determined by two codominant alleles, L^M and L^N. The following data were obtained:

Blood type	Number
M	1787
MN	3039
N	1303

Carry out a chi-square test to determine whether this population is in Hardy–Weinberg equilibrium at the MN locus.
The total number of individuals is 6129.
$f(L^M) = p = (1787 + 3039/2)/6129 = .54$
$f(L^N) = q = (1303 + 3039/2)/6129 = .46$

A population in Hardy–Weinberg equilibrium should have the following genotype frequencies:
$f(L^ML^M) = p^2 = (.54)^2 = 0.29 = 1777/6129$
$f(L^ML^N) = 2pq = 2(.54)(.46) = .50 = 3065/6129$
$f(L^NL^N) = q^2 = (.46)^2 = 0.21 = 1287/6129$

Now we set up a chi-square test:

Blood type	*Observed*	*Expected*	O–E	$(O–E)^2$	$(O–E)^2/E$
M	*1787*	*1777*	*10*	*100*	*.056*
MN	*3039*	*3065*	*6*	*225*	*.084*
N	*1303*	*1287*	*16*	*256*	*.20*

Chi-squared = $\Sigma (O–E)^2/E = .34$

The number of degrees of freedom is the number of genotypes minus the number of alleles = 3 − 2 = 1.

Looking at a chi-square table, we see the p *value is easily greater than .05, so we do not reject the hypothesis that this population is in Hardy–Weinberg equilibrium with respect to the MN locus.*

20. Most black bears (*Ursus americanus*) are black or brown in color. However, occasional white bears of this species appear in some populations along the coast of British Columbia. Kermit Ritland and his colleagues determined that white coat color in these bears results from a recessive mutation (*G*) caused by a single nucleotide replacement in which guanine substitutes for adenine at the melanocortin 1 receptor locus (*mcr1*), the same locus responsible for red hair in humans (K.

Ritland, C. Newton, and H. D. Marshall. 2001. *Current Biology* 11:1468–1472). The wild-type allele at this locus (*A*) encodes black or brown color. Ritland and his colleagues collected samples from bears on three islands and determined their genotypes at the *mcr1* locus.

Genotype	Number
AA	42
AG	24
GG	21

a. What are the frequencies of the *A* and *G* alleles in these bears?
f(A) = *(84 + 24)/174 = 0.62*
f(G) = *(42 + 24)/174 = 0.38*

b. Give the genotypic frequencies expected if the population is in Hardy–Weinberg equilibrium.
Expected genotype frequencies:
f(AA) = *(0.62)(0.62) = 0.384*
f(AG) = *2(0.62)(0.38) = 0.471*
f(GG) = *(0.38)(0.38) = 0.144*

c. Use a chi-square test to compare the number of observed genotypes with the number expected under Hardy–Weinberg equilibrium. Is this population in Hardy–Weinberg equilibrium? Explain your reasoning.

Genotype	*Observed*	*Expected*	O–E	$(O-E)^2$	$(O-E)^2/E$
AA	*42*	*33*	*9*	*81*	*2.45*
AG	*24*	*41*	*17*	*289*	*7.05*
GG	*21*	*13*	*8*	*64*	*4.92*

Chi-squared = $\Sigma(O-E)^2/E$ = *14.42*

The number of degrees of freedom is the number of genotypes minus the number of alleles = 3 − 2 = 1.

The p *value is much less than 0.05; therefore, we reject the hypothesis that these genotype frequencies may be expected from Hardy–Weinberg equilibrium.*

21. Genotypes of leopard frogs from a population in central Kansas were determined for a locus (*M*) that encodes the enzyme malate dehydrogenase. The following numbers of genotypes were observed:

Genotype	Number
$M^1 M^1$	20
$M^1 M^2$	45
$M^2 M^2$	42
$M^1 M^3$	4
$M^2 M^3$	8
$M^3 M^3$	6
Total	125

a. Calculate the genotypic and allelic frequencies for this population.
$f(M^1M^1) = 20/125 = .16$
$f(M^1M^2) = 45/125 = .36$
$f(M^2M^2) = 42/125 = .34$
$f(M^1M^3) = 4/125 = .032$
$f(M^2M^3) = 8/125 = .064$
$f(M^3M^3) = 6/125 = .048$
$f(M^1) = p = .16 + .36/2 + .032/2 = .16 + .18 + .016 = .356$
$f(M^2) = q = .34 + .36/2 + .064/2 = .34 + .18 + .032 = .552$
$f(M^3) = r = .048 + .032/2 + .064/2 = .048 + .016 + .032 = .096$

b. What would the expected numbers of genotypes be if the population were in Hardy-Weinberg equilibrium?
For population in Hardy–Weinberg equilibrium:
$f(M^1M^1) = p^2 = (.356)^2 = .127; .127(125) = 16$
$f(M^1M^2) = 2pq = 2(.356)(.552) = .393; .393(125) = 49$
$f(M^2M^2) = q^2 = (.552)^2 = .305; .305(125) = 38$
$f(M^1M^3) = 2pr = 2(.356)(.096) = .068; .068(125) = 8$
$f(M^2M^3) = 2qr = 2(.552)(.096) = .106; .106(125) = 13$
$f(M^3M^3) = r^2 = (.096)^2 = .009; .009(125) = 1$

Genotype	*Observed*	*Expected*	O–E	$(O–E)^2$	$(O–E)^2/E$
M^1M^1	*20*	*16*	*4*	*16*	*1*
M^1M^2	*45*	*49*	*–4*	*16*	*.33*
M^2M^2	*42*	*38*	*4*	*16*	*.42*
M^1M^3	*4*	*8*	*–4*	*16*	*2*
M^2M^3	*8*	*13*	*–5*	*25*	*1.9*
M^3M^3	*6*	*1*	*5*	*25*	*25*

Chi-squared = 30.65
Degrees of freedom = # genotypes – # alleles = 6 – 3 = 3
The p value is much lower than 0.05; this population is not in Hardy–Weinberg equilibrium for this locus.

22. Full color (*D*) in domestic cats is dominant over dilute color (*d*). Of 325 cats observed, 194 have full color and 131 have dilute color.

a. If these cats are in Hardy–Weinberg equilibrium for the dilution locus, what is the frequency of the dilute allele?
f(dilute) $= f(dd) = q^2 = 131/325 = 0.403; q = 0.635$

b. How many of the 194 cats with full color are likely to be heterozygous?
If $q = f(d) = 0.635$, *then* $p = 1 - q = 0.365$
$f(Dd) = 2pq = 2(0.365)(0.635) = 0.464; 0.464(325) = 151$ *heterozygous cats*

23. Tay–Sachs disease is an autosomal recessive disorder. Among Ashkenazi Jews, the frequency of Tay–Sachs disease is 1 in 3600. If the Ashkenazi population is mating randomly for the Tay–Sachs gene, what proportion of the population consists of heterozygous carriers of the Tay–Sachs allele?
If q *= the frequency of the Tay–Sachs allele, then* $q^2 = 1/3600; q = 1/60 = 0.017$.

The frequency of the normal allele = p = *1* – q = *0.983.*
The frequency of heterozygous carriers = 2pq = *2(0.983)(0.017)* = *0.033; approximately 1 in 30 are carriers.*

24. In the plant *Lotus corniculatus,* cyanogenic glycoside protects the plants against insect pests and even grazing by cattle. This glycoside is due to a simple dominant allele. A population of *L. corniculatus* consists of 77 plants that possess cyanogenic glycoside and 56 that lack the compound. What is the frequency of the dominant allele that results in the presence of cyanogenic glycoside in this population?
The frequency of the recessive allele = q; *if the population is in Hardy–Weinberg equilibrium (or if the population went through a round of random mating), then the frequency of homozygous recessives* = q^2 = *56/(77+56)* = *56/133* = *0.42;* q = *0.65.*
Then p = *0.35* = *the frequency of the dominant allele.*

*25. Color blindness in humans is an X-linked recessive trait. Approximately 10% of the men in a particular population are colorblind.
   a. If mating is random for the color-blind locus, what is the frequency of the color-blind allele in this population?
   For males, f(*color blind*) = p = *0.1.*
   b. What proportion of the women in this population is expected to be colorblind?
   For females, f(*homozygotes*) = p^2 = *0.01, or 1%.*
   c. What proportion of the women in the population is expected to be heterozygous carriers of the color-blind allele?
   For females, f(*heterozygotes*) = 2pq = *2(0.1)(0.9)* = *0.18, or 18%.*

*26. The human MN blood type is determined by two codominant alleles, L^M and L^N. The frequency of L^M in Eskimos on a small Arctic island is 0.80. If the inbreeding coefficient for this population is .05, what are the expected frequencies of the M, MN, and N blood types on the island?
If $f(L^M) = p = 0.80$; *then* $f(L^N) = q = 1 - p = 0.20$.
$f(L^M L^M) = p^2 + Fpq = (0.80)^2 + 0.05(0.80)(0.20) = 0.64 + 0.008 = 0.648$
$f(L^M L^N) = 2pq - 2Fpq = 2(0.80)(0.20) - 2(0.05)(0.80)(0.20) = 0.32 - 0.016 = 0.304$
$f(L^N L^N) = q^2 + Fpq = (0.20)^2 + 0.05(0.80)(0.20) = 0.04 + 0.008 = 0.048$

Section 25.3

27. Demonstrate mathematically that full sib mating (*F* = ¼) reduces the heterozygosity by ¼ with each generation.
For a randomly mating population, the heterozygote frequency = 2pq;
for inbreeding populations, the heterozygote frequency = 2pq – *2*Fpq.
If F = ¼, *then the heterozygote frequency* = 2pq – *2(¼)*pq = *1.5*pq.
The reduction in heterozygote frequency is ½pq, *or ¼ of the randomly mating heterozygote frequency of* 2pq.

Section 25.4

28. The forward mutation rate for piebald spotting in guinea pigs is 8×10^{-5}; the reverse mutation rate is 2×10^{-6}. Assuming that no other evolutionary forces are present, what is the expected frequency of the allele for piebald spotting in a population that is in mutational equilibrium?

Here, we use the equation $\hat{q} = \frac{\mu}{\mu + \upsilon}$; *where* $\mu = 8 \times 10^{-5}$ *and* $\upsilon = 2 \times 10^{-6}$; *the frequency at equilibrium is then* $8 \times 10^{-5}/(8 \times 10^{-5} + 2 \times 10^{-6}) = 8/8.2 = 0.98$.

29. For a period of 3 years, Gunther Schlager and Margaret Dickie estimated the forward and reverse mutation rates for five loci in mice that encode various aspects of coat color by examining more than 5 million mice for spontaneous mutations (G. Schlager and M. M. Dickie. 1966. *Science* 151:205–206). The numbers of mutations detected at the dilute locus are as follows:

	Number of gametes examined	Number of mutations detected
Forward mutations	260,675	5
Reverse mutations	583,360	2

Calculate the forward and reverse mutation rates at this locus. If these mutations rates are representative of rates in natural populations of mice, what would the expected equilibrium frequency of dilute mutations be?

Forward mutation rate $= \mu = 5/260{,}675 = 1.9 \times 10^{-5}$

Reverse mutation rate $= \upsilon = 2/583{,}360 = 3.4 \times 10^{-6}$

Equilibrium frequency $= \hat{q} = \frac{\mu}{\mu + \upsilon} = 1.9 \times 10^{-5}/(1.9 \times 10^{-5} + 3.4 \times 10^{-6})$

$= 1.9/2.24 = 0.85$

*30. In German cockroaches, curved wing (cv) is recessive to normal wing (cv^{+}). Bill, who is raising cockroaches in his dorm room, finds that the frequency of the gene for curved wings in his cockroach population is .6. In the apartment of his friend Joe, the frequency of the gene for curved wings is .2. One day Joe visits Bill in his dorm room, and several cockroaches jump out of Joe's hair and join the population in Bill's room. Bill estimates that 10% of the cockroaches in his dorm room now consists of individual roaches that jumped out of Joe's hair. What will be the new frequency of curved wings among cockroaches in Bill's room?

The proportion of migrants in the new population $= m = .1$. *The frequency of the allele for curved wings in the old population* $= q_{old} = 0.6$; *the allele frequency in the migrant population is* $q_{migrants} = 0.2$:

$q_{new} = mq_{migrants} + (1 - m)q_{old} = 0.1(0.2) + 0.9(0.6) = 0.56$

Now that we have calculated the allelic frequencies, we can calculate the new genotype frequency assuming random mating:

$f_{new}(cv,cv) = q_{new}^2 = (0.56)^2 = 0.31$

31. A population of water snakes is found on an island in Lake Erie. Some of the snakes are banded and some are unbanded; banding is caused by an autosomal allele that is recessive to an allele for no bands. The frequency of banded snakes on the island is 0.4, whereas the frequency of banded snakes on the mainland is 0.81. One summer, a large number of snakes migrate from the mainland to the island. After this migration, 20% of the island population consists of snakes that came from the mainland.

    a. Assuming that the mainland population and the island population are in Hardy–Weinberg equilibrium for the alleles that affect banding, what is the frequency of the allele for bands on the island and on the mainland before migration?
    Because banding is recessive, the frequency of banded snakes = q^2.
    On the island before the migration, $q^2 = 0.4$; $q = 0.63$.
    On the mainland, $q^2 = 0.81$; $q = 0.9$.

    b. After migration has taken place, what will be the frequency of the banded allele on the island?
    After the migration, $q_{new} = mq_{migrants} + (1 - m)q_{old} = 0.2(0.9) + 0.8(0.63) = 0.68$

32. Pikas are small mammals that live at high elevation in the talus slopes of mountains. Most populations located on mountain tops in Colorado and Montana in North America are isolated from one another because the pikas don't occupy the low-elevation habitats that separate the mountain tops and don't venture far from the talus slopes. Thus, there is little gene flow between populations. Furthermore, each population is small in size and was founded by a small number of pikas.
    A group of population geneticists propose to study the amount of genetic variation in a series of pika populations and to compare the allelic frequencies in different populations. On the bases of the biology and the distribution of pikas, predict what the population geneticists will find concerning the within- and between-population genetic variation.
    The small population sizes and the founder effects would cause strong effects from genetic drift. The geneticists will find large variation between populations in allele frequencies. Within populations, the same factors coupled with inbreeding will cause loss of genetic variation and a high degree of homozygosity.

33. In a large, randomly mating population, the frequency of the allele (*s*) for sickle-cell hemoglobin is 0.028. The results of studies have shown that people with the following genotypes at the beta-chain locus produce the average numbers of offspring given:

Genotypes	Average number of offspring produced
SS	5
Ss	6
ss	0

a. What will be the frequency of the sickle-cell allele (*s*) in the next generation?
The current allele frequencies are: f(s) = q = *0.028;* f(S) = p = *1* – q = *0.972*

The current genotype frequencies are:
$f(SS) = p^2 = 0.945$
$f(Ss) = 2pq = 0.054$
$f(ss) = q^2 = 0.001$
$W_{SS} = 5/6 = 0.83;\ s_{SS} = 0.17$
$W_{Ss} = 6/6 = 1.0;\ s_{Ss} = 0$
$W_{ss} = 0/6 = 0;\ s_{ss} = 1.0$
$\overline{W} = p^2W_{SS} + 2pqW_{Ss} + q^2W_{ss} = 0.784 + 0.054 + 0 = 0.838$

For the next generation, each genotype will have the following reproductive contribution:

$\frac{p^2W_{SS}}{\overline{W}}$, $\frac{2pqW_{Ss}}{\overline{W}}$, *and* $\frac{q^2W_{ss}}{\overline{W}}$ *for* SS, Ss, *and* ss, *respectively.*

$f(SS) = 0.936$
$f(Ss) = 0.064$
$f(ss) = 0$

The new allele frequency is then: $q' = 0.064/2 = 0.032$.

b. What will be the frequency of the sickle-cell allele at equilibrium?
We use the equilibrium equation for overdominance:

$$\hat{q} = f(s) = \frac{s_{SS}}{s_{SS} + s_{ss}} = 0.17/(0.17 + 1.0) = .17/1.17 = 0.145$$

34. Two chromosomal inversions are commonly found in populations of *Drosophila pseudoobscura:* Standard (*ST*) and Arrowhead (*AR*). When treated with the insecticide DDT, the genotypes for these inversions exhibit overdominance, with the following fitnesses:

Genotype	Fitness
ST/ST	0.47
ST/AR	1
AR/AR	0.62

What will be the frequency of *ST* and *AR* after equilibrium has been reached?

The equation for overdominance is: $\hat{q} = f(s) = \frac{s_{SS}}{s_{SS} + s_{ss}}$

$s_{ST/ST} = 0.53$
$s_{ST/AR} = 0$
$s_{AR/AR} = 0.38$
$\hat{q} = f(AR) = 0.53/(0.53 + 0.38) = 0.53/0.91 = 0.58$
At equilibrium, the frequency of AR *will be 0.58, and the frequency of* ST *will be 0.42.*

*35. In a large, randomly mating population, the frequency of an autosomal recessive lethal allele is 0.20. What will the frequency of this allele be in the next generation if the lethality takes place before reproduction?

If q =*0.2, then* p = *0.8. After one round of random mating, the genotype frequencies in the next generation will be:*

f(AA) = p^2 =*0 .64*

f(Aa) = 2pq = *0.32*

f(aa) = q^2 = *0.04, but these die.*

Adjusting for the death of all the homozygotes for the recessive lethal allele, the new genotype frequencies f′ *in the survivors are:*

f′(AA) = f(AA)/[f(AA) + f(Aa)] = *0.64/(0.64 + 0.32) = 0.64/0.96 = 2/3 = 0.67*

f′(Aa) = f(Aa)/[f(AA) + f(Aa)] = *0.32/0.96 = 1/3 = 0.33*

Here, we are simply taking the frequencies of the AA *and* Aa *genotypes and dividing each by the total survivors, those with* AA *and* Aa *genotypes.*

Then, the new allelic frequency q′ = f′(aa) + f′(Aa)/2 = *0 + 1/6 = 0.17; note that because the genotype* aa *is lethal,* f′(aa) = *0.*

36. The fruit fly *Drosophila melanogaster* normally feeds on rotting fruit, which may ferment and contain high levels of alcohol. Douglas Cavener and Michael Clegg studied allelic frequencies at the locus for alcohol dehydrogenase (*Adh*) in experimental populations of *D. melanogaster* (D. R. Cavener and M. T. Clegg. 1981. *Evolution* 35:1–10). The experimental populations were established from wild-caught flies and were raised in cages in the laboratory. Two control populations were raised on a standard cornmeal–molasses–agar diet. Two ethanol populations were raised on a cornmeal–molasses–agar diet to which was added 10% ethanol. The four populations were periodically sampled to determine the allelic frequencies of two alleles at the alcohol dehydrogenase locus, Adh^S and Adh^F. The frequencies of these alleles in the experimental populations are shown in the adjoining graph.

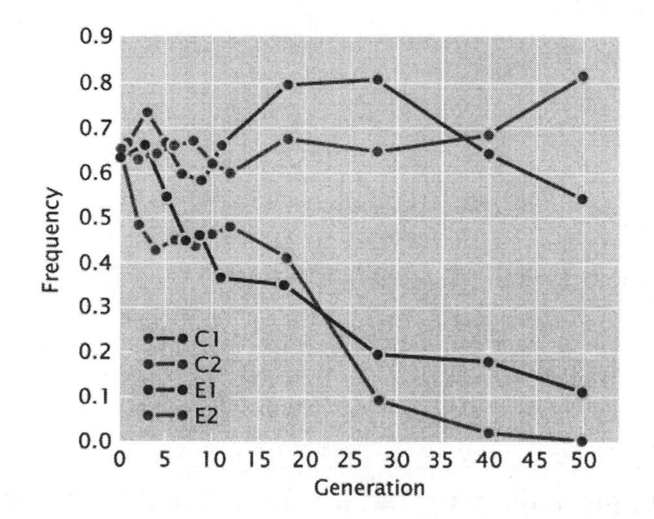

a. One the basis of these data, what conclusion might you draw about the evolutionary forces that are affecting the *Adh* alleles in these populations?
The populations raised on the ethanol-containing diet appear to be experiencing directional selection in favor of the Adh^F *allele, and against the* Adh^S *allele.*

b. Cavener and Clegg measured the viability of the different *Adh* genotypes in the alcohol environment and obtained the following values:

Genotype	Relative viability
Adh^F/Adh^F	0.932
Adh^F/Adh^S	1.288
Adh^S/Adh^S	0.596

Using these relative viabilities, calculate relative fitnesses for the three genotypes. If a population has an initial frequency of $p = f(Adh^F) = 0.5$, what will the expected frequency of Adh^F be in the next generation on the basis of these fitness values?
In the absence of data about relative reproductive rates, we use the relative viability data as a proxy for relative fitness.

$W_{FF} = 0.932/1.288 = 0.724$
$W_{FS} = 1.288/1.288 = 1.0$
$W_{SS} = 0.596/1.288 = 0.463$

The mean fitness $\bar{W} = p^2W_{FF} + 2pqW_{FS} + q^2W_{SS}$
$= 0.25(0.724) + 0.5(1.0) + 0.25(0.463) = 0.181 + 0.5 + 0.116 = 0.797$

We can then use the table method to calculate the frequencies after selection:

	Adh^FAdh^F	**Adh^FAdh^S**	**Adh^SAdh^S**
Initial genotypic frequencies:	$p^2 = (0.5)^2 = 0.25$	$2pq = 2(0.5)(0.5) = 0.50$	$q^2 = (0.5)^2 = 0.25$
Fitnesses:	$W_{FF} = 0.724$	$W_{FS} = 1.0$	$W_{SS} = 0.463$
Proportionate contribution of genotypes to population:	$p^2W_{FF} = (0.25)(0.724) = 0.181$	$2pqW_{FS} = (0.50)(1.0) = 0.50$	$q^2W_{SS} = (0.25)(0.463) = 0.116$
Relative genotypic frequency after selection:	$p^2W_{FF}/\bar{W} = 0.181/0.797 = 0.227$	$2pqW_{FS}/\bar{W} = 0.50/0.797 = 0.627$	$q^2W_{SS}/\bar{W} = 0.116/0.797 = 0.145$

After one generation, $p = 0.227 + 0.5(0.627) = 0.54$.

37. A certain form of congenital glaucoma results from an autosomal recessive allele. Assume that the mutation rate is 10^{-5} and that persons having this condition produce, on average, about 80% of the offspring produced by persons who do not have glaucoma.

a. At equilibrium between mutation and selection, what will be the frequency of the gene for congenital glaucoma?
The fitness of individuals with glaucoma is 0.8; the selection coefficient is 0.2.

For a recessive allele, $\hat{q} = \sqrt{\frac{u}{s}} = \sqrt{(10^{-5}/.2)} = 0.0071$

b. What will be the frequency of the disease in a randomly mating population that is at equilibrium?
Frequency of homozygotes $= q^2 = 5 \times 10^{-5}$

CHALLENGE QUESTION

Section 25.4

38. The Barton Springs salamander is an endangered species found only in a single spring in the city of Austin, Texas. There is growing concern that a chemical spill on a nearby freeway could pollute the spring and wipe out the species. To provide a source of salamanders to repopulate the spring in the event of such a catastrophe, a proposal has been made to establish a captive breeding population of the salamander in a local zoo. You are asked to provide a plan for the establishment of this captive breeding population, with the goal of maintaining as much of the genetic variation of the species as possible in the captive population. What factors might cause loss of genetic variation in the establishment of the captive population? How could loss of such variation be prevented? Assuming that it is feasible to maintain only a limited number of salamanders in captivity, what procedures should be instituted to ensure the long-term maintenance of as much of the variation as possible?
Genetic variation in the zoo salamander colony could be reduced because of a founder effect from the limited number of individuals used to establish a breeding colony. Genetic variation would be reduced further by inbreeding and genetic drift. Given that only a limited number of salamanders can be maintained in the zoo colony, regular introduction of wild salamanders from the spring into the colony will keep mixing in fresh genotypes. A continual influx of migrants from the spring will, over time, effectively increase the number of individuals sampled, keep the gene pool of the zoo colony close to the gene pool of the spring, and mitigate inbreeding. It is also important to maintain a 50:50 sex ratio, as deviations from 50:50 causes a reduction in the effective number of adults. Matings within the zoo colony should be carefully planned to avoid inbreeding.

Chapter Twenty-Six: Evolutionary Genetics

COMPREHENSION QUESTIONS

Section 26.1

1. How is biological evolution defined?
   Biological evolution is change in the genetic composition of a population over time.

*2. What are the two steps in the process of evolution?
   First, mutation causes genetic variation in a population and recombination creates new combinations of genetic variants. Second, the frequencies of the variants changes over generations, as a result of random processes or selection.

3. How is anagenesis different from cladogenesis?
   Anagenesis is change in a single group of organisms over time. Cladogenesis involves splitting of a group into two or more groups that become different from each other.

Section 26.2

*4. As a measure of genetic variation, why is the expected heterozygosity often preferred to the observed heterozygosity?
   The expected heterozygosity is independent of the breeding system and often represents the variation in a population better than the observed heterozygosity, especially for species that self-fertilize.

5. Why does protein variation, as revealed by electrophoresis, underestimate the amount of true genetic variation?
   Not all changes in DNA sequence cause changes in amino acid sequence, and not all changes in amino acid sequence cause a change in electrophoretic mobility of a protein.

6. What are some of the advantages of using molecular data in evolutionary studies?
   DNA and protein sequences are genetic; they can be studied and compared in all organisms; genomes provide a vast and growing amount of data in public databases; the molecular differences are quantifiable; comparisons of DNA and protein sequences provide information about evolutionary history and may be used to infer phylogeny.

*7. What is the key difference between the neutral-mutation hypothesis and the balance hypothesis?
   The neutral-mutation hypothesis proposes that most molecular variation is adaptively neutral. The balance hypothesis proposes that most genetic variation is maintained by balanced selection, favoring heterozygosity at most loci.

8. Discuss some of the methods that have been used to study variation in DNA.
   Three different types of DNA variation have been studied. (1) Restriction site variation has been analyzed by restriction endonuclease digestion of PCR-amplified fragments of DNA,

and analysis of resulting DNA fragments by gel electrophoresis. (2) Microsatellite length variation has also been analyzed, most frequently by PCR amplification of the microsatellite locus and gel electrophoresis to separate the PCR products of different lengths. (3) Variation in DNA sequences, such as base substitutions that do not result in changes in either restriction fragment length or PCR product length, may be characterized by sequencing of cloned or PCR-amplified DNA from the locus in question.

Section 26.3

*9. What is the biological species concept? What are some of the problems with it?
The biological species concept defines species as a group of individuals that can potentially interbreed with each other but are reproductively isolated from members of other species. One obvious problem is applying this concept to asexually reproducing organisms. Another is that, in practice, most species are defined by phenotypic or anatomical differences, and it may be difficult or impossible to determine whether individuals have the potential to interbreed if they existed at different times or their full life cycles have not been observed and studied.

10. What is the difference between prezygotic and postzygotic reproductive isolating mechanisms. List the different types of each.
Prezygotic mechanisms operate before fertilization of the egg by sperm (or fusion of gametes), and postzygotic mechanisms operate after fertilization. Prezygotic mechanisms include ecological isolation, behavioral isolation, temporal isolation, mechanical isolation, and gametic isolation. Postzygotic mechanisms include hybrid inviability, hybrid sterility, and hybrid breakdown.

11. What is the basic difference between allopatric and sympatric modes of speciation?
Allopatric speciation involves populations separated by a geographic barrier that precludes gene flow between the populations. Sympatric speciation takes place between populations occupying the same geographical area.

*12. Briefly outline the process of allopatric speciation.
First, a population is split by a geographical barrier that prevents gene flow between the two groups on either side of the barrier. The two groups then evolve independently; they accumulate genetic differences through various evolutionary processes such as mutation, selection, and random drift. If these genetic differences lead to reproductive isolation, meaning these individuals cannot or will not interbreed, then speciation has occurred.

13. What are some of the difficulties with sympatric speciation?
The main difficulty is that sympatric speciation requires genetic differences to accumulate between two groups that are initially freely exchanging genes. These genetic differences must then lead to reproductive isolation. However, interruption of gene flow requires reproductive isolation.In the absence of geographic separation, achieving this initial reproductive isolation appears difficult.

*14. Briefly explain how switching from hawthorn fruits to apples has led to genetic differentiation and partial reproductive isolation in *Rhagoletis pomonella*.
The flies that feed on apples mate on or near apples, whereas flies that feed on hawthorn fruit mate on or near the hawthorns. Moreover, apples ripen several weeks earlier than hawthorn fruits, so there is strong selection for flies that feed on applies to mate earlier than flies that feed on hawthorn fruits. These differences in location and timing of mating lead to reproductive isolation between these groups of flies and accumulation of genetic differences.

Section 26.4

15. Draw a simple phylogenetic tree and identify a node, a branch, and an outgroup.

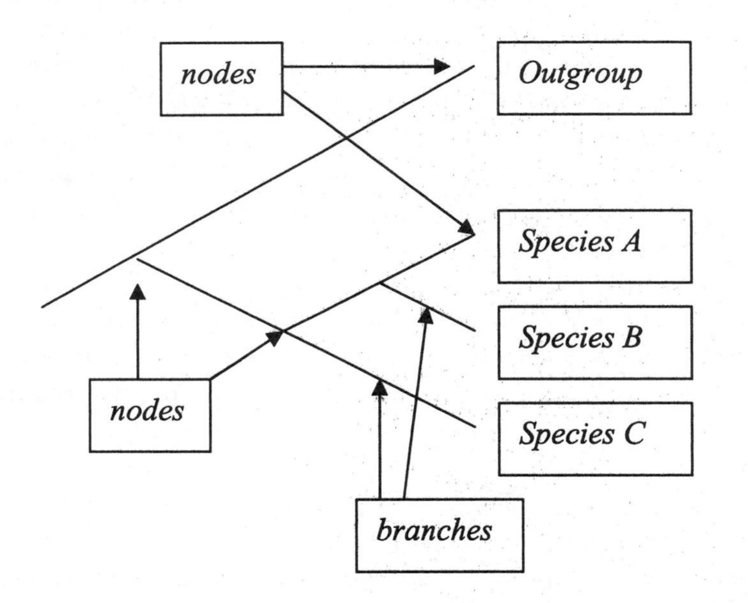

*16. Briefly describe the difference between the distance approach and the parsimony approach to the reconstruction of phylogenetic trees.
Distance approaches rely on overall similarity between species or gene sequences; most similar species are grouped together. Parsimony approaches try to reconstruct an evolutionary pathway that requires the minimum number of changes to arrive at the modern species or genes from a common ancestor.

Section 26.5

17. Outline the different rates of evolution that are typically seen in different parts of a protein-encoding gene. What might account for these differences?
Nucleotide substitutions that do not change the amino acid sequence occur far more frequently than those that do change the amino acid sequence. Changes also occur more rapidly in those parts of the protein sequence that are not essential for the function of the protein. Nucleotide changes occur most rapidly in intron sequences, except for those positions that affect splicing. Selective pressure to maintain the function of proteins

essential or advantageous for the survival and reproduction of the organism accounts for the differential rates of mutation and evolution.

*18. What is the molecular clock?
The molecular clock is the idea that the rate at which nucleotide changes take place in a DNA sequence is relatively constant over long periods of time, and therefore the number of nucleotide substitutions that have taken place between two organisms can be used to estimate the time since they last shared a common ancestor.

19. What is exon shuffling? How can it lead to evolution of new genes?
Exon shuffling is a hypothesis based on the observation that exons often correspond to functional domains of proteins. New genes with novel functions may arise through recombination events that bring together different exons encoding different functional modules of proteins.

20. What is a multigene family? What processes produce multigene families?
Multigene families are a group of genes in the same genome that are related by descent from a common ancestral gene. Multigene families arise through gene duplication events with subsequent diversification. Some members acquire mutations that render them nonfunctional and become pseudogenes.

21. Define horizontal gene transfer. What problems does it cause for evolutionary biologists?
Horizontal gene transfer, also called lateral gene transfer, is transmission of genetic information across species boundaries. Horizontal gene transfer occurs frequently in bacteria, through transformation and phage-mediated transduction. In eukaryotes, horizontal gene transfer may occur through endosymbiosis (e.g., mitochondrial and chloroplast genes), viral infections, parasitic infections, and human intervention (genetic engineering). Horizontal gene transfers confound phylogenetic inference using molecular sequence data because the evolutionary history of horizontally transferred genes differs from the evolutionary history of the rest of the genome.

APPLICATION QUESTIONS AND PROBLEMS

Section 26.1

22. The following illustrations represent two different patterns of evolution. Briefly discuss the differences in these two patterns, particularly in regard to the role of cladogenesis in evolutionary change.

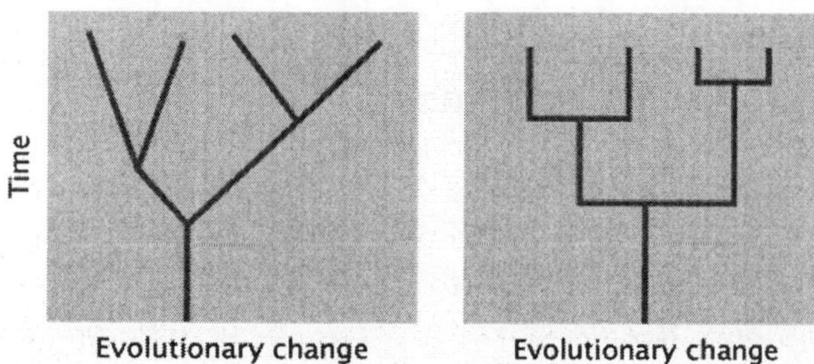

The two illustrations show the same pattern of cladogenesis. In each, an initial population splits into two groups with divergent evolutionary changes. Then, these two groups again each split into two other groups. The illustration on the left shows evolutionary differences accumulating gradually over time in each lineage after cladogenesis; cladogenesis occurs before diversification and may be the cause of diversification. The illustration on the right shows splitting of a population into two already divergent evolutionary lineages; cladogenesis occurs after, or as a result of, diversification of the population.

Section 26. 2

*23. Donald Levin used protein electrophoresis to study genetic variation in two species of wildflowers in Texas, *Phlox drummondi* and *P. cuspidate* (D. Levin. 1978. *Evolution* 32:245–263) *P. drummondi* reproduces only by cross fertilization, whereas *P. cuspidate* is partly self-fertilizing. The adjoining table gives allelic frequencies for several loci and populations studied by Levin.

a. Calculate the percentage of polymorphic loci and expected heterozygosity for each species. For the expected heterozygosity, calculate the mean for all loci in a population, and then calculate the mean for all populations.
For P. drummondii, *% polymorphic loci ranges from 50% in population 3 to 75% in populations 1 and 2. The mean expected heterozygosities in each population are:*

Pop 1: $[2(0.1)(0.9) + 2(0.02)(0.98) + 2(0.04)(0.96)]/4 = 0.296/4 = 0.074$
Pop 2: $[2(0.11)(0.89) + 2(0.31)(0.69) + 2(0.83)(0.17)]/4 = 0.906/4 = 0.227$
Pop 3: $[2(0.26)(0.74) + 2(0.14)(0.86)]/4 = 0.626/4 = 0.156$
Mean expected heterozygosity for all populations $= (0.074 + 0.227 + 0.156)/3 = 0.15$

For P. cuspidate, *% polymorphic loci ranges from zero in populations 1 and 2 to 25% in population 3.*
The mean expected heterozygosities are zero for populations 1 and 2.

Pop 3: $2(0.91)(0.09)/4 = 0.16$
Mean expected heterozygosity for all populations $= 0.16/3 = 0.055$

b. What tentative conclusions can you draw about the effect of self-fertilization on genetic variation in these flowers?
Self-fertilization reduces genetic variation.

Allelic frequencies

		Adh		**Got-2**		**Pgi-2**		**Pgm-2**	
Species	***Population***	**a**	**b**	**a**	**b**	**a**	**b**	**a**	**b**
P. drummondii	*1*	*0.10*	*0.90*	*0.02*	*0.98*	*0.04*	*0.96*	*1.0*	*0.0*
P. drummondii	*2*	*0.11*	*0.89*	*0.0*	*1.0*	*0.31*	*0.69*	*0.83*	*0.17*
P. drummondii	*3*	*0.26*	*0.74*	*1.0*	*0.0*	*0.14*	*0.86*	*1.0*	*0.0*
P. cuspidate	*1*	*0.0*	*1.0*	*0.0*	*1.0*	*0.0*	*1.0*	*0.0*	*1.0*
P. cuspidate	*2*	*0.0*	*1.0*	*0.0*	*1.0*	*0.0*	*1.0*	*0.0*	*1.0*
P. cuspidate	*3*	*0.91*	*0.09*	*0.0*	*1.0*	*0.0*	*1.0*	*0.0*	*1.0*

Note: a *and* b *represent different alleles at each locus.*

Section 26.4

*24. How many rooted trees are theoretically possible for a group of 7 organisms? How many for 12 organisms?

We use the formula:

$$\textit{number of rooted trees} = \frac{(2n-3)!}{2^{n-2}(n-2)!}$$

For 7 organisms, $11!/[(2^5)(5!)] = 39{,}916{,}800/[(32)(120)] = 10{,}395$ *rooted trees.*
For 12 organisms, $21!/[(2^{10})(10!)] = 13{,}749{,}310{,}575$ *rooted trees.*

25. Michael Bunce and his colleagues in England, Canada, and the United States extracted and sequenced mitochondrial DNA from fossils of Haast's eagle, a gigantic eagle that was driven to extinction 700 years ago when humans first arrived in New England (M. Bunce et al. 2005. *Plos Biology* 3:44–46). Using mitochondrial DNA sequences from living eagles and those from Haast eagle fossils, they created the adjoining phylogenetic tree. On this phylogenetic tree, identify **a.** all terminal nodes, **b.** all internal nodes, **c.** one example of a branch, and **d.** the outgroup.

a. *The terminal nodes are all the taxonomic groups listed on the right.*

b. *The internal nodes are all the branch points where lineages split.*

c. *The branches are the horizontal lines connecting nodes; the thick blue line illustrates one example of a branch.*

d. *The outgroup is the Goshawk, the bottom branch and node in the figure.*

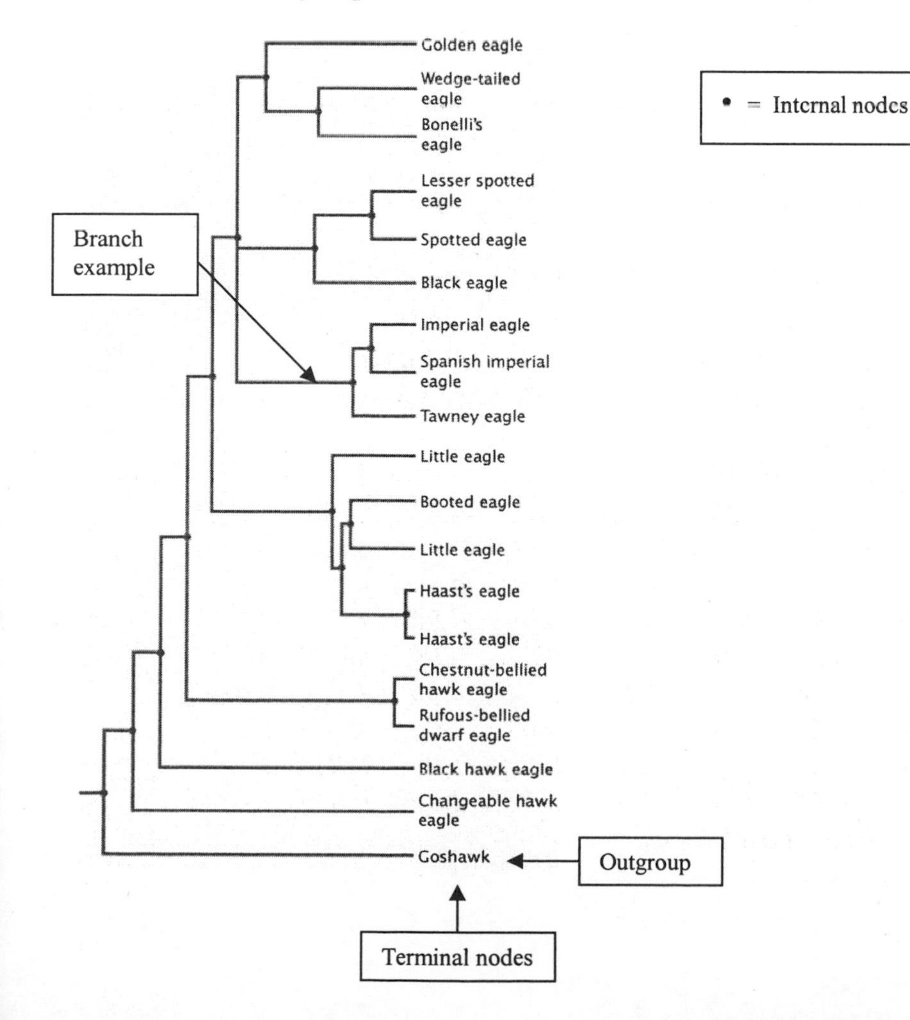

CHALLENGE QUESTION

Section 26.3

26. Explain why natural selection may cause prezygotic reproductive isolating mechanisms to evolve if postzygotic reproductive isolating mechanisms are already present, but natural selection can never cause the evolution of postzygotic reproductive isolating mechanisms.
If postzygotic mechanisms are already present, then matings between individuals from the different groups result in no progeny. Such matings are wasted opportunities. Individuals with mutations that prevent such matings (prezygotic isolating mechanisms) have a higher fitness because all their matings are productive and will result in higher relative reproductive success compared to individuals that can waste some of their opportunities in unproductive matings. These mutations have a good chance of spreading through the population.

    However, individuals with mutations that cause postzygotic reproductive isolation will have no progeny from some matings, whereas other individuals will have progeny from all their matings. Therefore, mutations that cause postzygotic reproductive isolation confer lower fitness and will tend to be eliminated from the population.